INTRODUCTION TO
CERAMICS

INTRODUCTION TO CERAMICS

GRAHAM FLIGHT

PETER LANE CONSULTANT EDITOR

Prentice-Hall

First North and South American Edition
published 1991 by
Prentice Hall Inc
A Division of Simon and Schuster
Englewood Cliffs, New Jersey 07632

Prentice-Hall Canada, Toronto
Prentice-Hall Hispanoamericana, S.A. Mexico
Editoria Prentice-Hall do Brasil, Ltda., Rio de Janeiro

ISBN 0-13-479247-5 paper
ISBN 0-13-479254-8 cloth

While every care has been taken to verify facts and methods described in this book, neither the publishers nor the author can accept liability for any loss or damage howsoever caused.

Introduction to Ceramics was conceived by
Thames Head, a division of BLA Publishing Limited,
TR House, 1 Christopher Road, East Grinstead
Sussex TH19 3BT

Extruded porcelain forms, with sprayed enamel color. The piece shown on the right is about 2 feet long (60cms); the piece on the left is about 3 feet long (90cms) by Oldrich Asenbryl (UK)

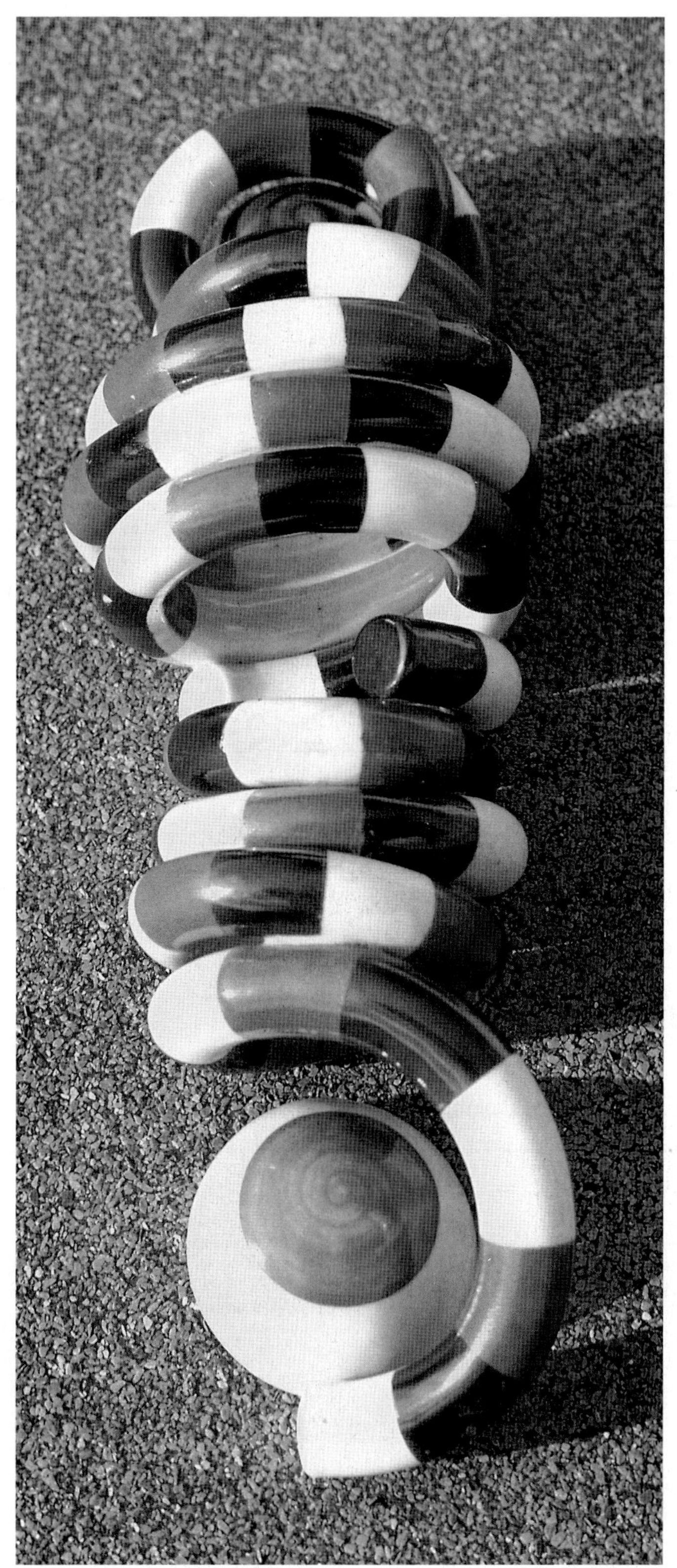

Designed, edited and illustrated by Playne Books, Gloucestershire, United Kingdom

Editors
Gill Davies
Alison Goldingham

Designers and Illustrators
Gail Langley
Marc Langley
Jane Moody
David Playne
Miles Playne
Melanie Williams
Karen Wilson

Picture Research
Diana Phillips
Pat Robertson
Christine Vincent

Typeset in ITC Gentleman on Scantext by Playne Books / Townsend Typesetters Limited

Printed in Italy.

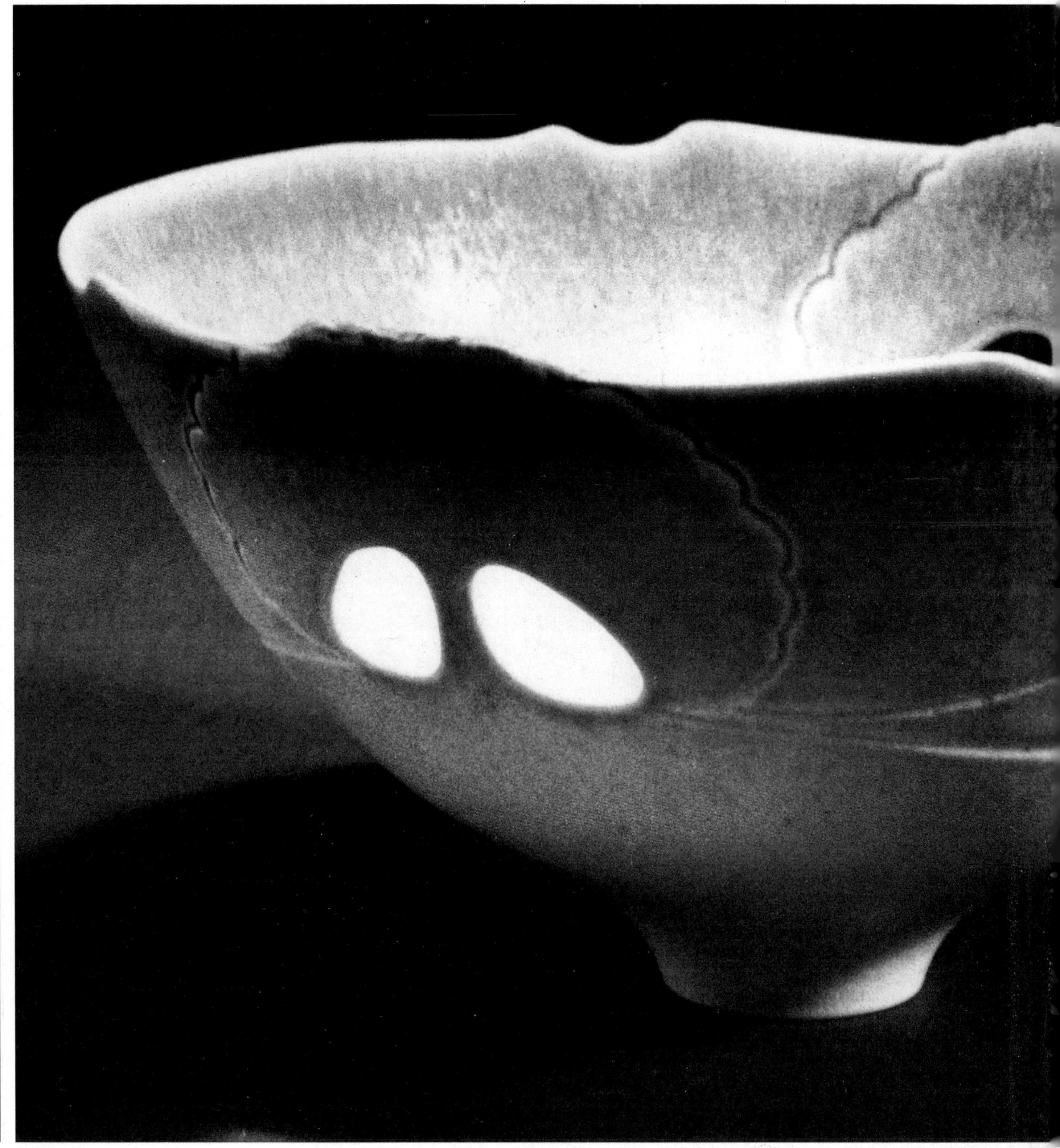

Contents

Porcelain bowl carved with a design of 'hills and trees'; 10 inches (25 cms) diameter, by Peter Lane (UK) 1979

For Rosie

The potters

G.E. Arnison
Oldrich Asenbryl
Paul Astbury
Rudy Autio
Maggie Barnes
Jenny Beavan
Tony Bennett
Maggie Angus Berkowitz
Sandra Black
Betty Blandino
Yvonne Boutell
Hilary Brock
Sandy Brown
Sally Bowen-Prange

Joan Campbell
Virginia Cartwright
Sally Cocksedge
Marianne Cole
Gordon Cooke
Delan Cookson
John S. Cummings

Greg Daly
Derek Davis
Sue Davis

Walter Dexter

Dorothy Feibleman
Ray Finch
Graham Flight
Tessa Fuchs

William Hall
Frank Hamer
Jane Hamlyn
Robin Hopper
William Hunt

Neil Ions

Roberta Kaserman
Lisa Katzenstein
Ann Kenny

Peter Lane
Jennifer Lee
Pauline Lurie

Jim Malone
John Maltby
Andrew McGarva

Peter Meanley
Eric James Mellon
Chris Myers

Susan Nemeth
Sidig El Nigoumi

Alan Peascod
David Pendell
Henry Pim
Ursula Morley Price

David Roberts
James Robison
Stuart Robyn
Ray Rogers
Jerry Rothman
Jill Ruhlman
Gail Russell

Jan Schachter
Karl Scheid
Peter Simpson
Ian Sprague
Petrus Spronk
Angus Suttie

Hiroe Swen
Geoffrey Swindell

Bryan Trueman

Sue Varley
Angela Verdon
Monique Vezina

Sasha Wardell
Robert Washington
John Wheeldon
Mary White
Lana Wilson
Mollie Winterburn
Andrew Wood

For further details about the potters and where examples of their work can be found in the book, see pages 336-7.

Introduction

The term, ceramics, encompasses a wide range of artistic achievements; from an ordinary homely bowl used daily for cooking or eating to large-scale ceramics used for architectural construction or embellishment, the possibilities are almost limitless.

To the beginner, who is perhaps handling clay for the first time, the range and scope of this new material are yet to be discovered; the initial difficulties of how to turn a lump of clay into a pleasing object is the main concern.

As a teacher who has introduced clay to many different age groups and to students who have approached the subject for many different reasons, it has always been fascinating to observe the varying reactions to this new medium. Some people react instinctively against the apparent 'dirtiness' of the material, regarding it distastefully as mud; others will begin straight away to experiment, squeezing it between their fingers and so becoming aware that it does not stick to the hands or make a mess—that it is pleasant to touch and malleable.

It is soon obvious which students will enjoy clay and be able to express their creative talent through this medium. Once the apt potter has established a relationship with clay, then the desire and the need to work in this art form will probably be retained for life. As students learn and develop their new skills with clay these new exponents of the art may gravitate towards a particular area of ceramics, perhaps choosing to explore one aspect that best matches individual temperament.

It is important that beginners follow their creative instincts first and foremost. Naturally there are some technical restraints but they should learn to work to the limits of the material. For instance, it is preferable to attempt to make a fine-walled pot, even though it may be in danger of collapsing, than to settle for a dumpy thick-walled version simply because it is 'safe'!

Ensuring that the creative urge is kept to the fore, even—in fact, especially—in the earliest lessons, it will then be appropriate to explore the technical aspects of the medium as and when needs arise

Like any other craft, pottery involves learning various techniques and basic skills and applying them to the making of a particular piece. One potter may find the spinning moving clay on the wheel inspires invention and leads to the development of satisfying curves and shapes, while another will welcome the slower process of handbuilding to shape a form that can be developed in any direction hands and eyes are inspired to take.

Whatever the process, the end result can produce a pleasing effect, from the basic beauty of a 'simple' food bowl to be used with pleasure daily, to the delight of a vase standing alone, or the constant interest of a piece of sculpture.

The need for activity that gives pleasure, excitement or peace of mind is within most people. To create something will often satisfy that need-be it a garden, a dress, or a poem. Making things is recognized as a fundamental need in children but is sometimes overlooked in the crowded days of adulthood. Pottery is just one way of fulfilling this need but it is a particularly rewarding one. To hold in your hands a pleasing form which has been produced from an amorphous lump of clay—and to have been personally responsible for that change—can give the potter a feeling of intense satisfaction.

Although this book must inevitably deal with the many technical aspects of the ceramic medium and evaluate the results, I have tried throughout the text to convey my own lasting delight in pottery in all its forms and to give the beginner some insight into the 'mystery' of clay and other related materials. It is at once a unique and exciting substance — unique in its potential and versatile in form and adaptation. It has inspired artistic creation throughout the centuries. It is hoped that this book will help to encourage further such creativity by introducing new potters to their art and helping more experienced potters to improve and develop their skills.

Graham Flight

Porcelain vase, height 5 inches (12 cms), by Lana Wilson (USA)

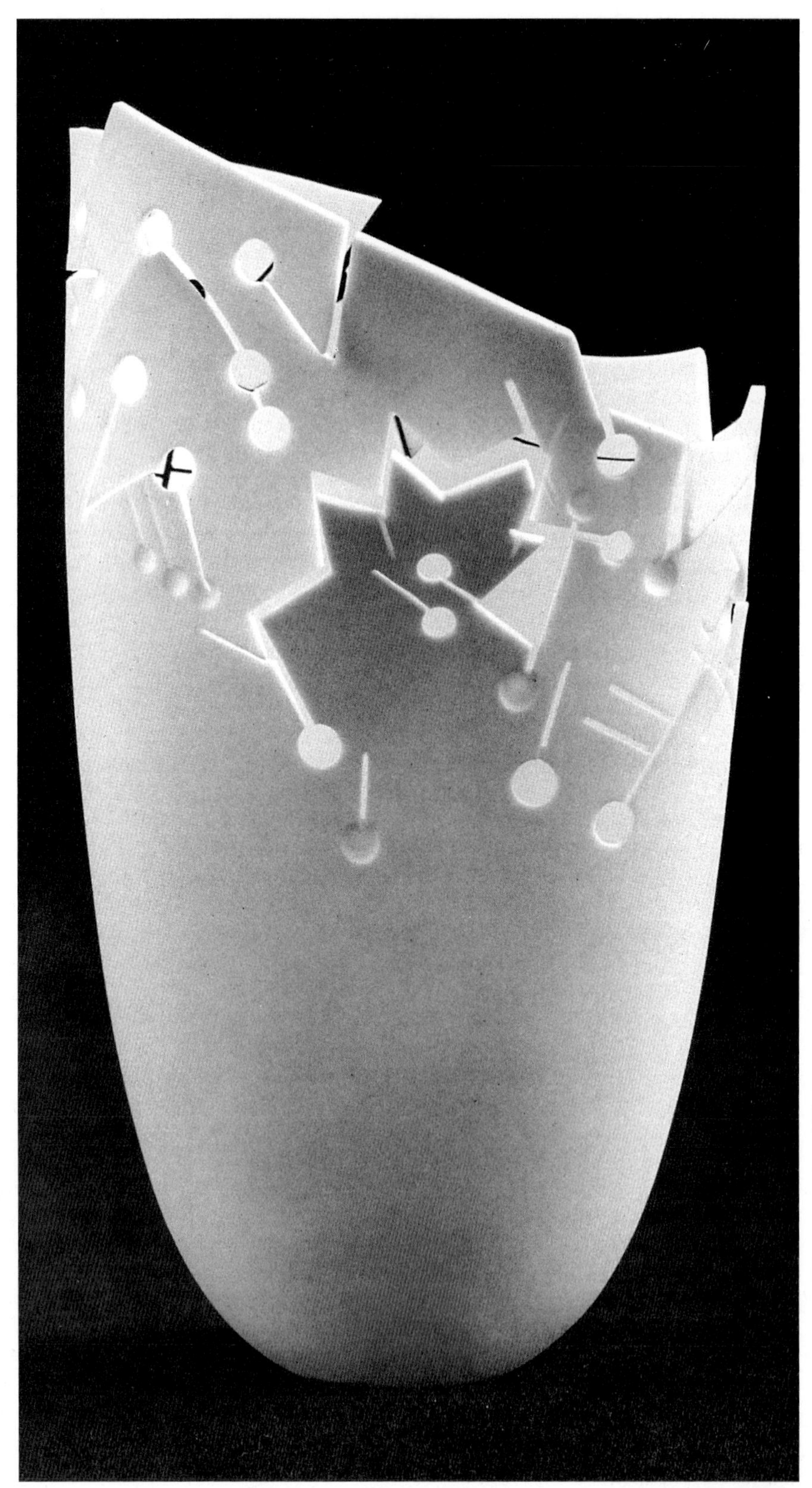

Asymmetric form by Angela Verdon (UK)
This piece is in fact only 4½ inches (11 cms) high. It has been decorated with incisions and piercings which exploit the delicacy of the bone china and enhance its translucency.

How to use this book

The Ceramics Manual has been designed to be used primarily by novice potters who are unfamiliar with this medium in order to help them explore both the techniques and the aesthetics of clay. The large number of illustrations and step-by-step descriptions of ceramic skills should enable even the complete beginner to explore three dimensional forms and understand how to tackle an enormous variety of work in clay.

Moreover, the book includes many examples of ceramic ware, both from historic sources and from work created by potters around the world today. In this way the novice potter is encouraged to both learn the skills involved and to appreciate the possibilities of each technique.

To those who teach pottery the Ceramics Manual should be a very useful tool; it provides a basic explanatory guide which shows through text and clear illustrations all that the beginner will need to learn—as well as discussing the work of current potters to demonstrate the techniques in action. It will act as an excellent supplement to the practical teaching in the studio or classroom.

The manual begins with an explanation of the medium and investigates the properties and preparation of clay bodies. This is followed by a succession of chapters which describe the techniques—including Handbuilding, Throwing, Glazing and Firing. Where appropriate, alternative methods are also explored to show how many different results can be achieved and just how innovative the potter can be. The chapter on Decoration is especially informative, with many illustrations to help stimulate creativity and imaginative ideas.

Inevitably many of the techniques are linked and the majority of skills can only be founded on those already acquired. None the less, it is hoped that each chapter can be read and digested as an individual course of instruction on that particular skill or area of ceramics.

The concluding chapter on History explores the development of ceramics through the centuries, country by country, and in particular examines its relevance to the potter today as a basis for the techniques that might otherwise be taken for granted and as an inspiration for style, form and further development.

The symbol **H** indicates that historical matter is relevant and cross-referenced. Similarly, the symbol **C** indicates a useful contemporary example on the page number stated.

The Useful Facts section (see page 312) is very comprehensive — almost a chapter in itself. It is full of interesting information, including glaze recipes, conversion tables, fault-finding guidelines, chemical knowhow, tools and equipment, health and safety, measuring tips and so on.

Throughout the text hazard symbols have been used to indicate clearly when a procedure is potentially dangerous. Readers are advised to watch out for these symbols and to proceed with due caution when embarking on any such activity and when handling toxic materials.

Every potter in time develops his or her own working methods. It is hoped that the techniques described and illustrated in the Ceramics Manual will provide a firm foundation on which to build a personal approach to this most stimulating art form—one that can provide practical functional ware, or which can be enjoyed as a relaxing tactile exploration of clay and which can become the vehicle for a wide range of creativity.

Moreover, the study of so many other potters' work, both in the History chapter and in the examples that have been used to elucidate the text, should enable us all to better appreciate the aesthetic contribution that ceramics have made, and continue to make, to the world in which we live.

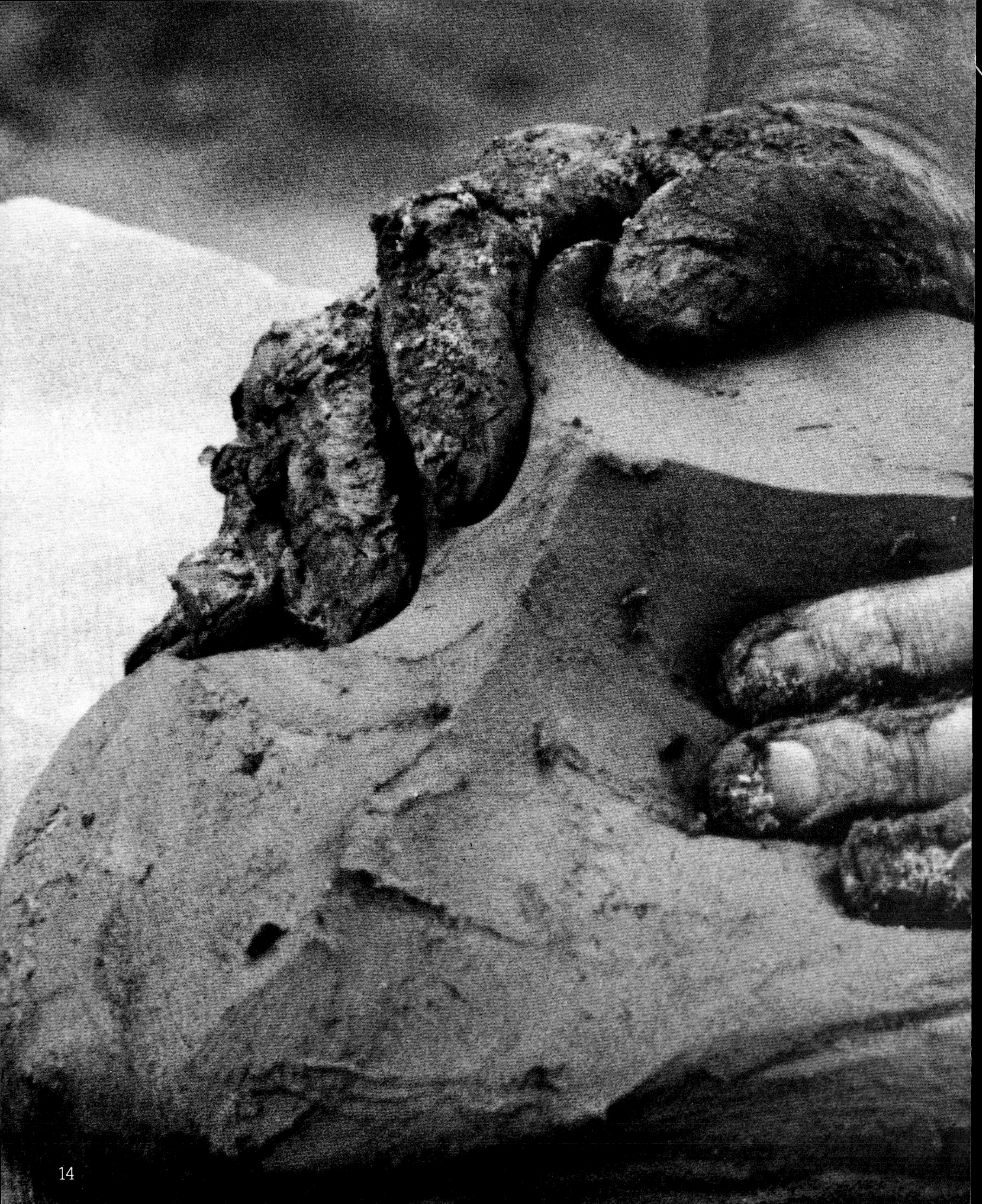

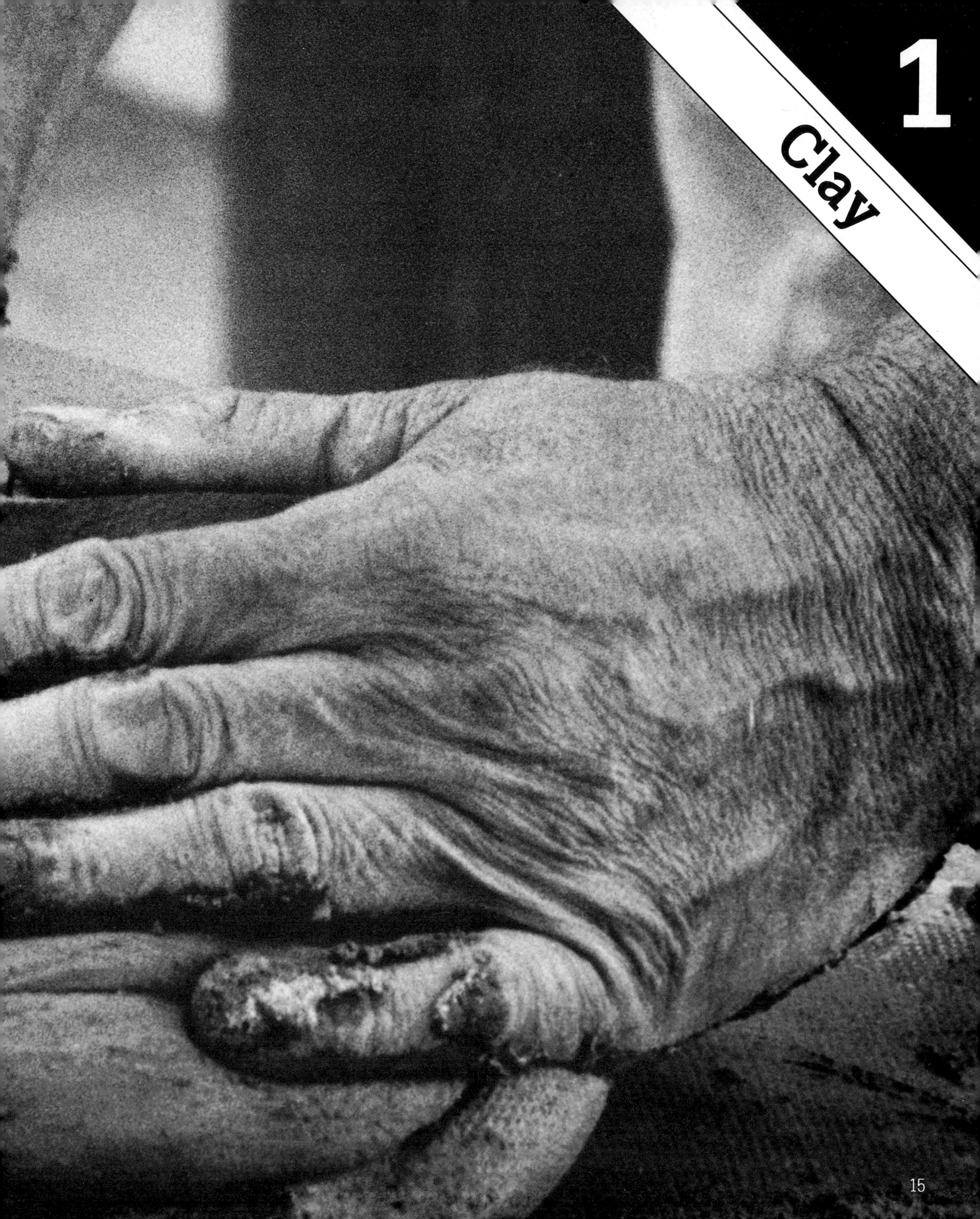

1 Clay

Clay

What is clay?

The main elements in igneous rock are silica and alumina which are the two essential elements in clay. The erosion and decomposition of the earth's surface is a continual process. Igneous rock is gradually eroded over hundreds of years to form fine grains. Cracked and broken by ice-cold water which seeps into the rock and turns to ice, constantly eroded by rainfall, washed away and ground down to tiny particles by the action of running water in streams and rivers; eventually the seemingly indestructible rocks become the minute 'seeds' of clay.

Clay is one of the cheapest and most abundant of all raw materials found throughout the world. The differences it exhibits in texture, quality and color depend on how it was deposited and what other minerals it has collected during its formation.

Clay is the only natural material which can be shaped directly by hand in its raw state. It will also, when wet and plastic, keep almost any shape into which it is formed and retain that shape even when the water content evaporates.

When dry, it is fairly solid and hard. However, when it is broken into small pieces and soaked in water again it becomes soft and sticky. Water lubricates the minute particles of its composition. The particles are thin and flat in section and cling together when wet, rather like a pack of playing cards. These flat particles can slide against each other and account for the unique property of clay; that is, its capacity for being reshaped. It can be pinched, stretched and rolled out. Moreover, unlike any other substance, its malleability allows it to be pulled and shaped on a potter's wheel. After a form has been made, the clay will continue to hold that shape although as the water evaporates away the particles of clay are drawn closer together; this means that the pot will shrink in size as it dries. Some clays shrink more than others, being finer-grained and more plastic.

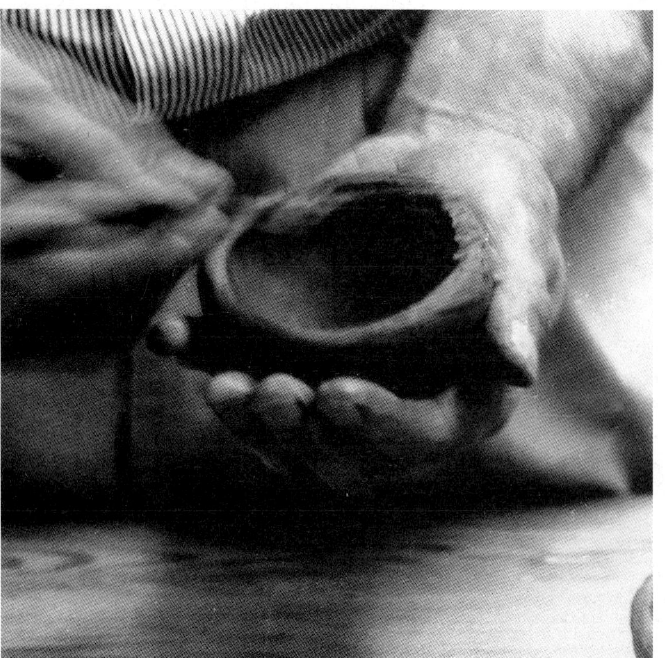

Composition and weathering

The plasticity of clay—that is, its ability to stretch and form curves without cracking or collapsing—is dependent upon its particle size and the decomposed organic matter present. Without this organic matter the clay is crumbly and loses its shape.

A few clays dug straight from the ground may be suitable for potting but others have to be mixed, usually because they are either non-plastic or too plastic.

Leaving raw clay outside to weather (or mixing old clay with new) will help to increase the plasticity. Chinese potters were said to use the clay dug and left by their fathers and grandfathers while the clay they dug themselves was left for their offspring.

A few days stored in an airtight bin will often improve a 'short' clay (one which breaks easily when stretched or bent into a curve). Small amounts of vinegar can also be added to improve plasticity. Some clays are, for economic reasons, used straight from the ground. Often, building materials such as bricks and drainpipes are made by a non-stop process of digging the clay, transferring it up a moving belt to a machine which extrudes the required shape, and then transporting these these pieces straight to the kiln for drying and firing. This is a continual process which reduces the cost of manufacture.

Exploring the medium

When a student first comes into contact with clay it may seem superficially to be a simple substance. Probably under instruction, the student will be given a wedge of ready-prepared clay (usually gray or brown) and instructed in the first stages of making something in this unfamiliar medium.

As the student's knowledge and natural curiosity

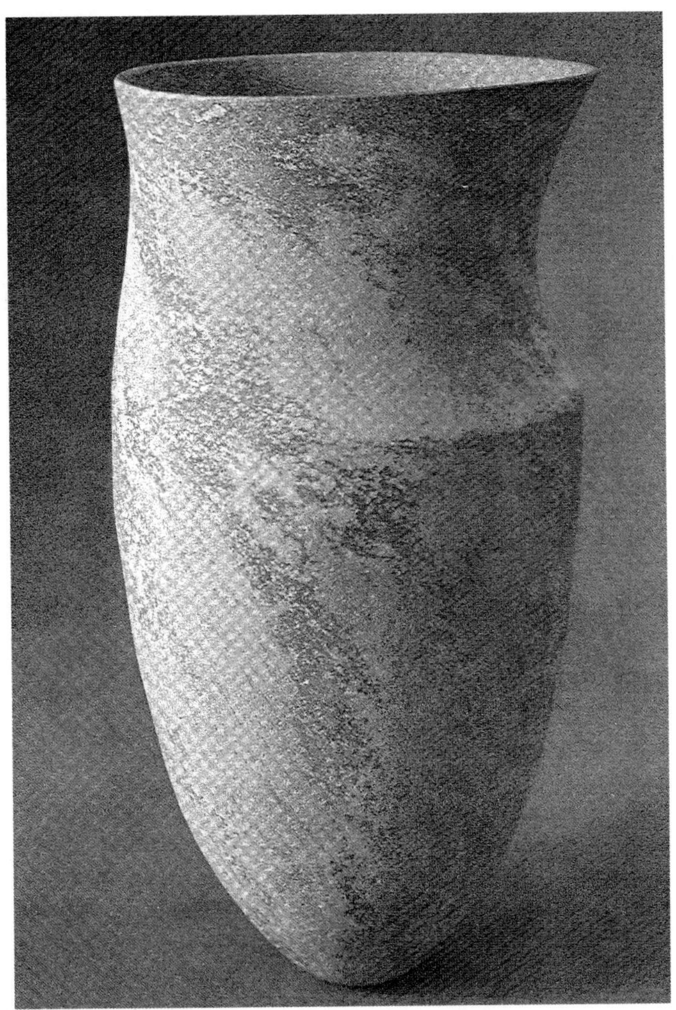

Handbuilt pot by Jennifer Lee (UK)

increases, so a greater variety of preparation methods will be developed. Moreover, a fuller consideration of the properties required will play an increasingly important part in the selection and preparation of the raw material.

To enable the potter to control the various stages of forming, drying, decorating and firing, the clay has first to be chosen and mixed with care.

It will be important to be fully conversant with the properties of the different clays. Is this one smooth? Will it need sand or grog mixed with it to open the texture? How much will it shrink when it dries? It may be useful to mix together different clays or add other materials. This mixture will then be known as a 'clay body'.

In other words the potter must know which materials will obtain the best results and most suit the demands of the particular task ahead. As experience grows, so the wish to learn more about the medium increase too. Understanding clay is fundamental to all aspects of the potter's craft.

How the potter chooses to use clay will depend largely on his or her own approach and instinctive feelings about this art form. Sometimes too, decisions may be influenced by the nature of the instruction received. Clay can be used to create a wide variety of forms and to reflect current art styles. Or it may be used in a far more primitive way as a simple material to form a handbuilt pot—a pot that will be finished, polished and fired as Neolithic potters did. There are so many alternatives. The clay can be formed and molded into a finely made piece of jewelry, decorated with precious metals, or used as a piece of industrial porcelain, so hard and heat resistant that it will survive the temperatures and pressures undergone by a spacecraft's outer surface. The possibilities are almost without limitation.

The primitive potters must have recognized the properties of clay when they dug it from a river bed—like some boys of six and seven I was recently watching at a school near my home. These youngsters had discovered a bank of clay in some man-made excavation on the school playing field.

They proceeded to excavate and create beaten clay ribbons of roadways on the bank for their toy cars. The labor was well divided; three boys dug the clay out with their fingers while the rest kneaded it and beat it out with stones until it formed a hard surface for the roadway. Thus it could have been that the people of a primitive settlement experimented, digging clay and shaping it and then eventually making cooking pots and water jars with this versatile material.

The value to the maker is immeasurable. Any potters who look back on the first pots they made will still remember with pleasure the thrill of creating those first pieces. Now looked at with a more critical eye, they may seem rather clumsy and misshapen but can still be regarded with affection by the maker. The pleasure of their making will be recalled and enjoyed once more.

Always try to be positive in your approach to the medium. Do not allow yourself to feel confined by the materials and the techniques. Try to use the material to its limits and practise as much as you can. Using clay is just one aspect of self expression and exploration, extending the levels of experience; the pleasure and satisfaction of working with clay are without equal for many potters.

Types of clay and their origins

Primary clay

Most decomposed and eroded material will be washed away from the site of the parent rock. Those clays that do remain near to their source are called primary clays. They are coarse grained as they have not been broken or ground down by movement through water and are usually non-plastic. They are white or light colored, being unstained by any other minerals. The most well-known and widely-used primary clay is kaolin or china clay.

Kaolin

China clay (or kaolin) is found in rock formations and not in easily dug beds. It is mined by washing it out of the ground with high-pressure hosepipes and is then left to settle in large settling tanks. China clay is used in many different industrial processes. It is especially important for the potter as it is one of the main constituents of high-firing clay bodies and glazes.

Kaolin is an uncommon primary clay. Uncontaminated by minerals and other clays, it is especially useful because of its whiteness (it is often added to a clay body to contribute whiteness) and its ability to withstand very high temperatures. It is the basis of most vitrified porcelain.

The discovery of kaolin in China (about 200 BC) made possible the first high-fired stoneware and eventually, with the improvements of high-firing kilns, led to the production of a transluscent porcelain in about 600 AD.

The discovery of this vital link in the making of porcelain was not made in Europe until the 18th century. Kaolin has a melting point of 3250°F (1800°C) and is seldom used alone but is mixed with other clays to improve its plasticity and lower its maturing temperature in the kiln.

Secondary clays

Secondary (or sedimentary) clays are those that have been transported, usually by water but sometimes by wind, to a bed far from the position of the parent rocks from which they were eroded.

The action of being transported by water has a vital abrasive effect on the eroded rock particles; they are ground against each other and reduced in size. Thus secondary clays have a fine particle size which makes them more plastic. On their journey they also collect various other materials and organic matter which color them. The minerals and carboniferous content affect their character and make the clay impure so far as the potter is concerned. The most common mineral is iron oxide which is why most common clays are brown, yellow or red.

Ball clay

Ball clay is one of the finest of secondary clays,and is so called because of the practice, when it was first mined, of transporting it in the form of large balls on the backs of pack horses. In this manner it travelled from the clay pits to the potteries.

This lighter colored clay has a smaller percentage of iron and is a smooth-textured very plastic material because of its fine particle size. It appears blue-gray to black because of the carboniferous material it has collected while being waterborne. Like kaolin, it will stand very high temperatures but it is too plastic to use alone and shrinks excessively. Often it is mixed with kaolin; each component then counteracts the other's throwing faults without losing too much whiteness. The clay can be used as a white slip for coating compatible dark bodies.

Fireclay

This is a secondary clay, so-called because it is often found in the region of coal seams. It is highly refractory; that is, it is resistant to high temperatures—because of the absence of iron oxide in the clay. Fireclays can be very plastic or completely non-plastic and coarse in texture.

Their main use is for the manufacture of firebricks and furnaces and lining crucibles, kilns, kiln shelves and supports.

They make a good addition to a stoneware clay if a little 'tooth' is wanted (that is, a grittiness and slight resistance to the fingers). They also provide a rough texture to the finished article.

Common clay

Common clay can be found almost anywhere in the world below the first few feet of top soil. It is a secondary clay that has collected minerals and organic matter before being compacted as a strata beneath a river bed or lake. Lighter shades contain fewer collective minerals and fire a buff color. Darker shades range from yellow to black and usually fire a terracotta red because of the iron content. Most have a maximum firing temperature of 1630-1830 °F (900-1000° C) after which they start to melt. Because it is so readily available and relatively cheap, common clay is used mainly for building purposes.

Primary and secondary clays

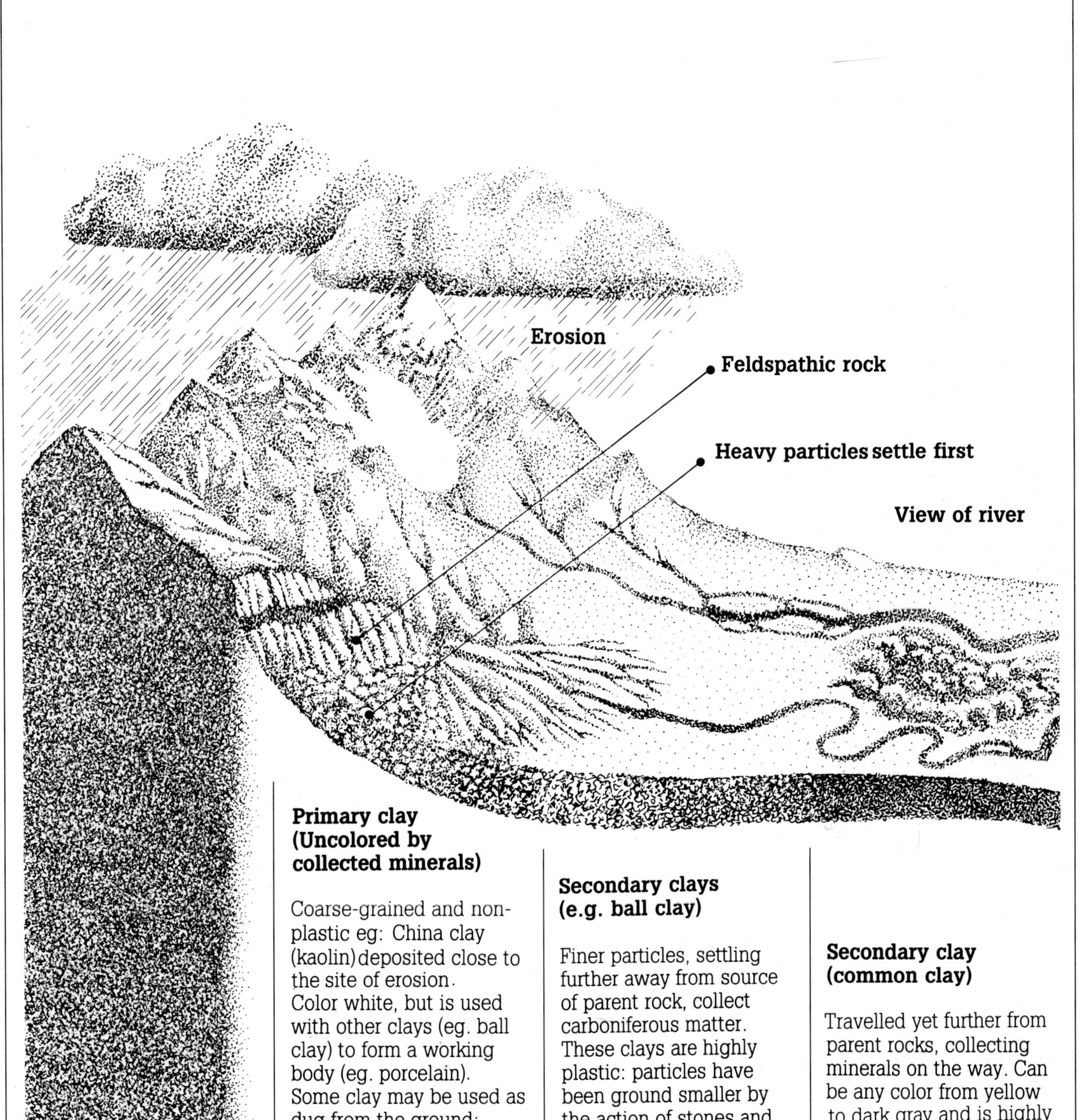

Primary clay (Uncolored by collected minerals)

Coarse-grained and non-plastic eg: China clay (kaolin) deposited close to the site of erosion. Color white, but is used with other clays (eg. ball clay) to form a working body (eg. porcelain). Some clay may be used as dug from the ground; most are mixed with other clays to form a better working body.

Secondary clays (e.g. ball clay)

Finer particles, settling further away from source of parent rock, collect carboniferous matter. These clays are highly plastic: particles have been ground smaller by the action of stones and water movement. They usually turn off-white when fired.

Secondary clay (common clay)

Travelled yet further from parent rocks, collecting minerals on the way. Can be any color from yellow to dark gray and is highly plastic. Usually turns orange, red or buff when fired.

Digging for clay

A potter can never know too much about the raw materials that he or she uses. It is therefore an interesting and valuable experiment to discover a local source of clay and learn how to unearth and prepare it. Attempting this yourself will not necessarily save a good deal of money and it is certainly not going to save any time—clay preparation is a long process—but it may well prove a particularly stimulating learning process.

Unfortunately the sticky clay lurking at the bottom of your garden is unlikely to be suitable for ceramic purposes. Surface clay is generally far too contaminated. Try instead to find some samples-a few pounds (about 2kg) of each will do—from clay beds deeper down. These may be exposed because of new building excavation, a road cut, a landslide, or where river action has scoured a deep channel and left a vein of clay exposed in its bank. You may be helped in your search by the local knowledge of geologists, miners, masons, brickworkers and so on.

Any impurities such as gravel, roots, stones and dead leaves will need to be removed so first dry the clay thoroughly, then break it up into small lumps and leave it to soak in a bucket of water for several days. (To achieve the best results, newly-dug clay should actually be left for another year or so to mature and remove the stratification but one assumes that this might try the beginner's patience and that the aim here is to experiment, rather than to achieve the highest quality clay!)

Skim off any floating matter. Stir again and then allow coarse heavy impurities to sink. Remove as much water as you can from the top by baling or syphoning. The clay will take on the consistency of thick cream in about half an hour. It should then be poured through a 20-mesh sieve to remove the coarsest impurities. A second pouring through an 80-mesh sieve refines it yet further. The slip should be dried in drying bats or on a plaster slab until it reaches a malleable plastic state and can

Extracting common clay at Littlethorpe in Yorkshire. Clay is sometimes hand dug and then loaded into trucks on a small-gauge railway.

be wedged. It can then be subjected to the various tests shown on pages 28-29. Store successful batches of clay in the usual way (for instance, in plastic). It will be particularly rewarding to be able to create ceramic pieces from 'your' clay. However, do not despair if for some reason a clay sample proves to be unsuitable as a formative medium; it may well be improved by adding flint, china clay, sand or grog. Or it may itself serve best as an addition to other bodies. Moreover, local clays often make excellent slip for slipware or a useful source of pigment which can be especially good on stoneware.

Commercial mining

Each year many millions of tons of clay are extracted commercially. The intention usually is to supply industry with the raw material for building bricks, insulating liners and so on. The clay may be dug out, mined, blasted, or washed from the surface with high-pressure hoses.

Different sources produce clays with distinctive qualities which serve varying purposes in the commercial world. The production of clay for potters generally plays a very small role in this.

Preparing the clay

Mixing a clay body

At some time most potters like to experiment with mixing clay bodies to produce just what is required for their own particular needs. The simplest way is to use dry ingredients and clay in powder form to which water must be added. Thus, even if you are unable to go back to the very beginning and excavate your own clay, mixing it yourself in this way can be very satisfying.

If you have one, a pugmill can be used. This will mix and perhaps help to aerate the clay body up to a point. However, the final work should be done by hand.

Clay does not deteriorate with age; it improves. If the clay is left outside to weather or kept wrapped in an air-tight bin this will increase its plasticity. As it sours, the organic matter in the clay decomposes to form a gel around its constituent particles. This process increases its plasticity.

An older clay, mixed with a new batch and left to mature, will improve the new clay in only a few days. For those who have neither the time nor the inclination to dig their own clay, there are many fine clay bodies available ready-packed. These can be bought and used immediately but the larger the quantity you can afford, the cheaper the over all price. Collecting it yourself from the suppliers in a van or trailer will reduce the costs still further.

Some potters buy and prepare large amounts of clay in advance, both to ensure the constant color and texture and also to allow the clay to mature that much longer.

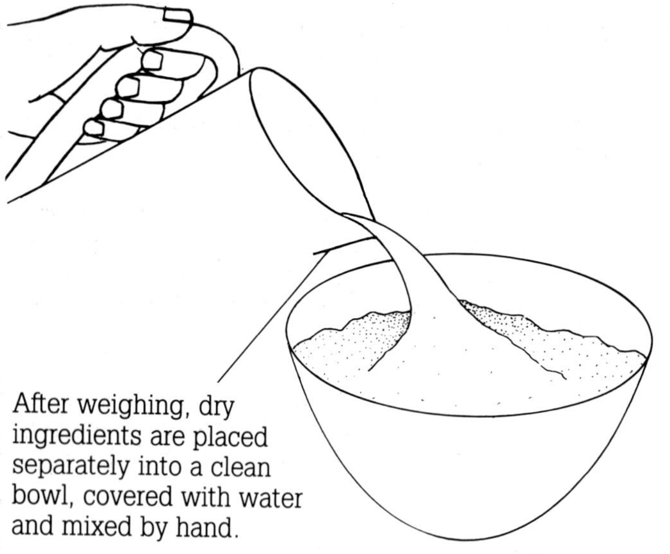

After weighing, dry ingredients are placed separately into a clean bowl, covered with water and mixed by hand.

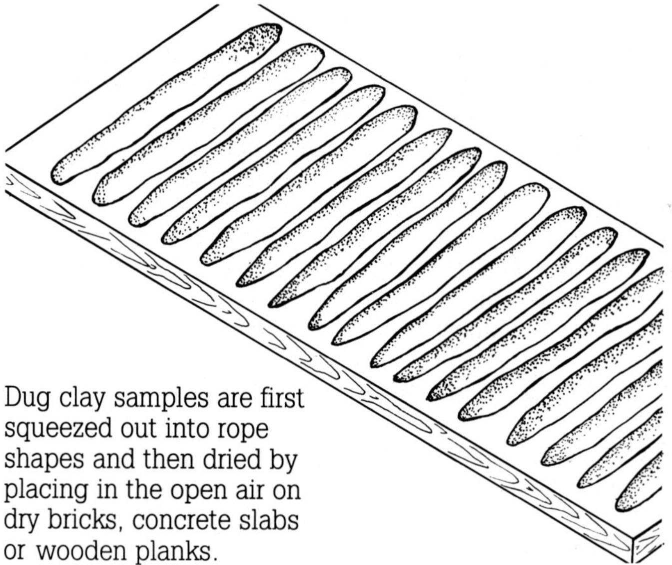

Dug clay samples are first squeezed out into rope shapes and then dried by placing in the open air on dry bricks, concrete slabs or wooden planks.

Soaking the clay

If you have dug your own clay, or if the clay you are planning to use is hard and rocklike, break it into small pieces ready for soaking. This may sound rather ridiculous—to dig a wet clay, dry it and then wet it again—but turning damp clay in to a liquid slip can take weeks. Dry clay, on the other hand, will accept water quickly and will be reduced to a workable slip in a very short time.

Mix the water and clay with a stick or by hand and then brush it through a coarse sieve. Use a 20-mesh at first and then repeat the process with an 80-mesh sieve.

Allow clay to settle and take off the clear water from the top.

Pour the thick clay slip into a plaster mold or on to a plaster slab (or a layer of clean dry bricks loose laid in a box form) and allow the water to soak away. Turn over the clay regularly to prevent uneven drying, each time placing it in small piles to let the air circulate around it.

If a fine-meshed nylon cloth is draped over the absorbent slab before the slip is poured over, this will help to keep it clean and the plastic clay can be collected economically.

Mixing the clay

1 Small pieces of broken clay are added to water and mixed with a stick.

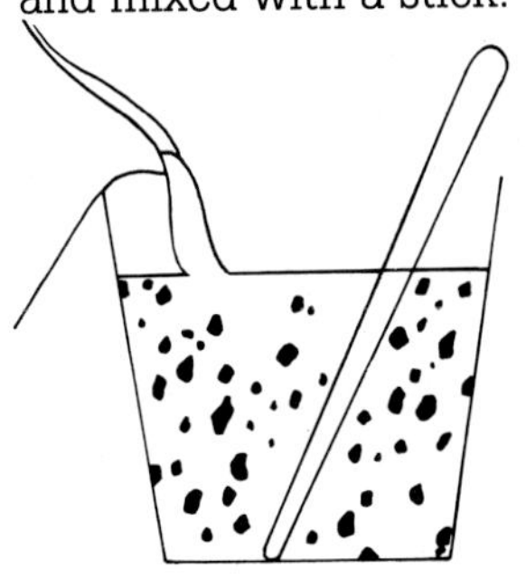

2 The mixture is then sieved through a 20-mesh sieve to remove any stones and impurities.

3 The mixture is sieved again through a finer sieve (80-mesh) to leave a fine slip.

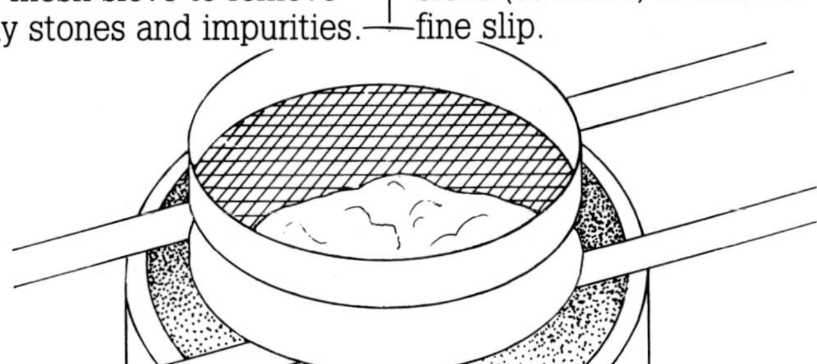

4 Finally the mixture is allowed to settle and the clear water is syphoned off from the top.

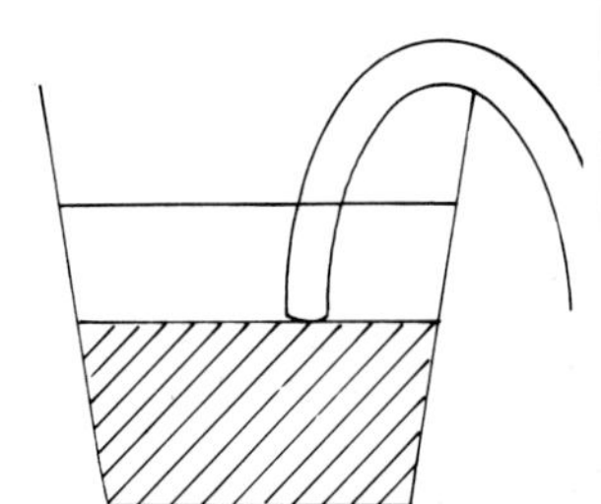

Grog

Grog is clay that has been fired once and then crushed to produce various grades. It is mixed with plastic clay to give texture, to improve its strength and to give 'tooth' to the clay. It will help the pot to dry evenly because the air can filter through it and shrinkage will be reduced. Grog can range from a fine powder to a coarse grit, giving anything from a sand-like texture to a coarse granular appearance. If made from a red clay it will add texture and surface color to a stoneware pot, burning through the glaze in dark speckles to produce an attractive finish.

◇ Adding grog at the slop stage

Adding grog to clay may be done at the 'slop' state (that is when the clay is in a mixture like mud) or at the wedging stage and the grog is usually added in the proportion of 10 to 15 per cent of the whole. The grog is graded according to the size of the particles which will pass through a particular size sieve—from the very fine dust which passes through a 100-mesh sieve, down to a coarse grit which will pass through a 10-mesh sieve and which is normally used for larger pieces of work. The grog and wet clay are mixed in a large bin. Be careful not to inhale clay dust and wear a mask if the grog is dry. ◇

Make as much as you can handle at once as the process is a long one and it is best to avoid having to repeat it constantly merely to produce small batches. Use your hands or a paddle of wood; add the grog slowly and stir really well.

If you want additional color, prepare this separately as a thin slop mixture and add to the clay and grog, mixing thoroughly. This ensures the color is distributed evenly through the clay.

1 Adding grog to a bucket of clay and water before drying.

2 Grog on the bench is being wedged into the clay during preparation.

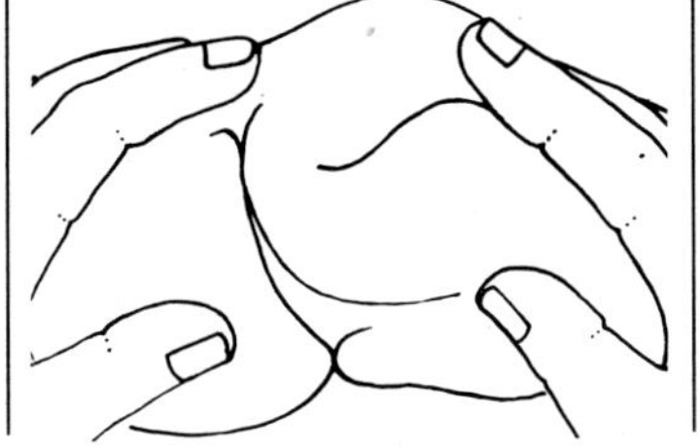

Adding the grog to soft clay

Grog may be added direct to a soft plastic clay body. If only small quantities are required, the grog can be sprinkled on to the wedging bench and rolled into the clay during the wedging. If the clay is soft, the dry grog will help soak up the moisture and stiffen the clay.

When all the grog is mixed in, the clay must then be left, to allow the grog particles to become saturated. Wrap the clay up or keep it in an airtight bin for two or three days. Take out the clay and re-wedge immediately prior to its use.

If you are mixing clay and grog it is best to first dry the clay and then break it up into small lumps, as described under 'Digging for clay' on the previous page.

Working the clay

Drying the clay mix

The mixture should then be set out to dry, somewhere that is neither too warm nor too windy; heat and wind will indeed dry the clay but will do so only on the outside, forming a hard crust with soft clay inside.

Construct a shallow tray of clean dry bricks, loose-laid in a convenient spot which is not too exposed to the elements, and then pour or scoop in the clay, to a depth of about two bricks. Repeat this if there is more clay than can be accommodated in one pan. Lay bricks over the top of the clay and leave it to the processes of evaporation.

The grog will soak up moisture and the clay will mature and increase its plasticity the longer the mixture is left. Keep an eye on it from day to day. Make sure it does not dry too hard. It should be in a fit state to mix and wedge. This drying may take some days to complete. The first batch is usually a slow one, and can seem especially so if you are impatient to start; subsequent batches can be left to mature while you are using the earlier ones.

1 Line the plaster mold with open-weave nylon mesh and pour in the mixture of clay and water. The water will be absorbed by the plaster through the nylon. When it is dry the clay can be lifted out cleanly on the nylon sheet.

2 For larger quantities of clay, shovel the mixture into a dry brick box constructed out of doors under a shelter of some form. The brick box can be made by loose-laying the bricks on a level surface. Surround the base with a two-brick high wall to contain the liquid. The bricks will absorb the moisture and the clay can be lifted out when it has dried.

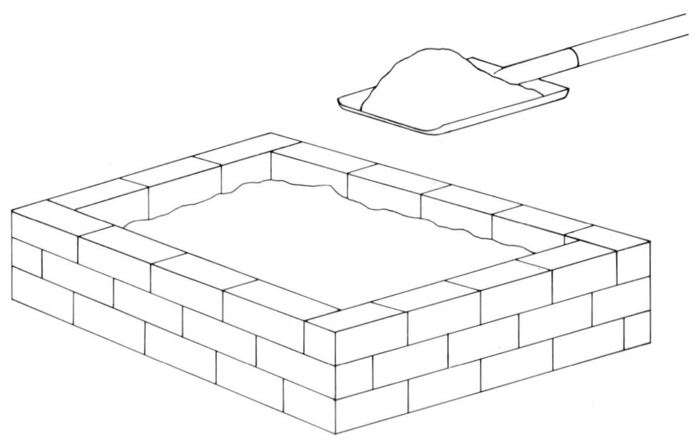

Working the clay

Although this is a fairly strenuous task, it is a satisfying one which will give you a feel for the clay you are going to use and will also help the beginner to understand some of the strange and varied qualities of this excellent material.

When the drying clay is in a soft plastic condition, remove it from the drying trays and mix and wedge it on a strong slab. Cut through frequently with a stout wire, slam the two halves together, and then mix and wedge again.

Make sure that the bench you use is not too high. A low bench, one about 2 feet (60 cm) high, will mean that it is possible to use the weight of the body when wedging rather than working with just hands and arms which would soon tire you out. Work with the largest piece of clay you can manage, until it is of an even texture and color when cut with a wire. Wrap the clay in a plastic bag or place it in an airtight bin until it is required.

Clay that is already mixed with grog of different grades can be bought ready-mixed from pottery suppliers, if you haven't the space, time or inclination to prepare your own.

Dog-head (or ball-head) wedging

1 Use both hands. Exert equal pressure downwards, using the heel of the hands.

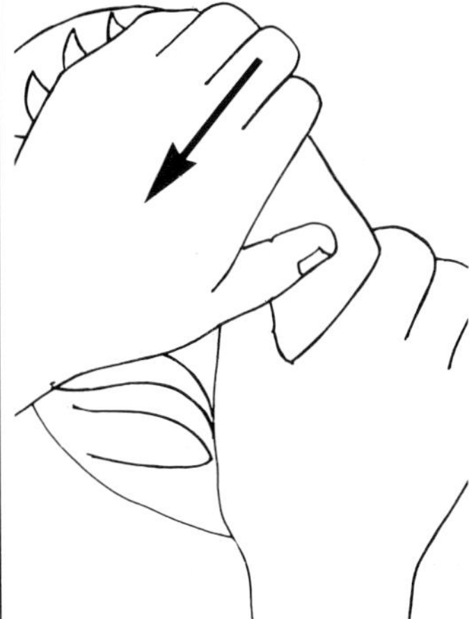

2 Clay is pulled from the back of the piece forwards and pressed back again into the clay with the heels of the hands; this is repeated. As you work it, the shape of the piece of clay will resemble the head of a dog with nose and eyes.

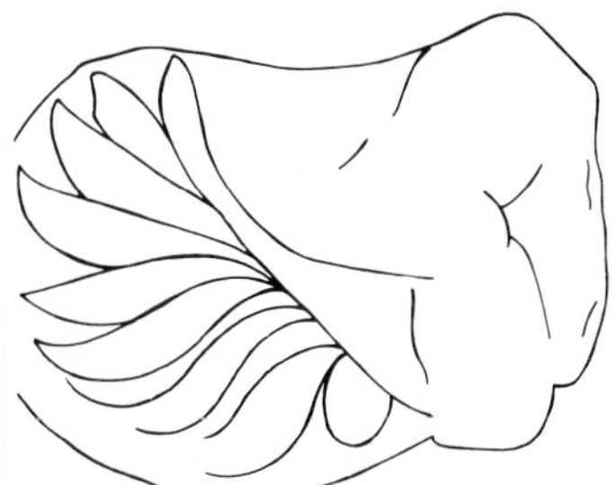

Spiral wedging

The clay is turned and, as the heel of the hand is pushed into the center of the clay, it produces a spiral cone-like shape.

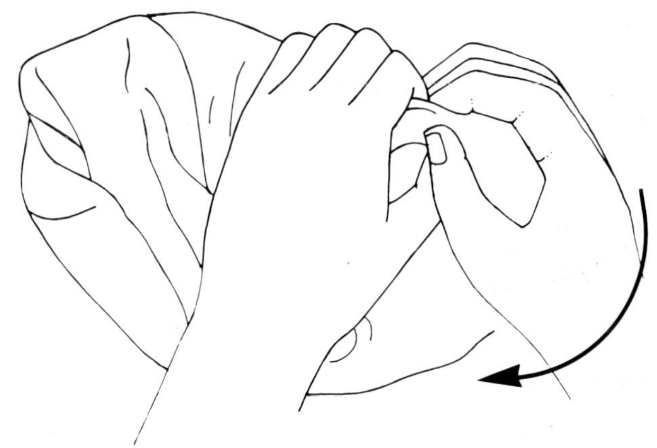

Lifting, beating and wiring

1 Clay is lifted high and beaten on a solid bench to drive out the air bubbles. This is repeated until all the bubbles have been removed.

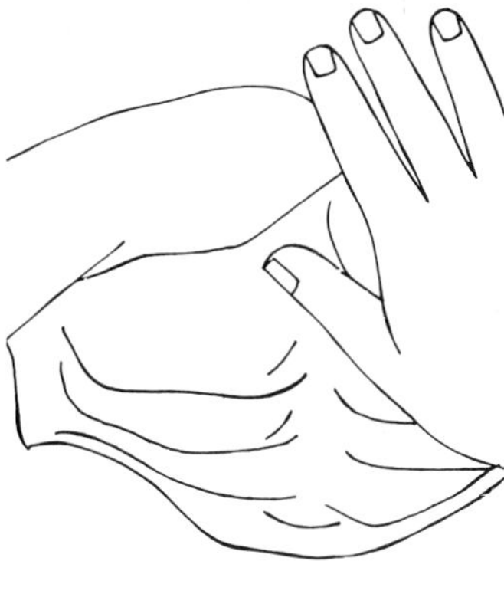

2 On the final throw down on to the bench, make sure one end of the clay is left free from the bench.

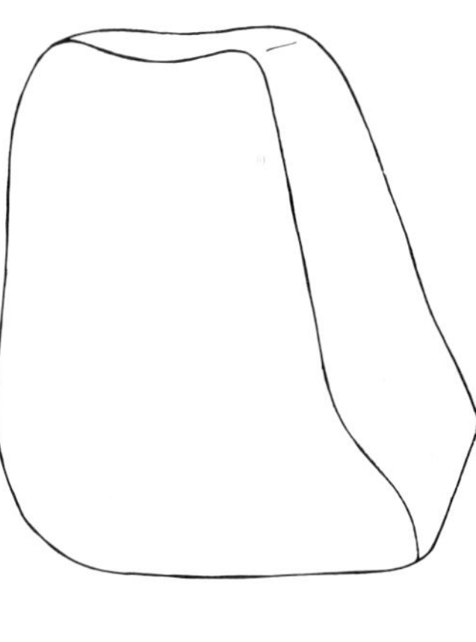

3 The clay can be cut through with a twisted wire from below, slicing upwards through the clay.

4 The two halves are turned and thrown together again, this process being repeated over and over until the clay appears smooth and even in texture.

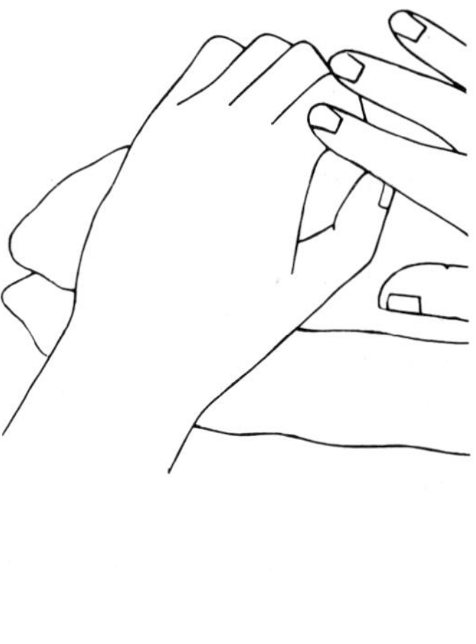

Testing the clay

Shrinkage

The shrinkage of a clay body occurs at three stages: first when the clay changes from wet to dry; the highest degree of shrinkage takes place in the first (bisque) firing; finally there is further shrinkage in the second (glaze) firing at a higher temperature.

Roll out a piece of clay and mark a fixed length (say, 6 inches or 15 cm), allow it to dry slowly until bone hard and then measure it again. Fire the test in a bisque firing and measure again. Finally fire the clay in a glaze kiln to test the overall shrinkage. Finer, more plastic clays with a small particle size will shrink the most.

If the shrinkage is excessive and deforms the piece, making some additions of grog to the clay or mixing in a little coarser clay will help to rectify the problem.

1 Roll out a strip of clay and mark a fixed length, for example 6 inches (15cm).

2 Allow the clay to dry naturally and measure the fixed mark again.

3 The clay shows the most marked shrinkage after a biscuit firing.

(You will notice that the main shrinkage has taken place at stage 3 during the firing of the clay.

Clay goes through several stages as it dries: plastic (see below) — damp/soft leatherhard (shape can still be modified without cracking) — leatherhard (almost rigid, can be cut, many decorative processes applied at this stage) — bone dry. The potter will soon learn to recognize the changing state of the clay.

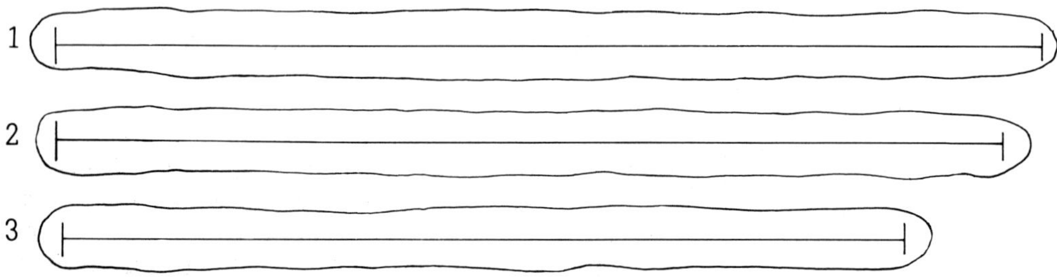

Test for plasticity

Roll a coil of clay and bend it into a circle. If the coils crack it is non-plastic and will need additions of small amounts of another clay that is older or more plastic (such as bentonite) to improve the plasticity Sometimes a few weeks' souring and maturing will suffice to improve its workability.

Experiment with the clay. Work it well; try coiling, pinching or throwing and judge how it stands up to being worked.

1 Roll out a rough coil from the clay sample.

2 Bend the coil into a fairly tight loop. If the clay cracks along the outer edge it will need additions of older clay or bentonite to improve it, or the clay can be left out in the elements to sour and mature until it improves.

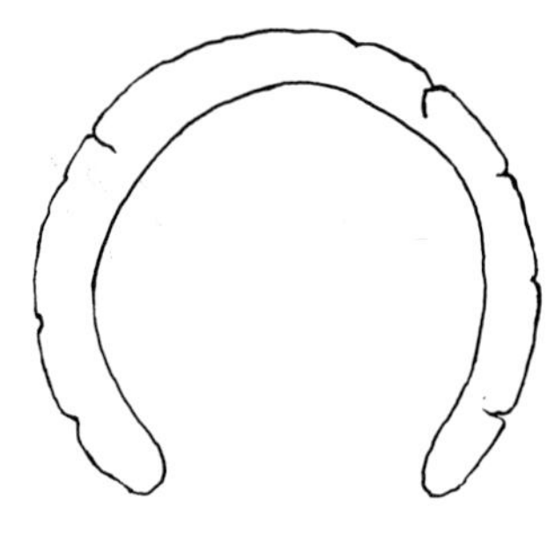

Porosity

Different clays absorb glaze at different rates—according to how open or fine a texture they have and the temperature at which they are biscuit fired. Most potters will fire to a specific temperature for a first firing, possibly about 1000°C (1830°F). If all samples of clay are fired to this temperature, then the absorption of glaze will vary only according to the type of clay chosen and this enables one to make a better judgment.

Test pieces of all the types of clay you may wish to use to the chosen temperature for a standard period of time (as appropriate to the selected glaze) and evaluate the results. You may find the glaze does not fit the body very well (it may 'crawl' or craze for instance). If this happens the bisque firing may need to be adjusted higher or lower to alter the porosity of the clay accordingly or the glaze composition altered (see Fault Finding on page 320).

Firing temperature

Some clays when fired to temperatures over 1830°F (1000°C) will melt; others will stand up to 2550°F (1400°C) or over before melting. (If a dug sample is dark red or yellow then this coloring indicates that it has a high concentration of iron oxide which acts as flux in the firing.)

Make several flat lengths of clay and dry them completely. Take one or two and place them on firing bars so they are suspended like a bridge, or place them over the rim of a biscuit-fired bowl. Place the whole contraption on a piece of old kiln shelf in case the clay test melts completely and runs. Fire to 1830°F (1000°C).

Repeat the process, firing to 2280°F (1250°C). The lighter colored the test clay, the more likely it is to fire to a high temperature. If the clay does melt at 2280°F (1250°C) it could become the base of a dark glaze.

Fire the lower-temperature test with an earthenware glaze (or glazes) to test both for color and the resulting fit between clay and glaze.

These tests involving glaze can vary enormously. Between the lowest bisque temperature and the highest glaze temperatures there can be a wide range of results—not to mention reversing the sequence. The possibilities are almost limitless!

1 Roll out a piece of clay.

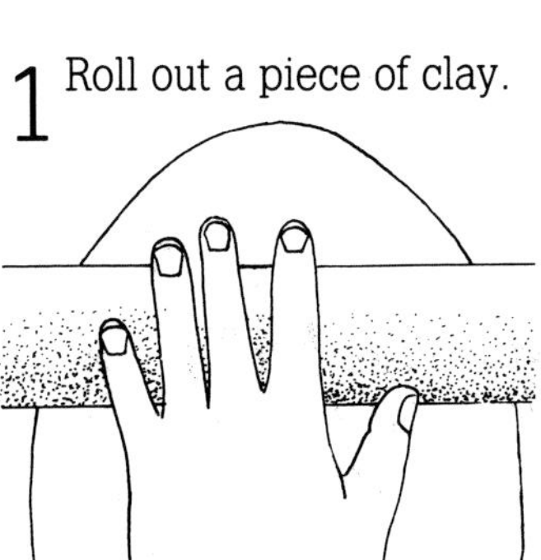

2 Cut into equal strips. Use a knife and a ruler.

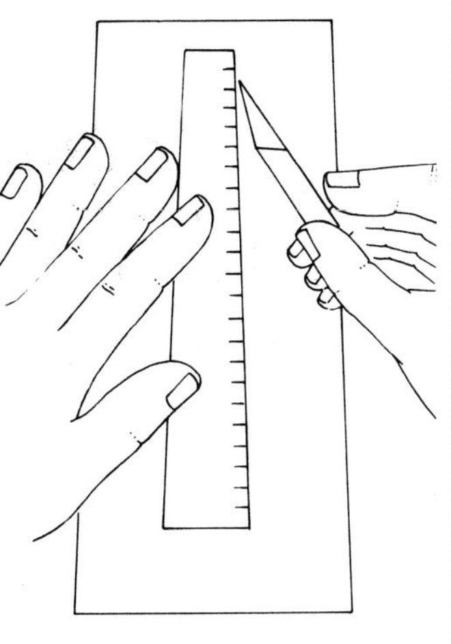

3 After drying, place the strips across two fire bars supported on a piece of kiln shelf. Fire to a biscuit temperature.

4 Fire again to stoneware temperature and check to see how the clay has stood up to the high temperature.

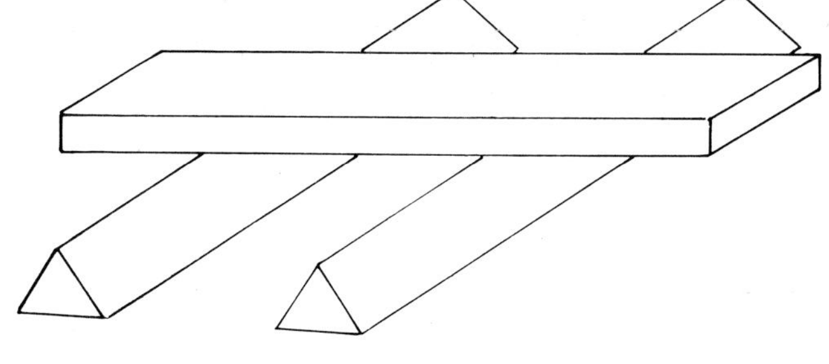

Clay bodies

Clay bodies are mixed from various clays with different properties to make a good working body.

Stoneware body

Stoneware bodies are made up from clays which normally fire to 2280°F (1250°-1300°C). They can be almost white through to a dark brown, depending on the mixture and the personal choice of the potter. They often have a percentage of fireclay (or grog or sand) to open the texture of the clay and to reduce shrinkage and warping. Stoneware clays may be dug straight from the ground as a fine plastic form of fireclay.

They become vitrified when fired to around 2370°F (1300°C) and are capable of taking and combining with the simplest form of feldspathic glaze into a buff or gray finish—depending on the atmosphere in the kiln.

Porcelain body

Porcelain body has a fine, smooth texture and fires pale gray to white. It is very pliable and, when used on a wheel, is capable of being stretched and potted very thinly so that it becomes translucent when fired. It dries swiftly if thin but will absorb moisture quickly to soften it again if required. When fired to a suitable temperature it vitrifies into a hard glassy material.

It is also used for industrial purposes to make tableware mechanically as a liquid slip poured and passed into molds. Increasingly during the last twenty years it has been used by studio potters for handmade pieces; in fact its whiteness, and acceptance of vivid colors has greatly influenced the style of studio potters.

Earthenware body

Most of the common clays around the world are classed as earthenware clays. When they are dug, most earthenware clays need to be screened to remove impurities such as gypsum and stones. They can be discovered almost anywhere the ground is dug and often have a high concentration of iron oxide. This makes the initial color vary from yellow to red-brown and black — but most of these colors fire to red terracotta. Earthenware clay will fire to around 2020°F (1100°C). Once above that temperature it will often start to bubble and turn black. Some earthenware clays melt well below 1830°F (1000°C) and can make quite a reasonable glaze for stoneware pieces.

These clays are quite soft in pottery terms when fired, and porous. If the pot created is to be used for holding a liquid, then it must be glazed all over to ensure it does not leak. Most housebricks, roof tiles and chimney pots were (and still are) made from earthenware bodies; their porous nature helps to keep the walls free from damp.

Raku body

Because of the thermal shock a clay body needs to endure during a Raku firing, its composition has to be open textured and coarse. Ground brick, sand or coarse grog are added to the clay in order to open up the texture; the different textures of the additions also impart an interesting quality to the finished pot. However, the firing temperature and the open texture will make the body more porous.

Sculpture body

If large pieces for modeling or thick slab construction are required, then the body has to be strengthened with perhaps 20 to 25 per cent of coarse grog or sand. This will enable the body to stand up to the drying, and will also help reduce shrinkage, warping and risk of collapse. The addition of the grog allows air into the clay and so assists the drying process; the resulting body will be far more even.

Clay for casting

Many potters today choose to use a clay body in a liquid form as a casting slip to make fine and delicate repeatable shapes. This is done by pouring the liquid body into a plaster cast, removing it when leather-hard and then cutting, drilling or squeezing it to suit requirements.

Some cast items (for example, large tiles) may need a groggy open-textured mixture to allow for even drying and shrinkage. Finer pieces will require a well-sieved smooth mixture which will create a delicate surface on which to decorate.

To make a body sufficiently liquid to pour and cast—without adding more water which could cause extra shrinkage and cracking—additions of sodium silicate will be required to act as a deflocculent and to improve the fluidity of the slip.

C see page 114

Examples of clay colors

Rhinoceros by Tessa Fuchs (UK)
A red earthenware piece, inspired by a safari in Kenya.

Dynamic serving dish by Sandy Brown (UK)
Oxides have been applied freely using brushes and fingers.

Asymmetric pot by Jennifer Lee (UK)
Stoneware pot in T-material impregnated with vanadium pentroxide to create a surface like lichen.

Porcelain bowl by Peter Lane (UK)
The incized design and delicate cut edge are emphasized by the simple white glaze.

Cylindrical vases by Graham Flight (UK)
These are made of porcelain and clear glazed with swirling bands of underglaze painting.

Test clays and firings

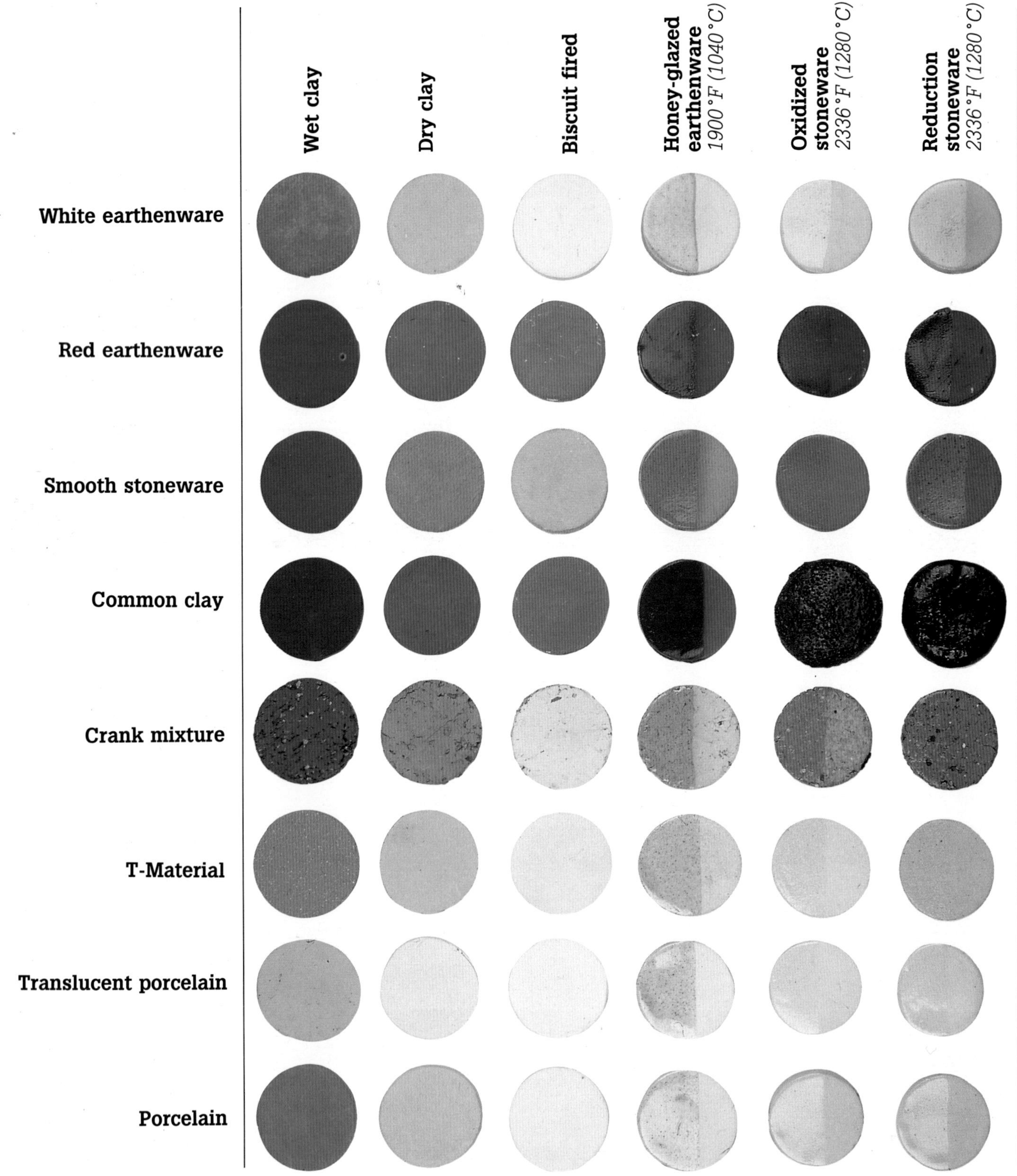

White earthenware

Smooth-textured earthenware is a highly plastic body used for throwing, handbuilding and casting. It fires ivory white up to 2120°F (1160°C) and shows colored earthenware glazes to advantage.

Red earthenware

This is a fine-textured rich red body that fires up to 2156°F (1180°C), creating a classic terracotta color. It is excellent for both throwing and handbuilding purposes, or for modeling.

Smooth stoneware

A blend of fireclay, china clay and ball clay, this is a light gray color that fires to a warm buff in an oxidized kiln or gray in a reduction firing. It is good for all types of throwing and handbuilding, firing at up to 2336°F (1280°C).

Common clay

Smooth-textured clay can be dug directly from a clay bed and used after cleaning, weathering and pugging. The color is dark gray to black when dug. It has a heavy iron content, firing rich red at 1940-2010°F (1060-1100°C).

Crank mixture (Chamotte clay)

A heavily grogged open clay, this is mainly used for handbuilding, modeling, sculptural pieces and Raku. The coarse-grained texture withstands shrinkage and thermal shock; it fires at over 2370°F (1300°C)

T-Material (an English china stoneware clay)

This is a beautifully plastic grogged clay that is ideal for large or sculptured pieces. It can be used alone or mixed with other bodies to improve their strength or texture. It fires white—up to 2370°F (1300°C).

Translucent porcelain

This has similar properties to porcelain but will allow light to show through when thrown or modeled very finely, giving an added beauty to the finished pieces.

Porcelain

Pure white, fine textured and dense, porcelain is good for handbuilding and modeling detail. It can be thinned to a fine edge, is smooth and creamy when thrown but needs to harden slowly. It fires up to 2370°F (1300°C).

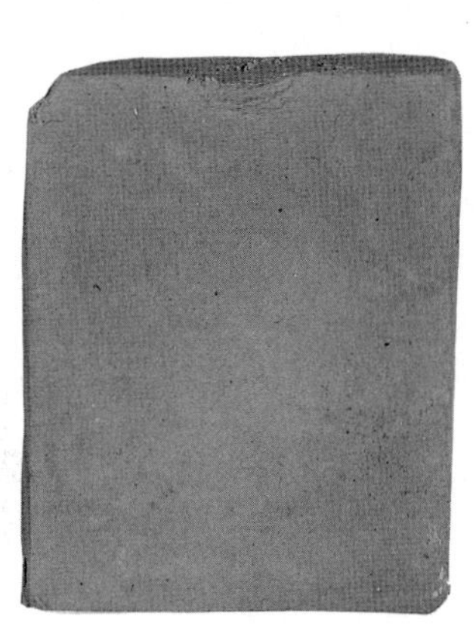

Handbuilt forms by
Ursula Morley Price (UK)

2

Handbuilding

Handbuilding

No committed potter would deny the sheer pleasure of handling plastic clay for it is a material which responds instantly to the slightest pressure. Most will also confess to experiencing both delight and disappointment in their work from time to time. Many will claim that they remain students of their craft throughout their lives. Whether working on the potter's wheel or producing three-dimensional forms by one or more of the handbuilding techniques described in detail later, you are urged to keep an open mind in all aspects because few things are absolute in so complex a subject as ceramics. The variables are legion. Learn, especially, to respect your tools and materials. Deepen your understanding and widen your appreciation of ceramics at every opportunity by visiting exhibitions, reading around the subject, and talking with other potters.

Although individual handbuilding methods are described and illustrated in this book you should always remember that the boundaries between techniques are not fixed. A single pot or ceramic sculpture may require fashioning by a variety of processes. It may be built up from a number of separately made sections—some may be thrown, others modeled—but, however produced, those individual parts must be thoroughly joined together if they are to survive the stresses of drying, shrinking and firing. Beginners are often frustrated when they find they have not understood this basic principle.

Clay can be worked at any stage from wet to dry but joining normally takes place while moisture is still present. Parts to be joined should be in the same condition wherever possible or uneven shrinkage will cause cracking and force them apart.

The most familiar pottery vessels are those having a round section. Many have been wheelthrown while others will have been constructed from individual coils or strips of plastic clay. Some potters find such forms inhibiting to creative thought and prefer to alter thrown pieces away from the round with additions and/or subtractions to or from the initial piece. Some further possibilities are discussed later but, although the different techniques involved are often illustrated here by symmetrical pots, in most cases, they apply equally to sculptural and asymmetrical forms. Whichever methods appeal to you most it is important to remember that is all they are: processes by which you can give tangible reality to your ideas.

Early experimentation

In this chapter the various techniques of handbuilding will be explored and explained, bearing in mind that this is the method by which most beginners are introduced to ceramics. These techniques include:

Modeling
Pinched or thumb pots

Envelope form by Ursula Morley Price (UK)
Here the coiled and pinched form has been made of stoneware and its striated finish achieved with wood-ash glazes. (Oil fired 2264°F 1240°C)

Coil, rope or strip building
Forming clay slabs
How to use hard slabs to make flat-sided boxes
How to use soft (or stiffer) slabs to make curved shapes

These basic handbuilding methods are the first ones that novice potters are likely to use. Through practice with these new techniques they will develop their skills and begin to understand the properties of clay.

Presented with clay, young children are natural handbuilders, creating animals, birds, monsters and innovative pots; they have no inhibitions or preconceived ideas. All beginners would do well to give vent to their imaginations in the same way. They should feel themselves to be restricted by only a few ground rules—such as the need to hollow out the clay if the form is too thick, the obvious necessity for firm adhesion of any additions and how to avoid explosions! (For instance, these can occur if the clay is too thick to allow the release of steam during firing, if air is trapped in a section of the pot, or if the piece is still too damp to be safely fired.)

The free flow of ideas should be the prime consideration when first handbuilding with clay. Beginners will then develop an understanding of the material and their own creativity—as well as being able to produce a finished article which can be kept and fired.

However, simply because it is the method by which most potters are first taught their skills, handbuilding should not be regarded as the lesser art form. It is by no means merely the stepping stone to the wheel. Often it is quite the reverse, since the potter does not suffer the initial restriction of the circular section. After all, every pot, vessel or ritual figurine was made by hand in the days prior to the invention of the wheel.

Moreover, throughout the world, many potters today prefer to use traditional handbuilding methods to produce beautiful and original work. Handbuilding techniques allow the work to progress more slowly and many potters prefer this relaxed pace to the immediacy of the wheel.

It can be a very pleasurable activity and it is hoped that, by following the guidelines in this chapter, many new potters will be able to discover all it has to offer.

Choosing clay for handbuilding

Clays and their preparation have already been discussed in chapter one (pages 22 -27). However, the choice of clay is very important when handbuilding. In particular the texture plays a vital part. Not only does groggy clay create an interesting surface on the finished pot but also allows a thick piece of sculptural handbuilding to dry and shrink evenly as air is allowed to penetrate the body. The choice of a fine body for a delicate pinched or modeled form is equally important.

A small thumb pot or pinched pot may demand the fine texture of a porcelain clay, whereas a large pot may need a coarse, grogged open clay. Make a note of the proportions of the materials used so that they can be repeated if and when required.

The qualities of a particular clay will dictate both the color and texture of the finished result. Clay can range from a creamy smooth body through to a coarse-grained, open-textured heavily grogged body. (Most clays are described as 'bodies' because they are composed of more than one ingredient.)

Therefore the selection and careful preparation of clay must relate to the finished article with regard to texture, color and firing range.

The purpose of the finished pot must also be borne in mind. For instance a high-fired piece for food preparation will require stoneware or porcelain clay. An earthenware clay would bubble or melt at these temperatures. Lower-fired earthenware, however, may be ideal for making flameproof cookware. A fine-textured clay can be used to produce a smooth-surfaced eating bowl, whereas a clay with a coarse grain would be ideal for, say, a planter or a decorative panel—or perhaps a cooking pot which has to withstand the thermal shock of heating in an oven. Cooking pots need not all be of high-fired ware; many countries produce rough-textured, low-fired earthenware that is used for cooking. Some of these heavily grogged pots will survive when placed over an open flame, but using a heat diffuser or mat will reduce the possibility of cracking.

All this must be considered before commencing the building process.

Preparing the clay for handbuilding

The mixing of clay is an important first step in handbuilding. It may take some time for the clay to be in a fit state to use, as the clay needs time to mature and soak and then to dry into a workable plasticity. The time taken over these tasks will be reflected in the finished result. (See pages 22 -25.)

Handbuilding pinch pots

Handbuilding techniques such as pinching and coiling are popular methods of producing functional and decorative forms. They both allow great flexibility and can produce pots of remarkable symmetry.

Pinching requires only the use of the fingers with properly conditioned clay. Small sculptural pieces can be pinched from one mass of clay, or built up from many units. Functional items such as ancient cooking vessels had tripod feet added to them to allow them to be placed above hot coals. Japanese tea bowls show pinching at its most artistic and today many fine porcelain pieces illustrate just how delicate the results can be.

A ball of plastic clay simply invites the fingers to make a mark or hollow, or to squeeze and pull it in various ways. From this first handling has evolved the thumb or pinched pot. Making a vessel by pinching the clay between the thumbs and fingers allows the beginner to explore the texture and strength of this strange new material. By handling the clay and gradually squeezing the shape the novice potter can begin to understand and judge the qualities of the clay. It is heavy for its size; it will dry in the warmth of a hand; it must be kept damp to stop it crumbling. If the sides of the pot are too thick it will feel and look too heavy for its size. If the walls of the pot are pinched too hard, the clay will be too thin to support its own weight—the pot will sag and look floppy and formless.

An overall evenness must be achieved or there will be problems later when the pot dries. The thinner parts of the walls dry first; if there is too great a variation of thickness the pot will be pulled out of shape and cracks may appear. The evenness of the walls cannot be measured but must be judged by feeling the thickness between thumb and fingers as the pot is formed.

These first experiments with clay help to develop judgment of the clay's performance and to better understand the relationship of the thickness to the size and weight of the growing pot. A shape which is left thick might *look* like a fair example of a finished pot but once it is handled it will *feel* unfinished; it will remain just a lump of clay with a thumb hole in the middle and give the maker no satisfaction at all.

Developing the skills

The urge to improve should encourage the budding potter to try again and be more courageous; eventually, with growing confidence and appreciation of the medium, the potential of the thumb-pot form can be properly exploited.

C see pages 32 33 45 46 163

A pinched bowl by Sue Varley (UK)
Natural landscape outlines make a useful source of inspiration.

These small pots do not require elaborate pattern or coloring. A simple polished finish, or a surface of natural gritty clay scraped to reveal the texture, can be quite pleasing. This is another skill to develop—the creation of minimal surface decoration, learning not to over-decorate. Some of the best results can be achieved by the simple addition of a complementary color rubbed into the surface and then polished with a smooth pebble or the back of a spoon. Or try beating the pot gently into the shape of a seed head and burnishing it so that the shiny surface resembles African or Peruvian pots. The hard compressed surface could be given a contrasting pattern of scratched lines.

Working with fine clay

If the clay is a fine-textured one (porcelain, for example) it can be pinched out to the delicate thinness that will become translucent when it is fired. Both the clay and the hands will have to be dampened for this operation, as warm hands will dry out the thinner clay. The shape will probably have to be supported as it becomes thinner. This can be done by using any suitable vessel lined with paper to stop the clay sticking; a mug or bowl of the right shape and size will do.

The clay will have to be left to dry and stiffen for a while until it can be handled again without the risk of distortion so a second or third pot may be started—the pots being worked in rotation as they dry in between each spell of work. Sometimes this thin floppy state can be very useful as it allows the clay to be formed into clothlike folds or shell-like forms; the folds of clay can be overlapped so that they look like flower petals or folded tissue paper. Try to make a pattern of light and shade in the clay itself without relying on the use of added decoration and color.

Forming a pinch pot

1 The thumb is pressed into a ball of clay held in the other hand. Push firmly—well down into the clay. Work with the clay held in the hands, not on the table. If the pot is placed down on a table or bench it becomes flattened and loses its shape.

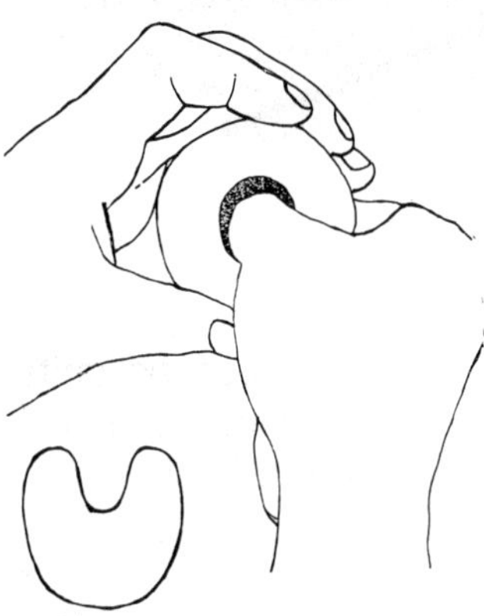

2 Turning the pot slowly in the hands, the clay is squeezed between thumb and fingers all the way around and across the base to thin it evenly. This is vital. The clay can be thick or thin but it must be consistently so to avoid distortion and cracking.

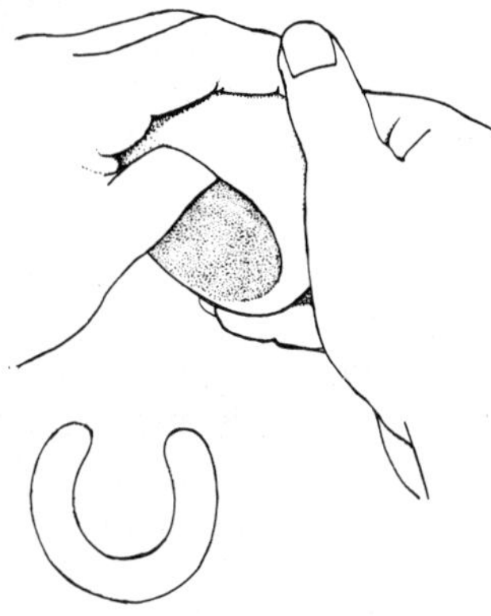

3 While thinning and squeezing the clay, the thickness can be felt between fingers and thumb; make sure it is even. Unless intending to make an open bowl, try to keep the top narrow; it will have a tendency to flare open and become bowl-shaped.

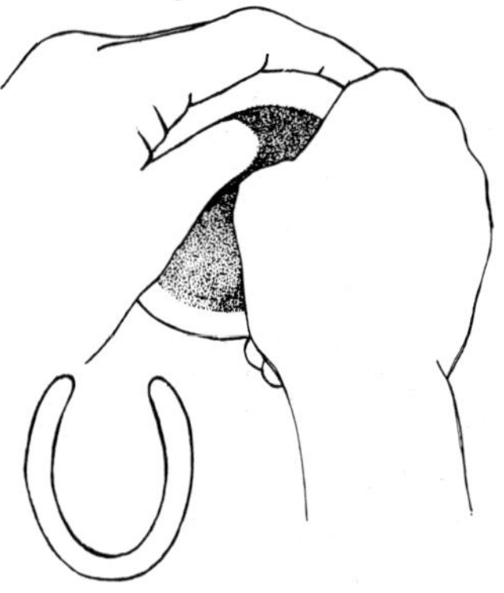

4 Pinch and thin the neck of the pot, smoothing the sides and perhaps forming a small foot with a coil of clay. You should now be generally finishing and improving the shape. After the pot has dried slightly the clay can be scraped, textured, polished and smoothed with the hands.

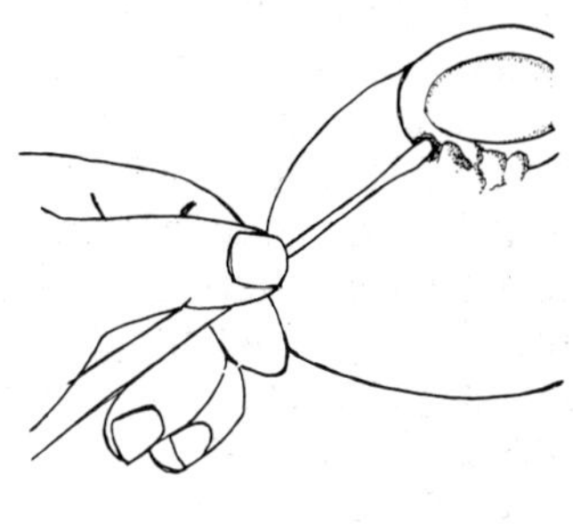

Smoothing

5 Smooth the surface by hand, first with a sponge and then with a rubber kidney.

Alternatively, apply a coat of iron oxide or manganese oxide and then polish the surface with a spoon. This will give a hard rock-like surface — especially if it is fired by a smoky sawdust firing (see page 243) which will enhance the finished polish.

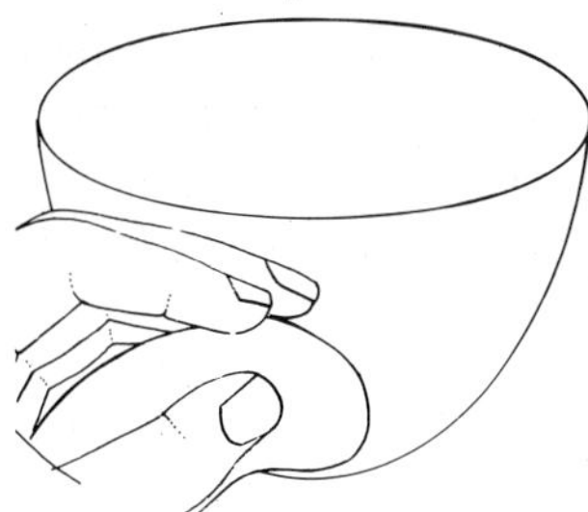

Textures

6 Patterns can be scraped into the surface of the pot with a saw blade, a fork or a fine needle. Either texture the surface all over or in bands of scraped and cut decoration.

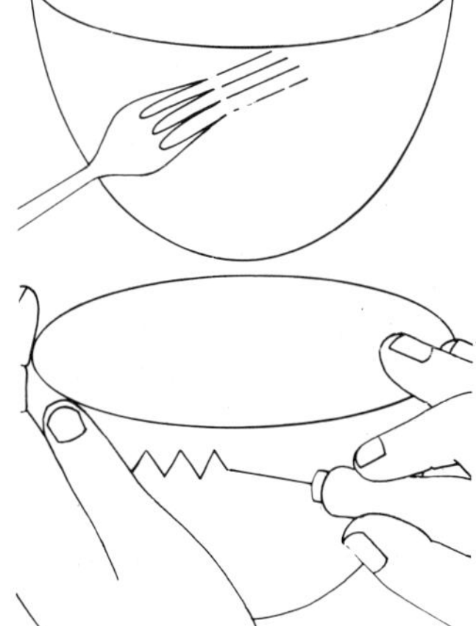

Impressed decoration

7 This allows for a greater control of design. Press patterns into the surface with a variety of tools—a pencil point, a ball-point pen, needles, pins, drills and drill pieces. This method can be used to create patterns or haphazard shapes that may result in a pleasing stone or rock-like 'eroded' effect.

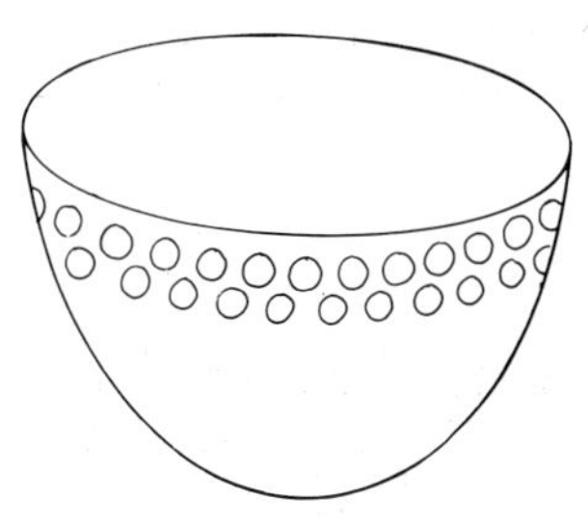

8 Small additions of clay pellets or fine coils applied to the surface and beaten or pressed in will give a variety of smooth and textured patterns. These catch any glaze or color applied when the pot is fired.

Joining pinch pots

Extending the pinch pot and attaching a collar

1 The simplest and most versatile method of extending the pot is to add strips or coils of clay.

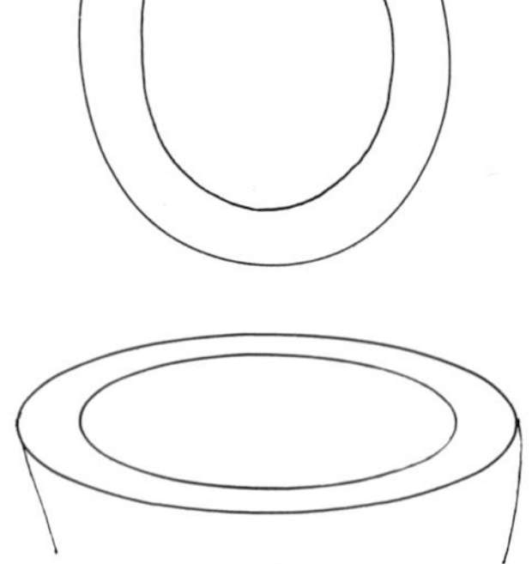

2 Score the rim of the existing pot with a fork and coat with slurry.

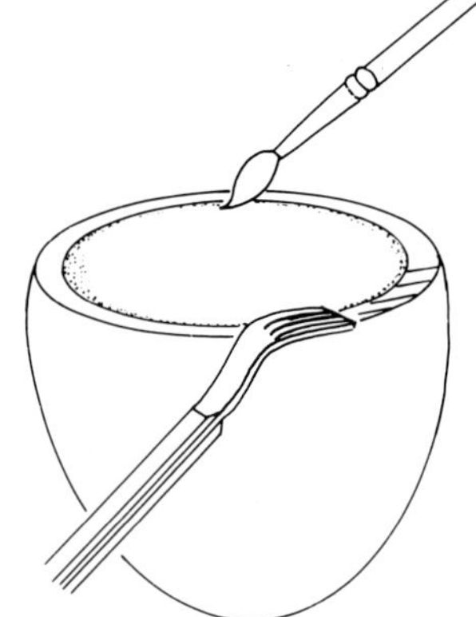

3 Roll out an appropriate rope or coil of clay with flattened palms.

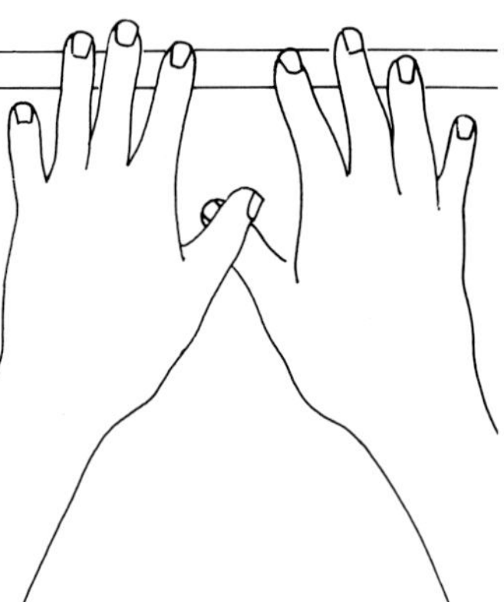

4 Position coil carefully on rim of pinch pot. Blend in and smooth join if required.

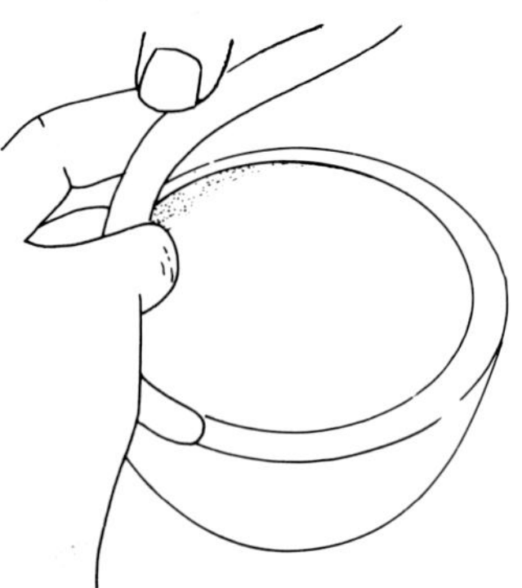

5 Successive bands or coils of matched, decreasing or increasing lengths will build walls into required shape.

6 Finish smoothing the coils.

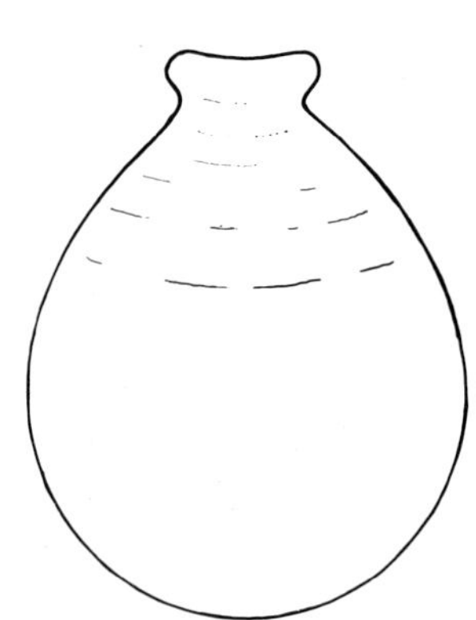

7 In order to achieve a finer roundness or eradicate any dents, blow through the neck of the pot, while it is still soft.

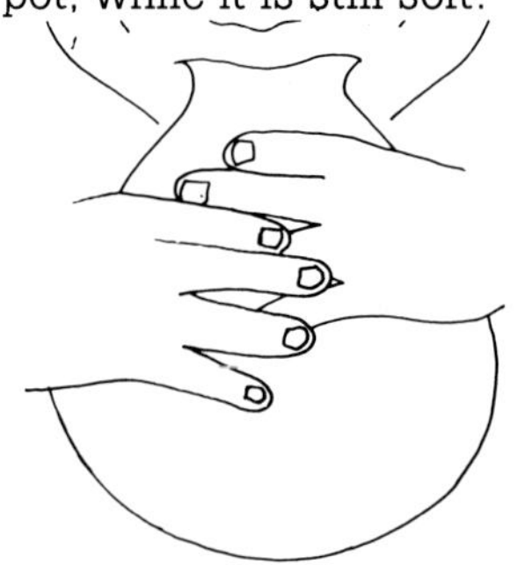

8 The choice of final decoration or finish should complement the design.

A further development of the pinch pot method is to use two open bowl-shaped pots, and then join them together to make an airtight ball of clay. This can be beaten and shaped to make the main form of a pot, or perhaps to form the body of a bird or animal. The method of joining two thumb pots which is described here is of basic importance because it applies to the joining together of any two clay pieces.

Two pots are formed so as to have the same width at the edge and the same thickness. Each edge to be joined is then scored with an old fork (or some such tool) and a slurry is made of the same clay. When the two edges have both been given a good coating of slurry, press them firmly together; then, using the fingertips, pull the clay across the join to seal the gap. The joint is now reinforced by a thin coil of clay fixed around the join, and smoothed over to make an airtight seal.

The shape can be gently beaten or squeezed into many forms, because the air trapped inside acts rather like that in an inflated cushion and thus prevents the ball collapsing. This is an ideal base for a bird shape for instance, as it stays full and round, supported by the trapped air.

Before the firing,a small hole must be pierced through the clay to allow the air to escape.

Joining two pots to make a ball

1 Pinch two similar-sized pieces of clay to form two half-round open-topped pots. Leave the rims fairly thick.
Score these with a fork and then cover each rim with thick slurry.

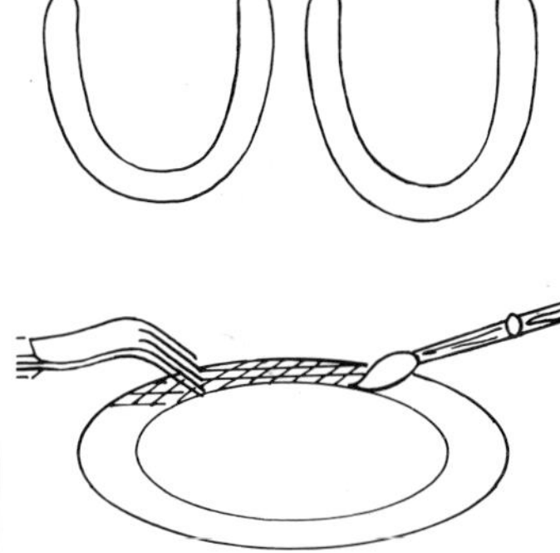

2 Join both pots rim-to-rim, pressing firmly together to make a good seal. Knit the joint together by pulling clay from one half to the other (alternately) to make a good airtight seal.

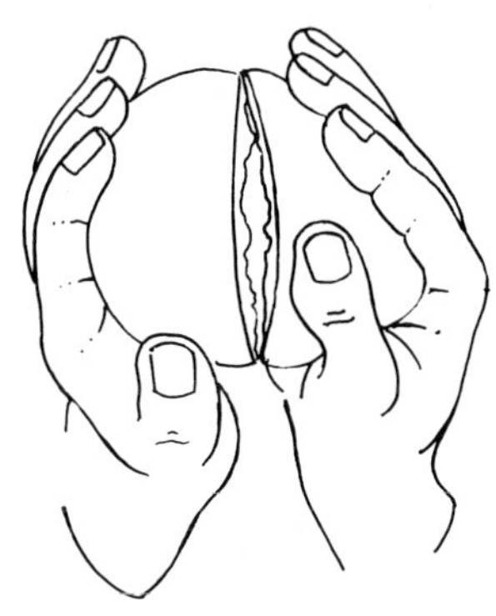

3 To cover the joint and complete it, put a rolled small coil around the joint, smoothing into the surface until it disappears.

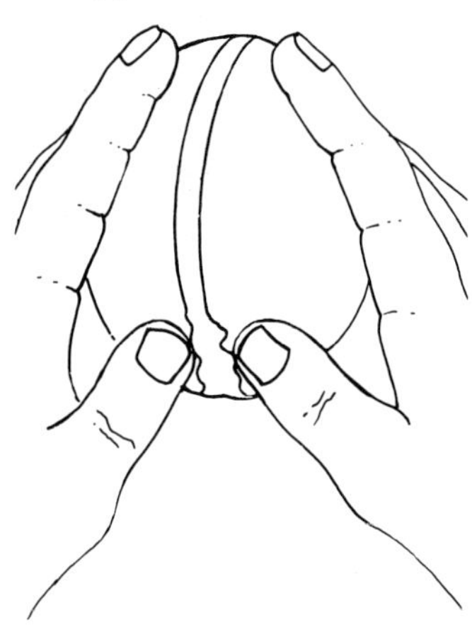

4 The complete joined pot becomes an airtight ball of clay ready to be shaped into the form required.

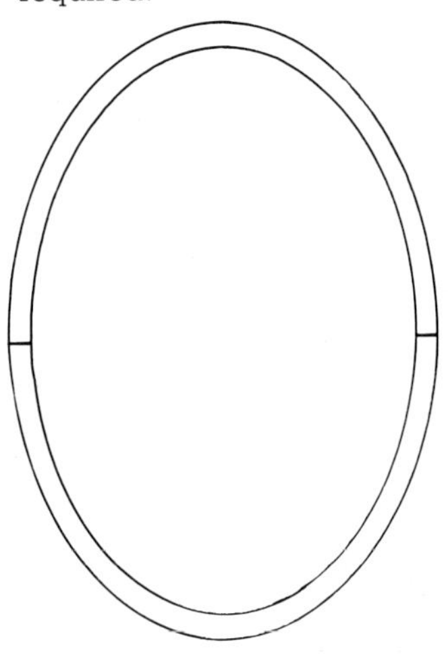

Developing a shape from the joined pinch pots

1 An egg-like shape could be the basis for a modeled head.

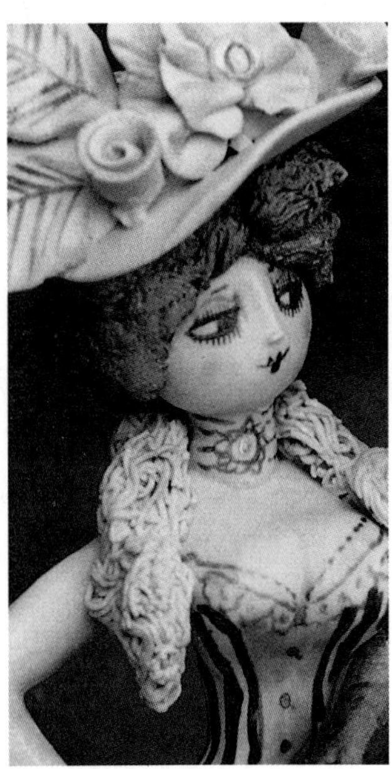

2 The ends of the joined pinch pot can be eased, squeezed and pulled out into a lemon-like shape which can form the basis for a bird.

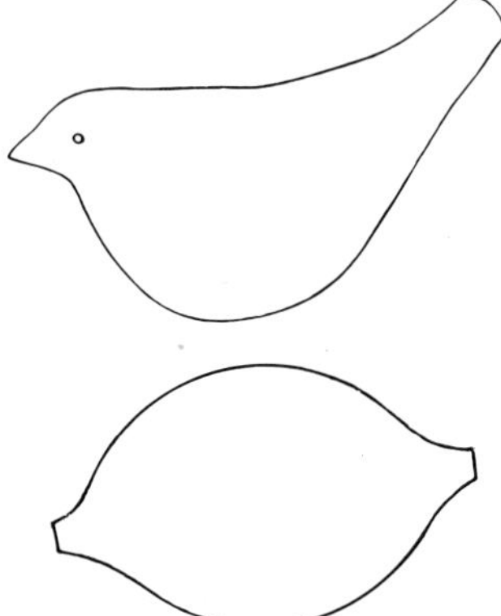

Modeled figure by Hilary Brock (UK)

C see page 41

3 If tapped on the bench, one side becomes flattened and will make a domed form which could be the basis for a mouse body, hedgehog, a cat or some form of crouching animal.

4 If the pot is beaten gently (still preserving the trapped air inside) using a flat piece of wood —squares, cubes, and other flat-sided objects can be constructed. Remember to drill a small hole somewhere as the pot dries to allow air to escape before firing.

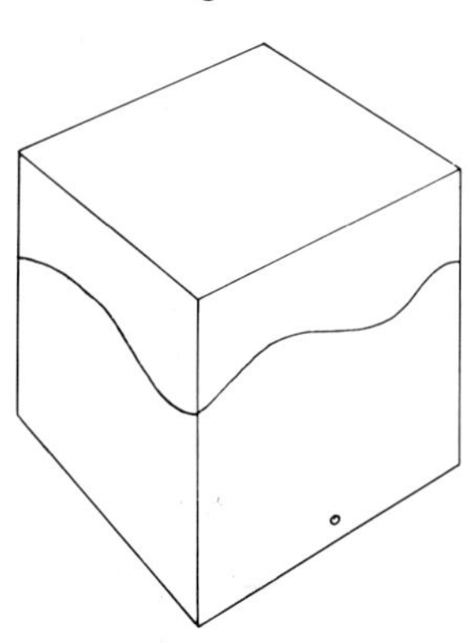

Pinched pots

Poppy-seed pot

Natural objects will often be a useful source of inspiration for form and shape. The poppy-seed for example is an attractive subject from which to draw ideas.

1 Using both hands, gently squeeze the double pinch pot into a slightly pointed shape, rather like a pear.

2 Gently beat the top and the bottom flat; the wider end will form the base of the pot, the narrow end will become the top.

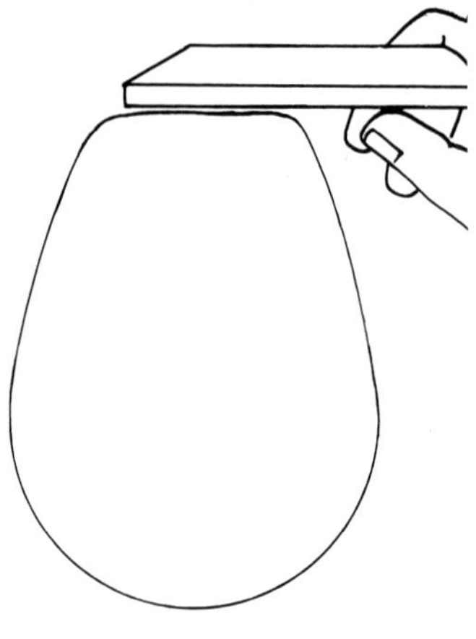

3 Roll out a piece of the same clay and cut out a rough circle that is approximately the width of the base, or slightly wider.

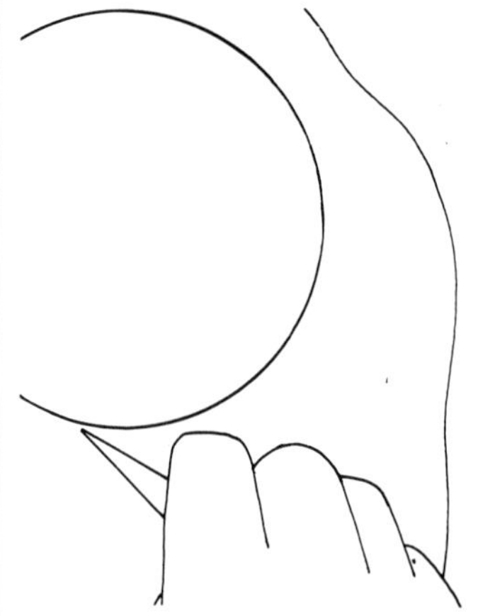

4 Score both edges and then use slip to attach the disc of clay to the top of the flattened pot.

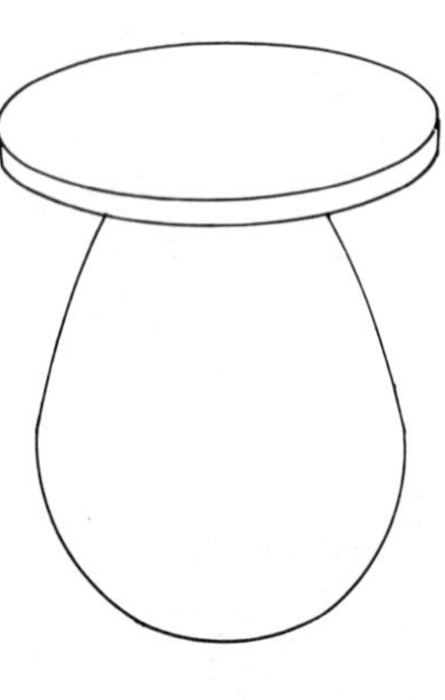

5 Using the fingers and thumb, score outwardly radiating ridges in the disc, thinning the edges until they are jagged and uneven so as to break the line of the circle.

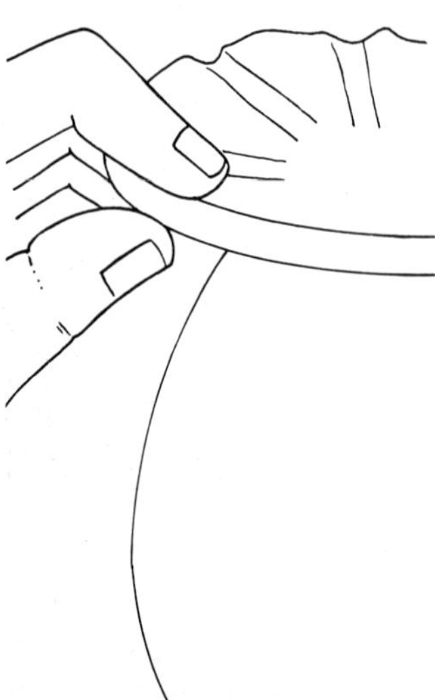

6 Score textures and patterns in the body of the pot. Make slight indentations with the fingers. Use any method to create a poppy-like head. You can even try rubbing dry oxide into the texture to create various colors. Finally, cut a hole in the center of the disc which will be the neck of the pot — this will also release the trapped air before firing.

The finished poppy-head pot made from pinched pots has textures that are filled with oxides, glazed with a woodash glaze and fired to 2268°F (1260°C).

Making a bird from a double pinch pot

1 Beginning with a lemon-like shape, both ends of the pot should then be squeezed and eased out to form the rudimentary head and tail of the bird.

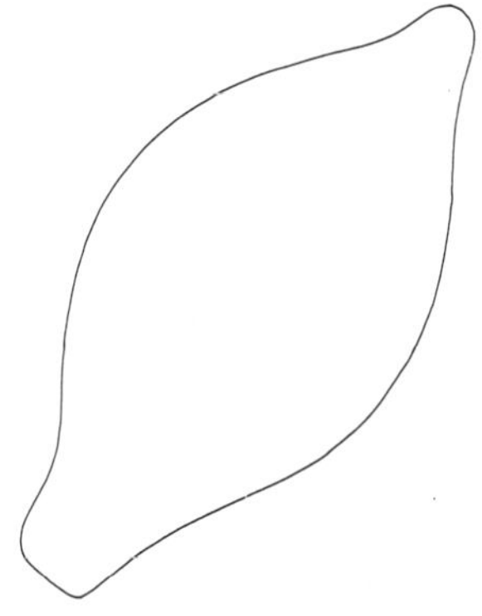

2 The head is made by pulling and squeezing the clay with one hand and shaping it into a neck and rounded head. Applying a little water helps this process.

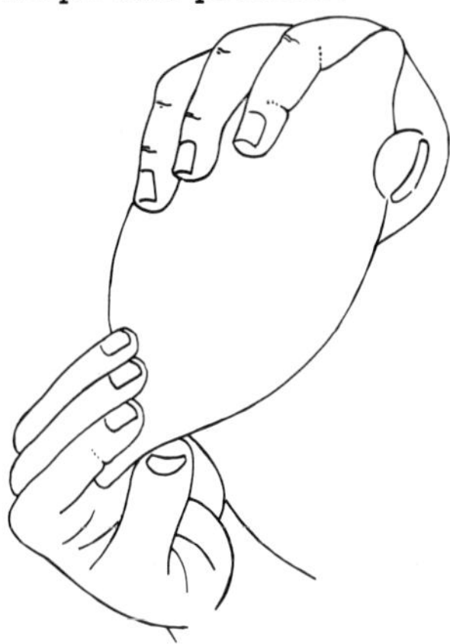

3 The tail is also squeezed and pulled, using water, but this time the shape is slightly flattened by the fingers and thumb to form a flat spade-like tail.

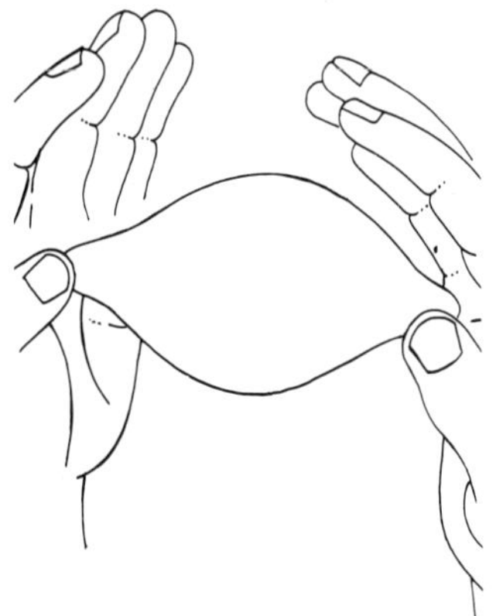

4 When the basic shape is made, angles and planes can be beaten into it with a flat slat of wood. This will give a little more vitality to the shape and will help to lift the head and tail from the base.

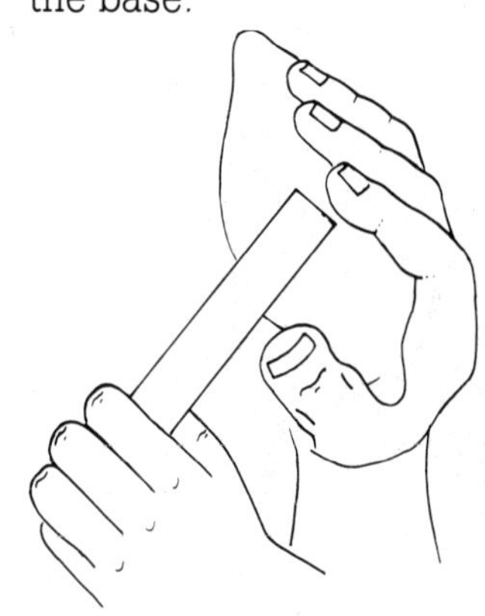

5 Wings can be added by flattening a small piece of clay in the palm of one hand and beating it with the heel of the other hand until it becomes a thin plate of clay. Leave the textures of the palm on the wing to create a pattern of fine lines.

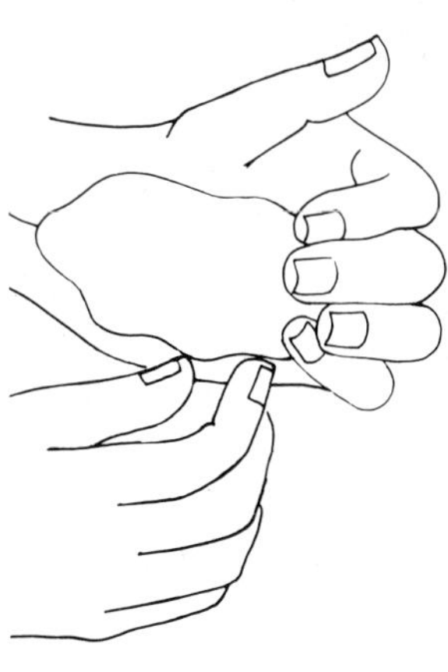

6 Attach the wings to the sides of the bird by the scoring and slurry method. Smooth the bottom edge into the shape and then pinch and thin the top edge into a suitable shape to create a wing-like effect. A feathered effect is difficult to achieve but it is simple to suggest the image of a bird with folded wings.

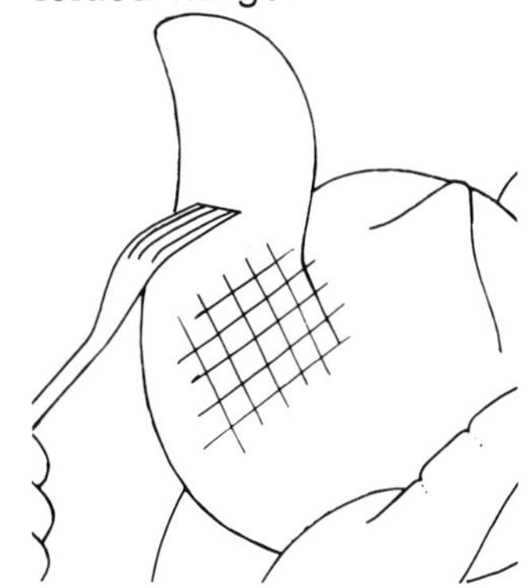

7 Now add the final touches to the shape. Use a pencil point or pen to make two eyes in the head. With a fine blade, fork or similar tool, add texture to the surface of the bird just below the neck by creating patterns which can be highlighted with oxides or may be glazed later in the process.

The bird without wings has heavy scored patterns filled with green slip and has been glazed with a clear stoneware. The bird with wings has been made with porcelain clay. Its texture patterns have been filled with a body stain in blues and greens, glazed with a woodash glaze.

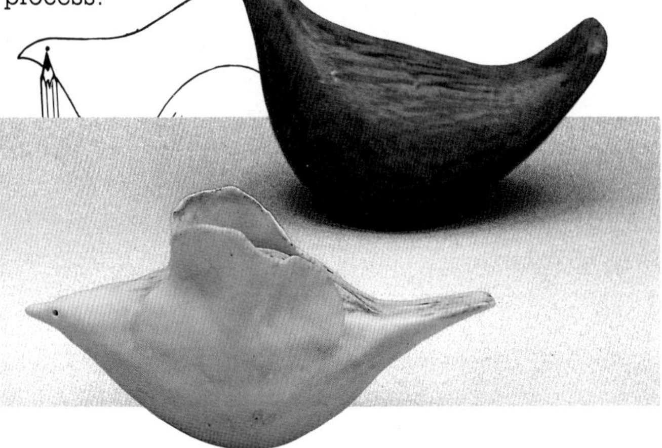

Pinched pots

Porcelain pinched pot by Mary White (Germany)
It is 4 inches (12cms) high and has an ocher/white feldspar glaze.

'Seed' bottle by Mollie Winterburn (UK)
Inspired by a seed from the garden this is made of stoneware clay cut from slabs. Black with yellow threads, it is 12 x 7 inches (30 x 18cms) in size.

Handbuilt pots by Jennifer Lee (UK)
Stoneware pots have been coiled and pinched, using colored clay to create bands both inside and outside the pots. Simplicity, balance and color are all important.

A pinched bowl by Sue Varley (UK)
Colored clays have been worked in to emulate landscape paintings of the Welsh border counties. After painting some areas in wax, the pot was glaze fired and then smoked in a sawdust kiln.

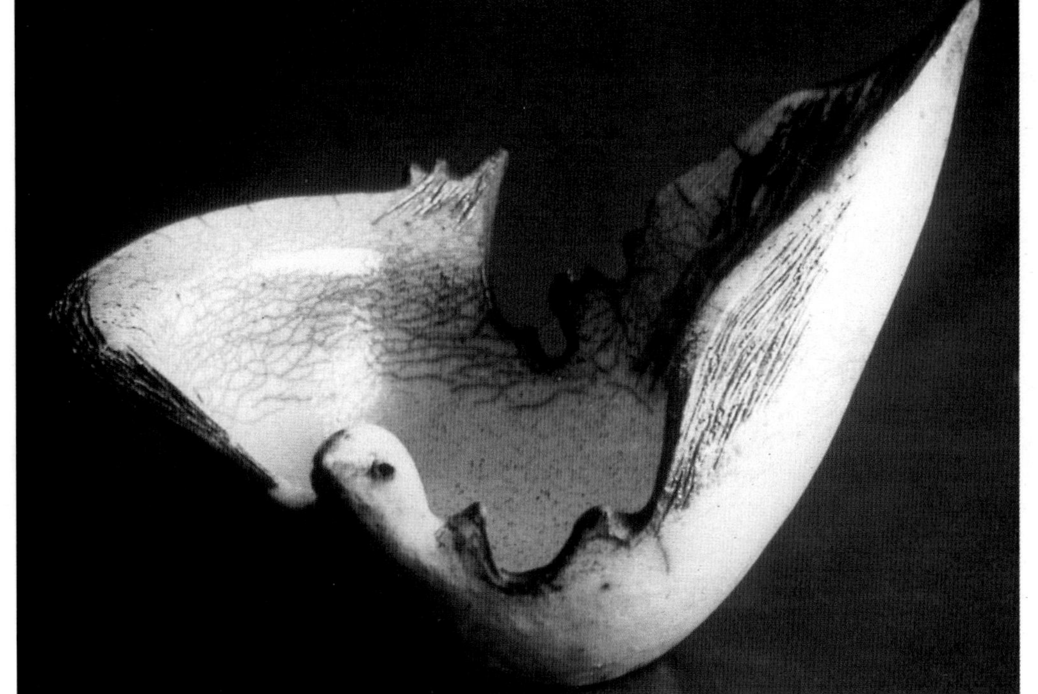

Bird bowl by Monique Vézina (Canada)
A handbuilt white sculptural bird shape has been used to make an unusual bowl that is raku glazed and fired.

Spherical pots by Ray Rogers (NZ)
These pots 7 inches (18cms) and $8\frac{1}{2}$ inches (22cms) with global swirls of color and a 'fungoid' effect have been made from coils and then pit fired.

Clown and friend by Sally Cocksedge (UK)
Simple figures have been sensitively modeled by hand and brightly painted with slip.

Bottle flange by Ursula Morley Price (UK)
This basically simple flanged bottle has been deeply grooved and shaped to create an exciting surface. It is made of red stoneware.

Porcelain bowl by Yvonne Boutell (USA)
Porcelain has been dyed with different oxides and then inlaid into this delicate bowl in a bold design.

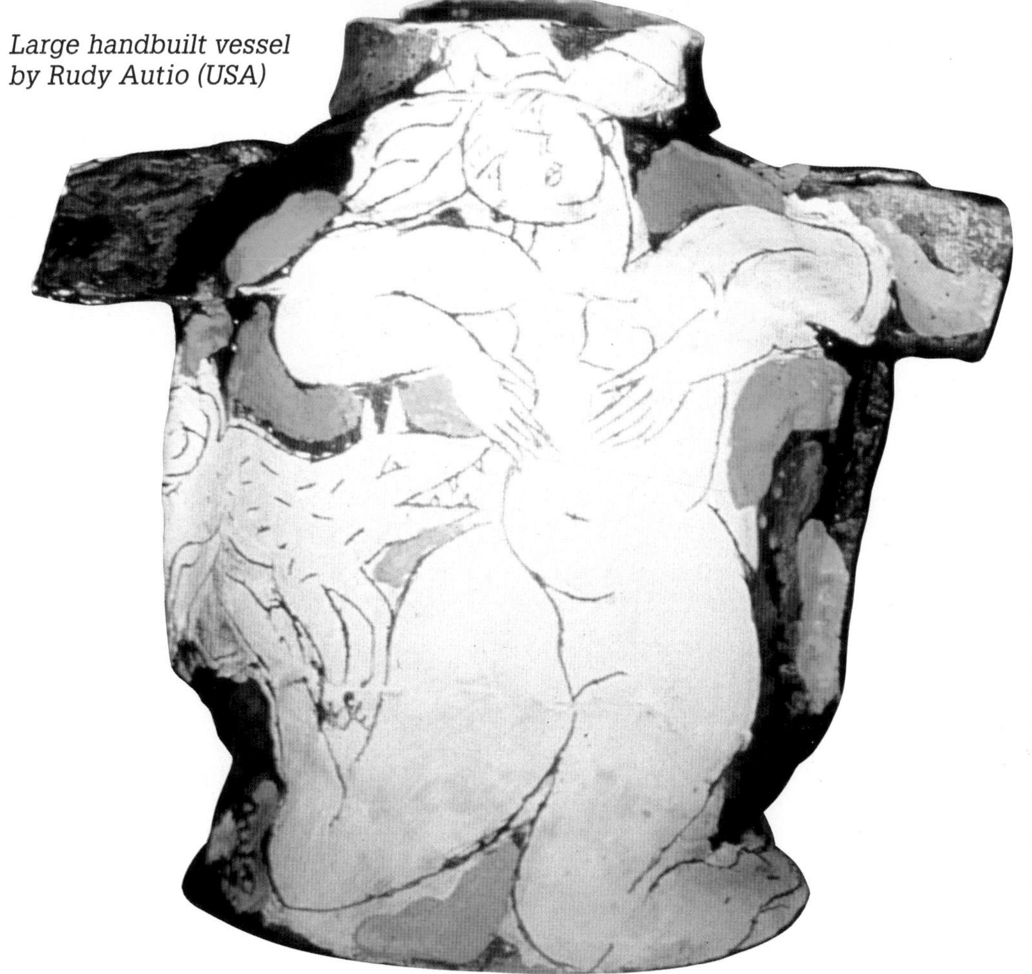

Large handbuilt vessel by Rudy Autio (USA)

Porcelain figure by Jill Ruhlman (USA)

Handbuilt pot
by Betty Blandino (UK)
Large thin-walled vessel in oxidized stoneware: mottled blue color has been painted and scraped on a white dolomite glaze.

Porcelain jampot
by Jenny Beavan (UK)
Here thrown forms have been finished with fine hand-modeled decoration and delicate tenmoku and gold glazes.

Ocarinas
by Neil Ions (UK)
These delightful Grebe and chick ocarinas are ceramic musical instruments inspired by American Indian artefacts. The rich designs were painted with earthenware slips.

Coil-built bowl
by David Roberts (UK)
This is a large open vessel 22 inches (56cms) diameter. It has been Raku fired to create an interesting surface effect on a simple form.

Coiling

From these first trials using the pinch-pot method, small pots may be made. Then perhaps comes the need to develop a shape, to make it larger than the pinching method will allow. The natural progression is to add more clay to the basic shape to increase the size. The most versatile way of doing this is to add ropes of clay to the top edge of the pot. This is coiling. Sometimes coils may be added to a base cut out from a sheet of clay or preformed in a mold. Potters develop their own styles and methods for increasing the size of a pot, choosing a way which suits individual needs. One method is for the clay to be rolled out on a table and cut into strips. Then it can be placed on edge around the base already formed and smoothed into base, thus raising the height of the walls.

Another method is to squeeze the clay into a rough rope, then lead it round the edge of the base and join it to the pot. This is a fairly quick method of increasing the size. The pot soon grows but unless the pot and coils can be dried quickly, the newly acquired height may be uncontrollable. This method is perhaps best suited to a warm climate where the potter may be working outside and the pot is hardening quickly in the sun and the wind. A potter working in a cooler climate may prefer to work on a number of pots in succession, one being allowed to harden while the potter is working upon the next. This will prevent the weight of the new coils pulling the pot out of shape or causing it to collapse. Alternatively pots can be stiffened by blowing warm air evenly across the surface with a hair dryer or hot-air gun. Some potters are known to use the flame from a blow-torch for this purpose. Ensure drying is even and do not leave pots unattended in front of hot-air blowers.

The first attempts at coiling may not be very good; the coil may become flattened or oval shaped as a result of too much pressure—or it may be unevenly thick and thin along the length. If the coil is unsatisfactory you can roll the clay up and start again, but it will be drying all the time and may need to be mixed with softer clay if more than one or two attempts are made. Alternatively, dampen the bench to keep coils moist. A short time spent practising rolling coils will result in a round even coil which can be produced almost automatically, leaving the potter free to concentrate on the shape of the pot and on how it is developing.

In order to achieve a good result when making a coiled pot, there are just a few essential requirements. It will, for instance, be necessary to have a suitable table or board, but in general the tools that are required will all be fairly unsophisticated.

Tools for coiling

The best tools the potter has are his or her fingers which with the addition of one or two hand-made wooden implements may be all that is needed. A flat board or a wooden-topped table makes the best working surface. (A plastic-topped table, although easy to clean, can make the clay stick.) The clay for making coils needs to be fairly soft. A wooden table top will absorb moisture as the coils are made and the clay should not stick to it.

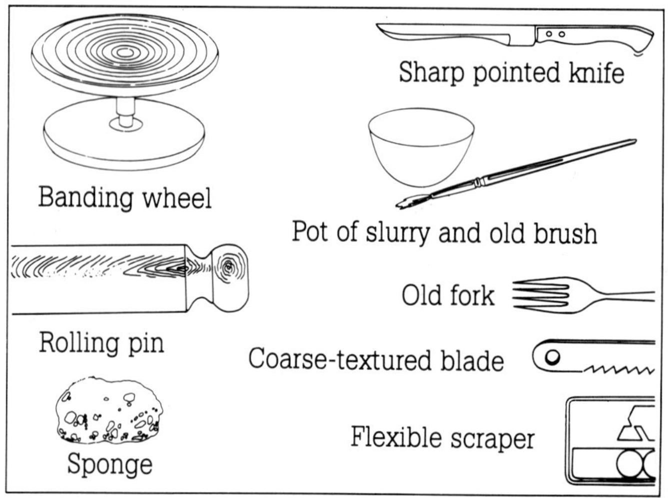

As the wood absorbs the moisture, it will form a thin coating of drying clay. DO NOT BRUSH THE DRY CLAY FROM THE BENCH; ◇ instead sponge it away. This stops the dry particles being spread into the atmosphere of the studio where they may be inhaled by the potter. It also dampens the bench top, and helps to stop the rolled coils from drying out and cracking as they are made.

Give yourself plenty of room; it is impossible to roll good coils in a cramped space.

One piece of equipment which is a great help is a banding wheel. This will stand on the table top and allow the pot to be turned around without having to be handled. The banding wheel needs to be heavy enough to rotate easily without tipping over. This item can be bought, or made in wood or metal, and should spin smoothly so that the pot may be examined from all sides while it is being made. If no wheel is available, the pot can be placed on a small board or piece of card, which can be turned while the pot is being constructed.

Household tools

Although hands and fingers are the best basic tools, there are other everyday articles which can be very useful when handbuilding. Most bought tools are expensive but many ordinary household items can make ideal tools. These will cost next to nothing and have the advantage of being readily adapted to personal needs. Gradually the potter may develop an acquisitive eye for any old implements which might be used for claywork—and almost any object can be used! These should be kept together in a box and may include the following:

Various pieces of wood whittled and sanded to suit the potter's needs, a cheese grater; old wooden spoons sanded down for beating; worn-down old kitchen knives; twist drills for making holes; old forks and spoons; a needle in a cork, or a hatpin; old paintbrushes; a rolling pin; broken hacksaw blades, ends sharpened, for cutting and scraping; a fine natural sponge; wire of different thicknesses, some fitted with toggles; metal strips for scraping, fixed to a handle; pastry cutters for cutting discs of clay; lollipop sticks; metal strips shaped for specific needs; metal and plastic tubes or pieces of pipe; the list is endless.

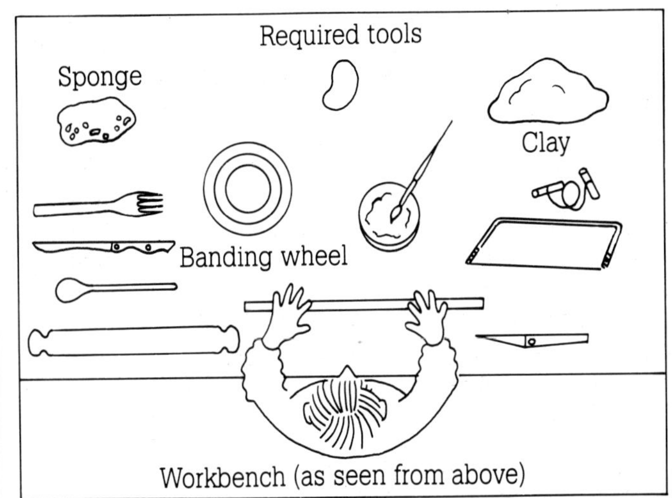

Give yourself plenty of room to roll out the coils. Make sure the bench remains damp throughout; this will help prevent the coils cracking and stops dust flying everywhere.

Making and using coils

1 First you will need to roll the coil. Cut a piece of clay from that which is already prepared. Take a piece that is just large enough for the hands to fit around side by side. Squeeze it out roughly into a rope shape and place on bench top.

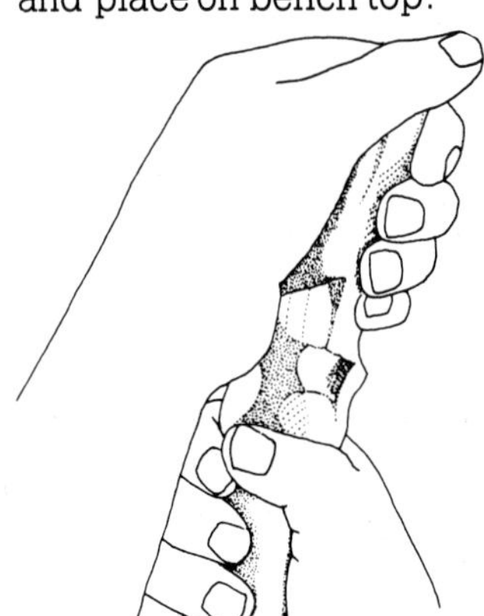

2 Put the hands on the ends of the coil and roll them to a point. (This stops uneven ends

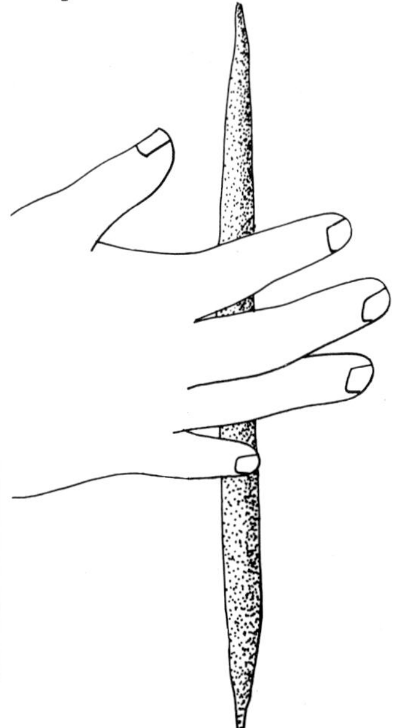

3 When the ends are rolled, place the hands side by side at the center of the coil, with the fingers spread as wide as possible to cover as much of the coil as is practical. Exerting a steady pressure, roll the coil back and forth from fingertip to wrist, rolling it over as many times as possible. It is this action and not downward pressure that produces the best coils.

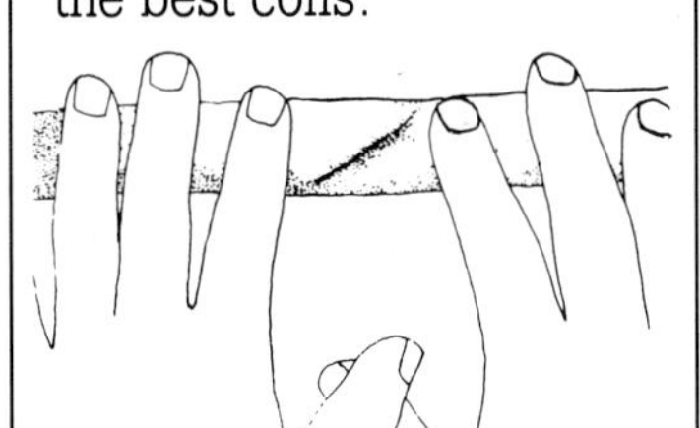

4 Change the position of the hands after each movement so that all the coil is rolled evenly. The coil will lengthen and become thinner but be careful it does not become thinner than the base to which it eventually will be fixed.

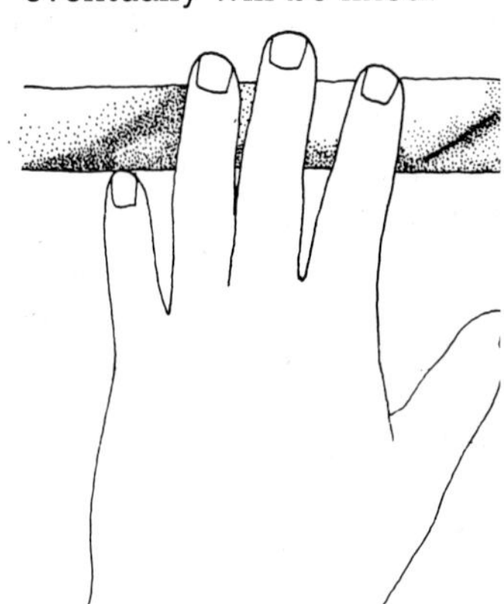

Coiling

Once you have decided on the shape to make, (perhaps, like the thumb pots, it might be based on a natural form), the first requirement is a base. This can be any shape you choose. The coils will follow the base shape but can quickly be altered to change the contours of the pot. For example, a pot with a square base may contain a round or oval middle section and revert to a square shape for the neck. Hollow forms of animals, birds and fish can be coiled, or a pattern of lines and swirls created when coils are placed in a dish mold.

To control the shape of the pot — to make, for instance, the sides spread out or the top narrower — the coils must be placed in the right position. To make the shape wider, the coils are placed on the outside edge of the coil below; to bring the shape in again, the coils are placed on the inside edge. For a straight-sided pot, make sure each coil is placed as evenly and upright as possible on to the coil below.

If the pot becomes soft and floppy while it is being made, it will have to be left to harden slightly before any more coils are added. Place the pot into a bowl or tin which is big enough to support the sides. Alternatively, pieces of clay can be placed around the outside of the pot to support its walls. If the pot has to be left unfinished, wrap it in plastic sheeting, making sure this is airtight, and the pot will then stay in the same condition until you are ready to work on it once again.

The process of handbuilding with coils is a quiet relaxing method of making a large pot of any shape. It is very satisfying to feel the coils being made and see the shape grow, and later scraping the surfaces over with a shaper or leaving the finger-marks to be part of the finished work. The freedom of choice is limitless when using coils rolled or cut from strips or squeezed out roughly. The pot may grow to any dimension unimpeded by the symmetry of a wheel-made pot and the surface treatment at all stages—wet, leather-hard or dry – gives endless scope for decoration. The fingers control the soft clay, the wide surfaces inviting a brush of slip engobe or glaze to add a sweeping design. Whatever the finish, the pleasure derived from the process will more than compensate for the effort of preparation and the more laborious early stages.

Do remember coil pots need not necessarily be round. The advantage of handbuilding (as opposed to throwing on the wheel) is that shapes can be angled and asymmetrical and coiling is one of the earliest and most versatile techniques.

H see pages 255, 300-1

Edwardian woman by Mollie Winterburn (UK)
This coiled stoneware clay pot was inspired by the swirl of necklaces on the bust of a lady from the 1900s. Colored black, off-white and green (using copper oxide and a tin glaze), the pot was given a slow firing. It measures 16 x 14½ inches (40.5 x 37cms).

Making a coiled pot

1 Flatten a round ball of clay on the turntable, placing a piece of paper on the wheel first. This will stop the base sticking to the wheel while you are flattening the clay. When the base is flat and about ½inch (1cm) thick, remove the base from the wheel.

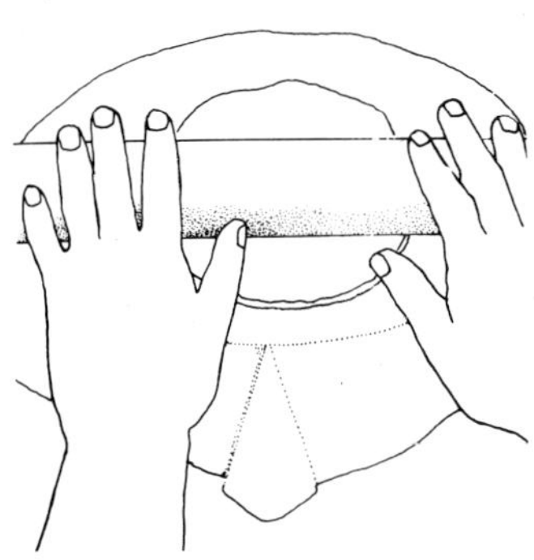

2 The shape will not be completely round and needs to be cut to shape. A jar or tin with the right diameter can be used. Cut around this to make a base, using a template made of paper or card. Alternatively, mark the circle with a needle while the clay is turned on a banding wheel.

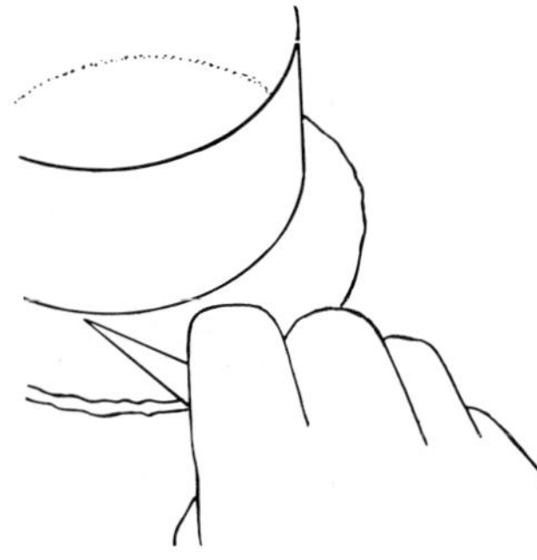

3 Cut the base out with a blade-edged tool and score the outside top edge of the base with an old fork or wooden tool. This helps to provide a 'key' (a surface which aids adhesion) so a good join can be made when the first coil is attached to the base. Dampen this roughened part slightly.

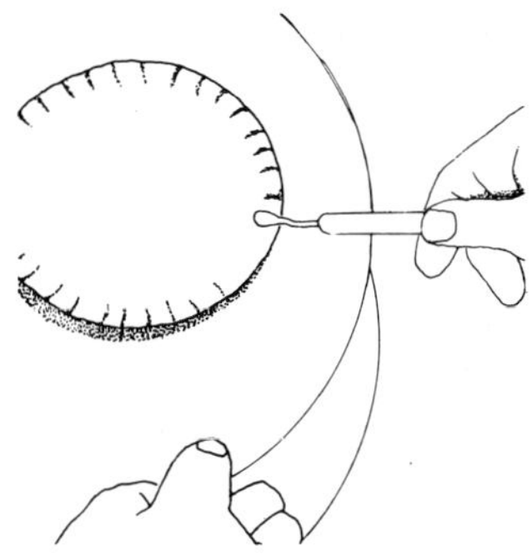

4 Place the first coil around the edge of the base where it has been roughened. Make one coil at a time, with each one long enough to go around the pot once. Join the ends where they meet. If the coil is longer than required, do not be tempted to try and use it all. Cut off the spare clay and make just one coil at a time. This will enable the top edge to be kept even and level.

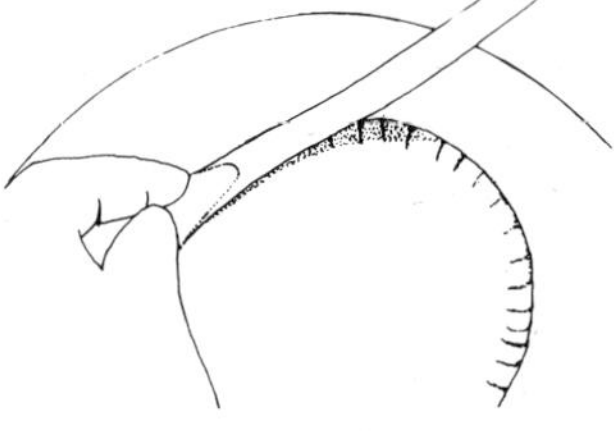

5 The second and successive coils are made in the same way. They must be about the same thickness as the base, and are then placed on top of the first coil. All the coils should be joined on the inside and outside by scraping the clay from the coil into the wall of the pot. In this way a firm join will be achieved, with no gaps showing through the wall.

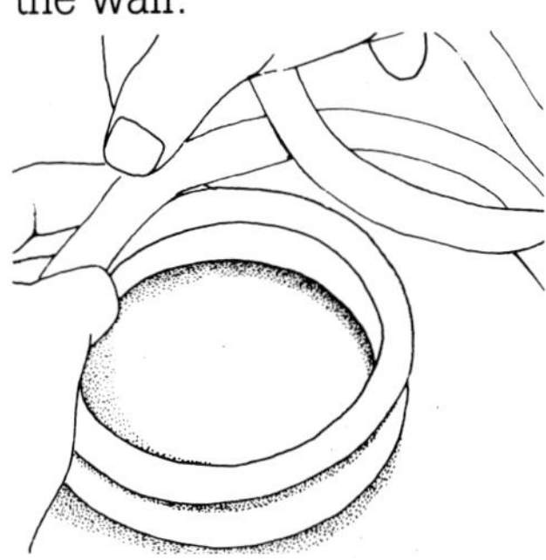

6 The joining of the coils on the inside of the pot will ensure a tight seal, while the joining of the coils on the outside can also serve as the start of a textured finish; the finger-marks might become part of a pattern, or the surface might be smoothed over in readiness for a form of decoration to be applied at a later stage. Much depends on the intended use of the pot when it is finished.

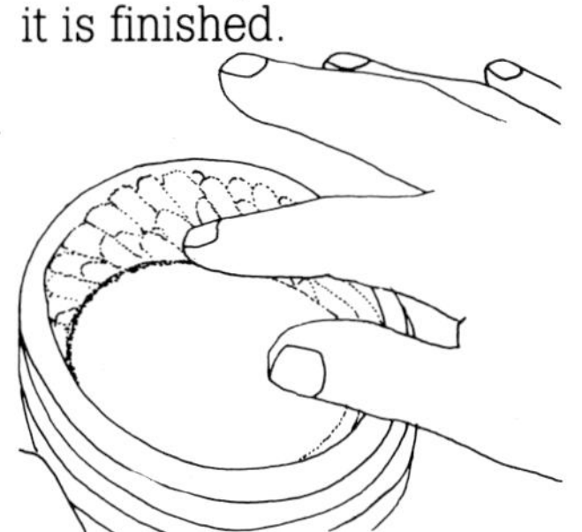

7 Coils placed on the outside edge lead the shape outwards. When placed on the inside edge, they will have the opposite effect and lead the shape inwards. If a straight pot is required, then the coils must be placed directly on top.

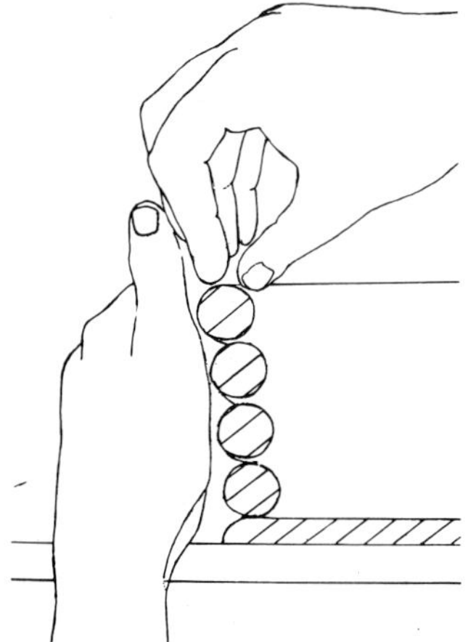

8 Once joined with smooth strokes or with controlled finger patterns, the finished effect may be exactly what the potter requires. If, however, a smoother surface is preferred, the pot can be gently beaten both inside and out with a paddle (a flat slab of wood) or with a wooden spoon. The outside may be smoothed with a flexible scraper to trim the shape and thin the walls.

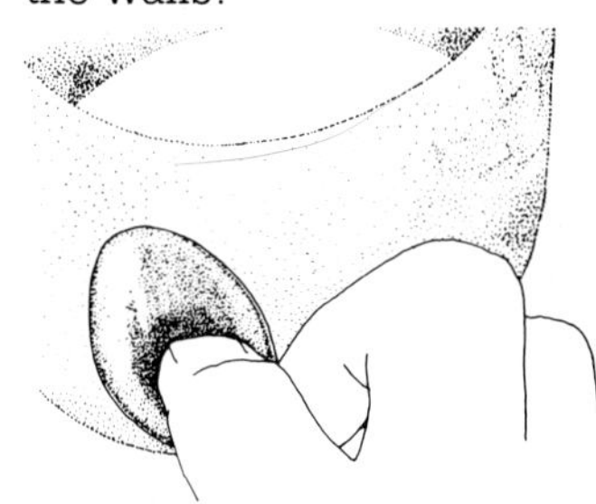

Slab building

Clay which has been rolled or sliced into a sheet is the basis for a wide range of objects which can be made by the slab method. (Although the example here is a simple box, shapes need not be rectangular or have straight lines.) As with all clay products, it is vital to have enough material to complete the job properly. So once the clay has been chosen and mixed (before actually embarking on a new creation), make sure there is enough clay to provide good-sized slabs.

Slabs can be cut from a lump of stiff clay with a wire stretched across a harp.

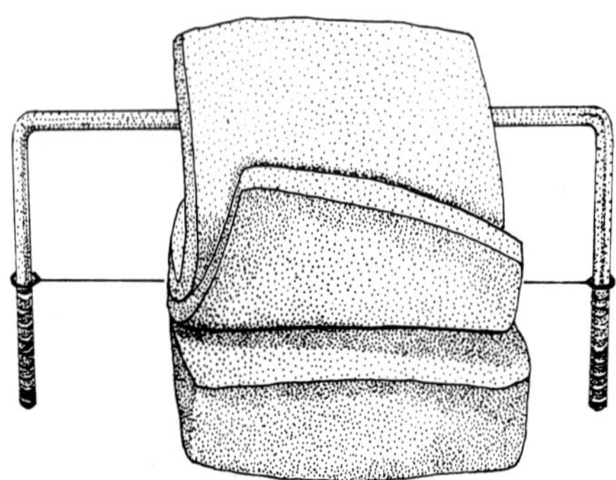

Preparing the clay.

Beat and roll out the clay on a clean piece of canvas; lift it constantly to release the tension and make the rolling easier. Finish by using a clean wooden roller running between two pieces of wood of equal thickness. Lay cardboard templates on to the slab. These will enable you to cut out accurately the sides and base of a box construction. Use a T-square and a sharp thin knife, as shown.

Using an old fork or similar tool, score the edges to be joined. Coat them liberally with slurry mixed from the same clay. When both the surfaces to be joined are ready press them firmly together.

Place a thin coil of clay on the inside of the join and smooth this in to give extra support. Check with a set square that the two pieces are at right angles (90 degrees).

Unless it is intended for press molding the slab will not be hard enough to use at once so it will have to be left to harden on a clean flat board. It must be allowed to dry slowly. Do not try to hasten this process as the slab will warp if it is dried too quickly. The slab should be of a leather-hard firmness when it is ready to use, such that it will stand up on edge without flopping over at all.

Tools for slab building:

A straight-edged measure
A set square to ensure accurate angles
A knife
A needle in a handle for trimming spare clay
An old fork for scoring the edges to be joined
A fine sponge for finishing
A pot of slurry (of the same clay as the pot) with a stiff bristled brush for its application
Metal and wooden scrapers for finishing

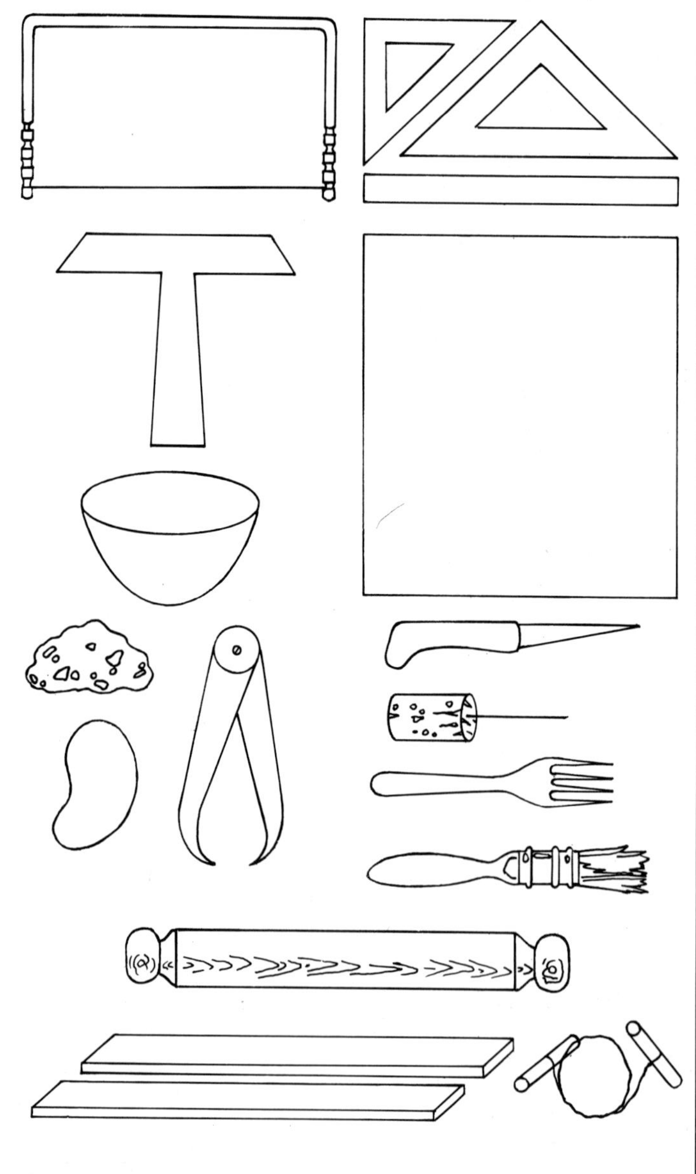

Making slabs

1 Place the clay on a flat table or board, which is usually covered with a piece of linen or similar cloth so that the slab can easily be lifted on to a board or into a mold. Beat the clay with the flat of the hand, working from the center outwards and gradually thinning the piece. Lift the slab frequently, to relieve the tension in the clay and allow it to stretch and grow without sticking.

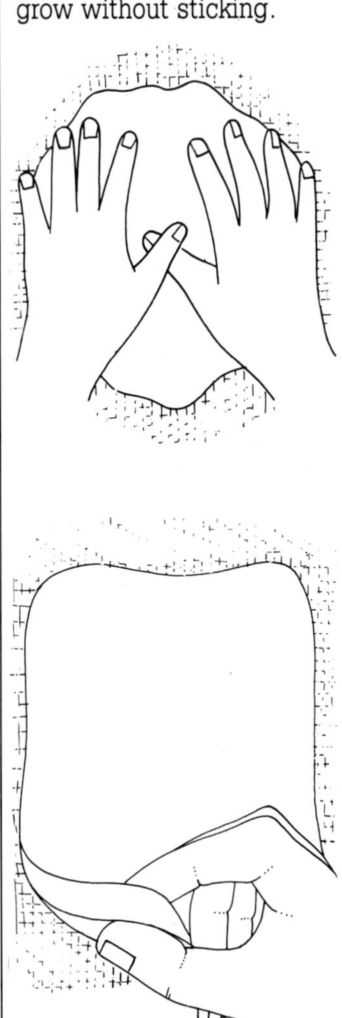

2 When the slab has been beaten out, place one slat of wood on each side of the slab. These slats must be of equal thickness and will dictate the finished thickness of the slab as the beaten slab is then rolled with a wooden rolling pin.

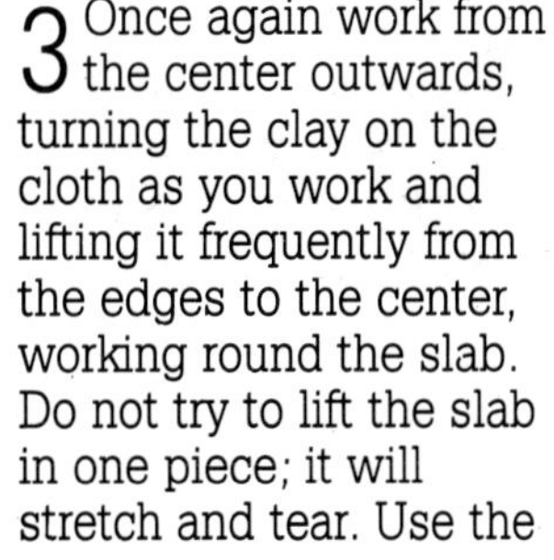

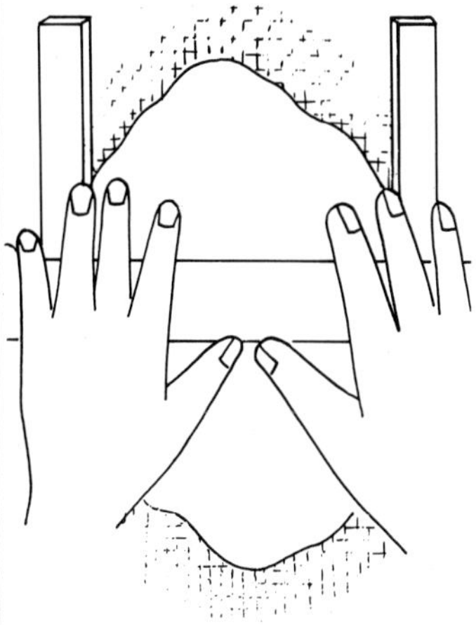

3 Once again work from the center outwards, turning the clay on the cloth as you work and lifting it frequently from the edges to the center, working round the slab. Do not try to lift the slab in one piece; it will stretch and tear. Use the rolling pin to support the rolled slab when moving it or lifting it. Eventually the roller will run along the slats of wood, and this will indicate that the slab is of an even thickness, equal to the slat height.

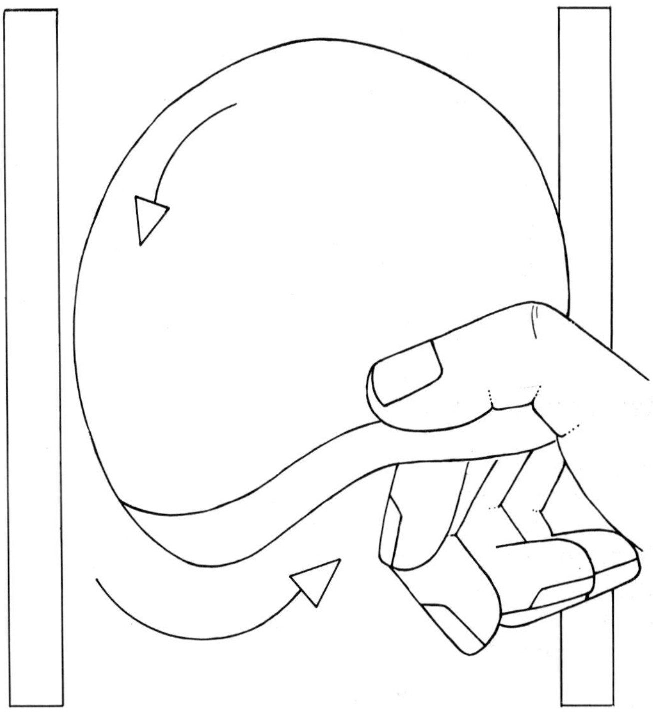

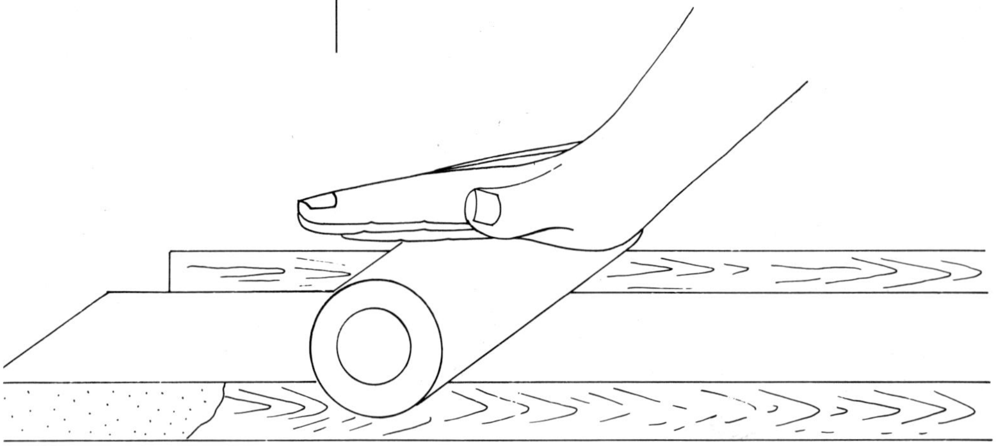

Making slab pots and boxes

Making a slab pot

1 First place the slab on a clean, dry, absorbent surface—a wooden board or table. Then position the template on the slab. Place the straight edge over the template and cut out the shape. Templates cut into curving profiles with scissors can produce some interesting shapes, especially three-sided pots.

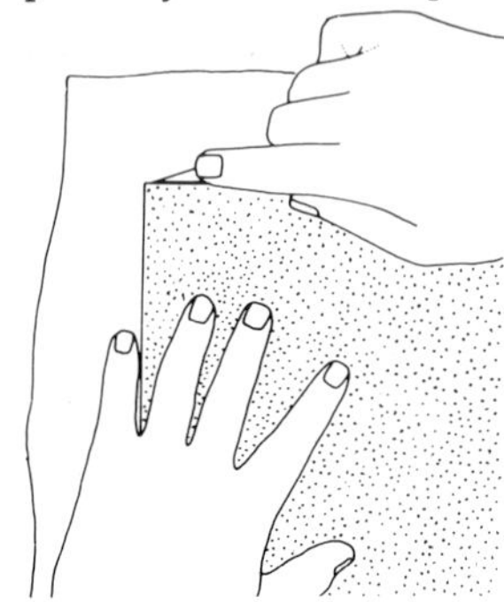

2 Ensure that the knife is kept at a constant angle while cutting, usually at ninety degrees to the bench top. Cut out all the sides, putting the base carefully aside until the sides have all been joined together.

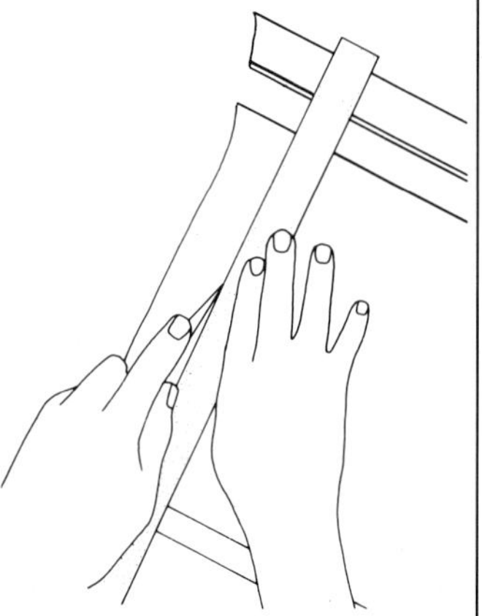

3 Take two of the sides to be joined and score down the edges that will touch each other, using an old fork. Coat the scored parts with the slurry.

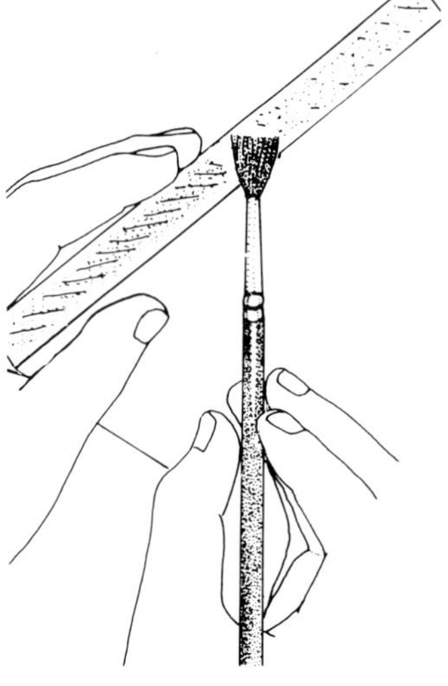

4 Press the two sides together, tapping them lightly to make a firm join. Roll a thin coil and place it on the inside of the join, smoothing into the angle to give extra support.

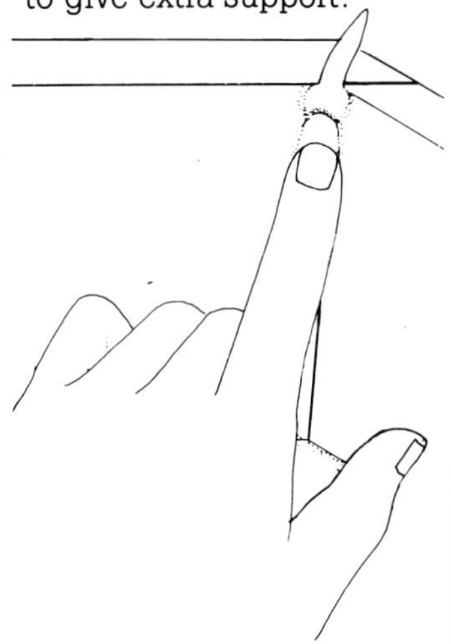

5 Use the set square to check that the sides are fixed at right angles to each other. Continue in the same way to join on the other two sides. Take great care when cutting and handling the slabs and when joining them. Carefully smooth the joins and the surface with a sponge and scraper.

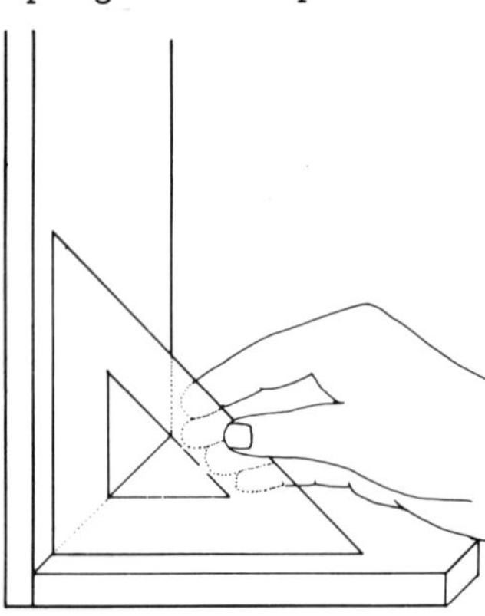

6 Lift the assembled form and place it carefully on the base to join in the same way; the weight of the pot will help the base to stick and keep it flat. If the pot is fragile, support it with blocks while it hardens. Slow down the hardening by covering the pot with plastic sheeting to prevent the pot warping or cracking open at the joins.

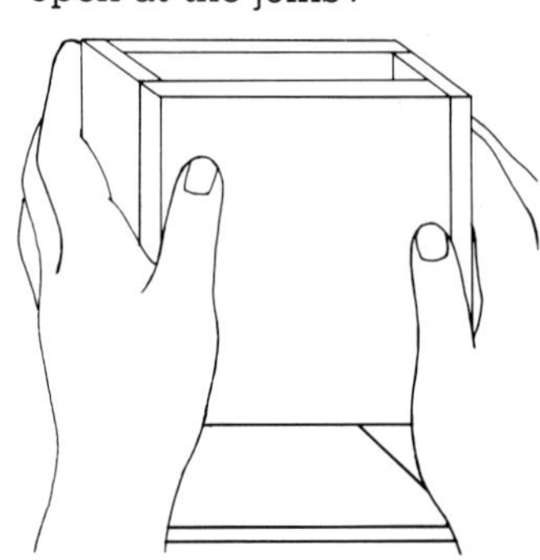

Cloud reclining woman by Lana Wilson (USA)

The square box

1 Each side is equal in size; the ends butt neatly into each other.

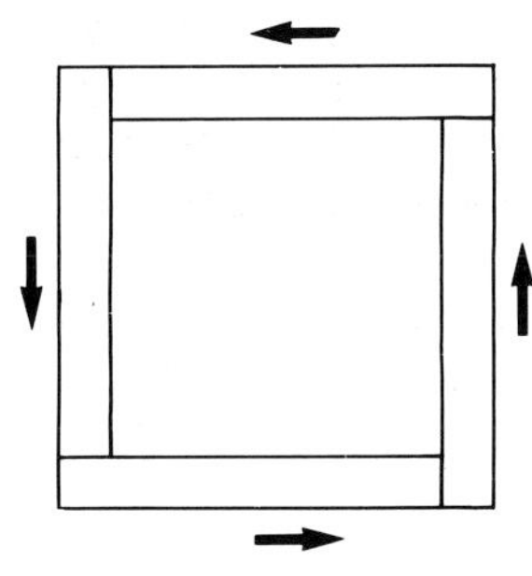

2 The rectangular box: this has two short and two long sides neatly butted together.

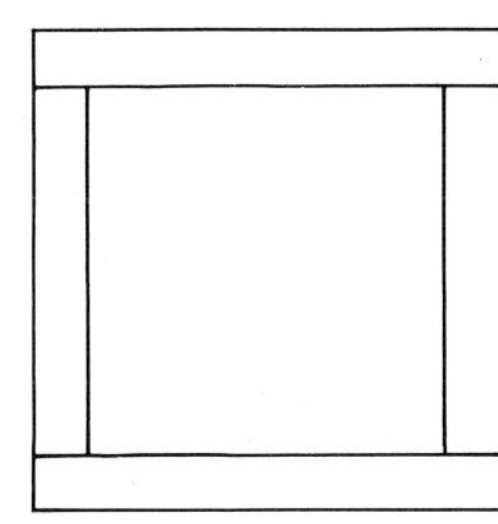

3 Angles may be cut to allow for more than four sides.

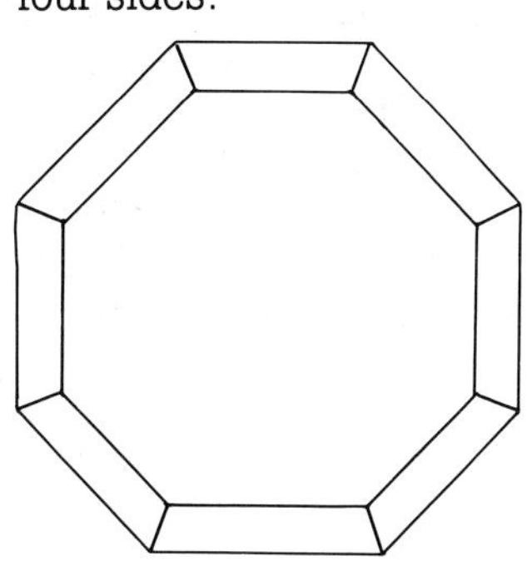

4 When the sides are added, these may be fitted to sit on top of the base or be joined on to the sides of the base.

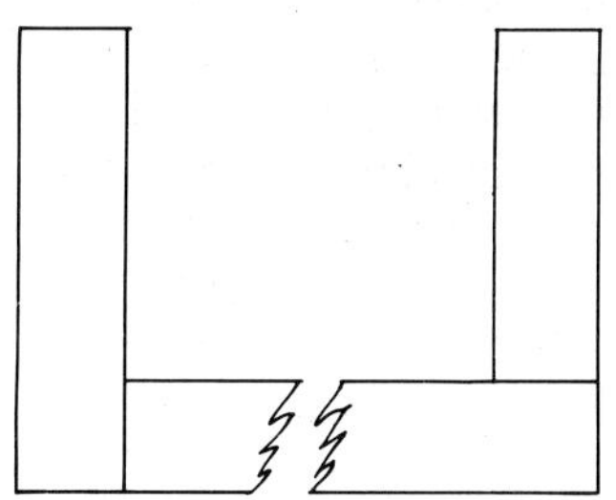

Construction of a lidded box made from slabs.

1 First construct the four sides. Place these on to the base and *then* cut out the base—not before. Check at all times to make sure the angles are square.

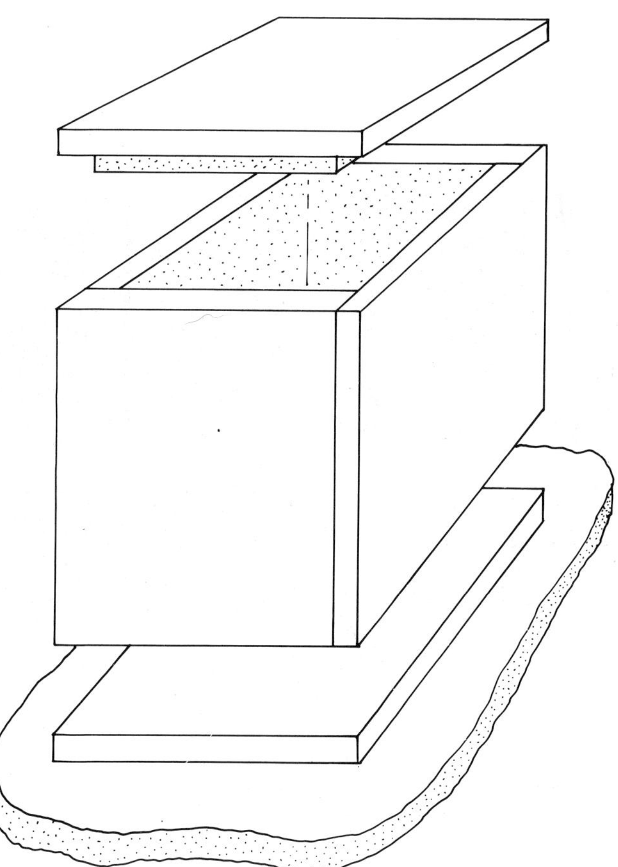

2 Lids require some form of 'locking' device. Here they are shown upside down to illustrate various ways in which they can be made to fit neatly on to the pot.

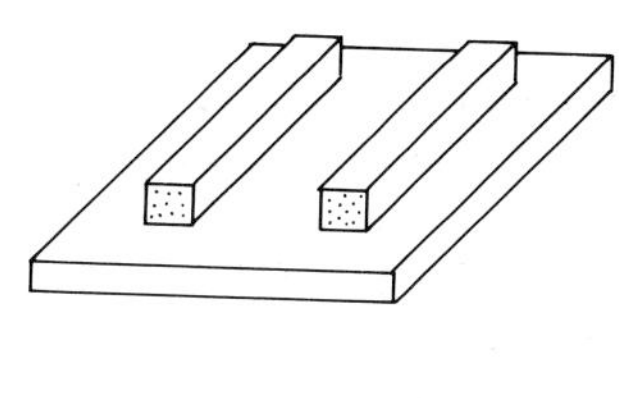

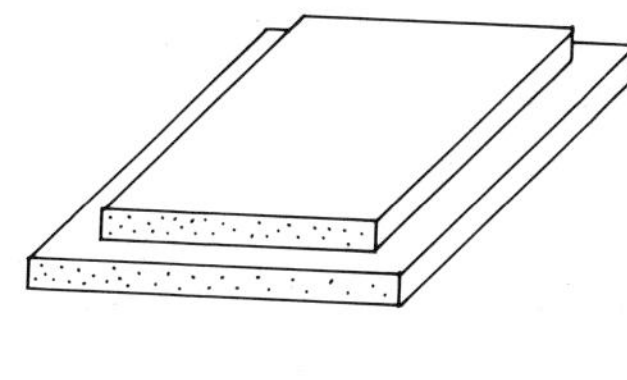

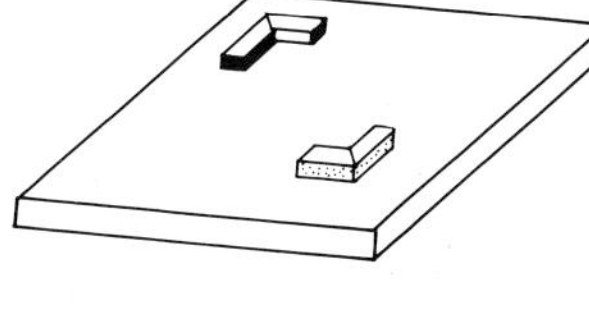

3 Sometimes the lid may be a piece cut from a sealed cube.

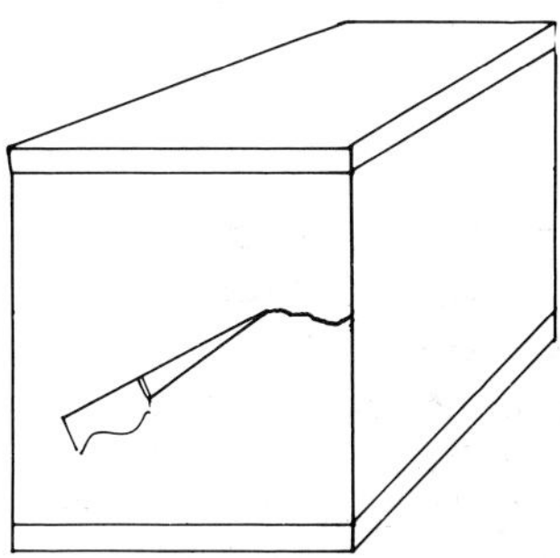

Top part of box cut through to form a close-fitting lid.

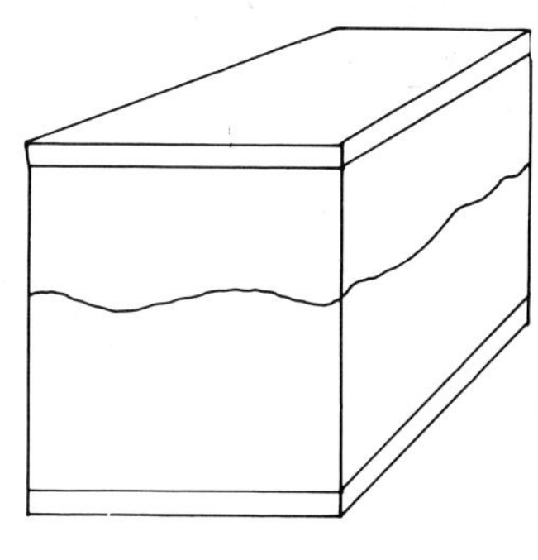

Soft slabs

Using soft slabs

Soft clay slabs will take on the shapes of objects they are draped on to or over or into, such as a stone or a recess on a brick surface, or an indentation in the ground. Cover the surface to be molded with a layer of damp newspaper to allow the clay to be detached easily when hard.

A free-form curved dish can be made by tying the four corners of a piece of cloth to the legs of an upturned stool. The cloth will form a natural curve which will be adapted by the clay placed in this hollow. A flat base or a foot ring may be attached when the leather-hard dish is removed (see page 60).

For a flat-sided shape, such as a box or trough, templates of card or paper can be cut around and then used to form the sides and base.

Soft slab pot

Using slabs that are soft and pliable it is possible to create pots by draping or rolling these slabs around a former. Useful formers can be found in everyday objects such as cardboard tubing, bricks and blocks of wood.

Soft slab box by Sandy Brown (UK)

1 Soft slabs may be formed around a tin or plastic container or a tube of cardboard. Soft clay tends to stick to tin or plastic so this should be covered with paper first. Cut a straight edge along one side of the slab, place the paper-covered form on the clay and then roll the clay around it.

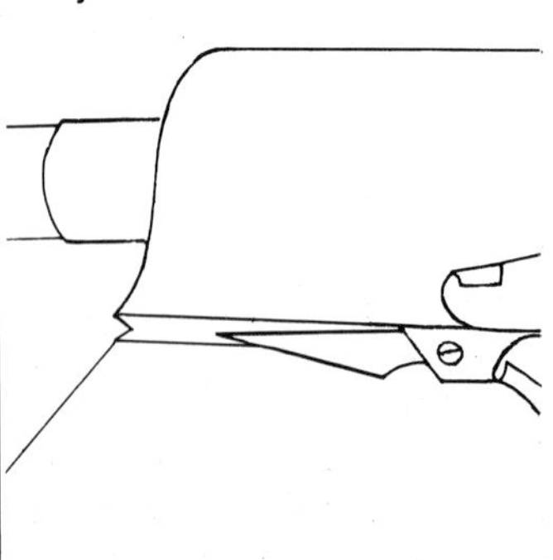

2 Join the two ends by cutting them carefully to butt together, and fix with slurry. The join may also be made by overlapping the ends and then scraping the spare clay away with a steel scraper. Alternatively the ends may be pinched together and the spare clay trimmed away.

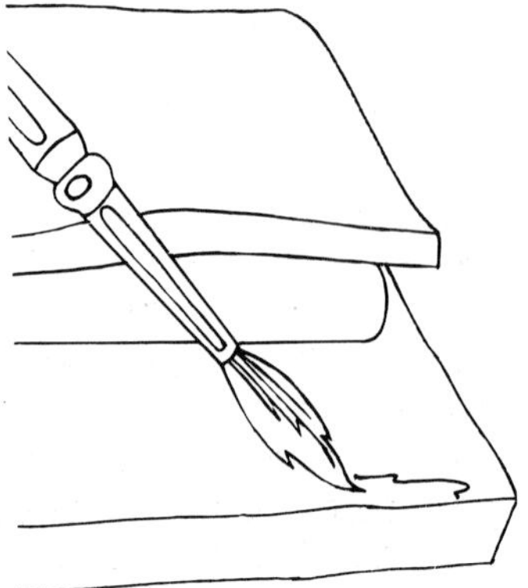

3 The pot (with the former in place to help retain the shape) is then fixed to the base in the same way as before. Remove the cylinder that has been used as a former straight away so that it does not become trapped when the pot dries and shrinks.

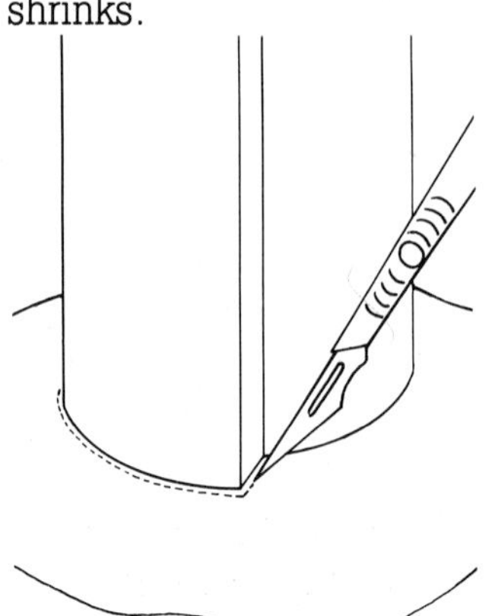

4 The edges can be butted together and then smoothed over. Alternatively, they can be overlapped and the extra thickness incorporated as part of the design.

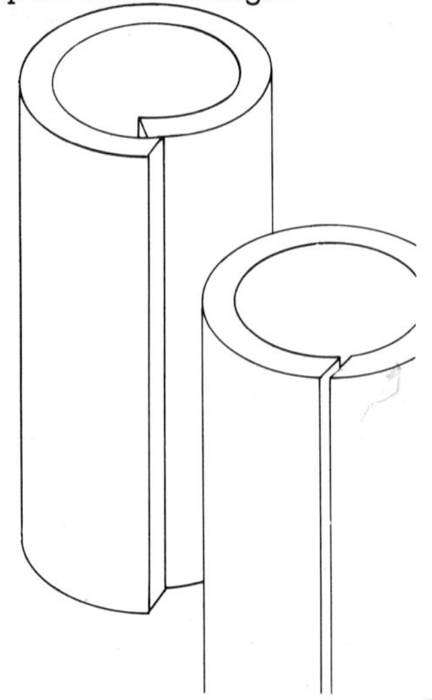

1 First the slab is rolled around a tube. Join the ends by butting them together. Alternatively, overlap the ends and use the overlap as a decorative feature, as shown.

2 Mugs can be made using the soft-slab method.

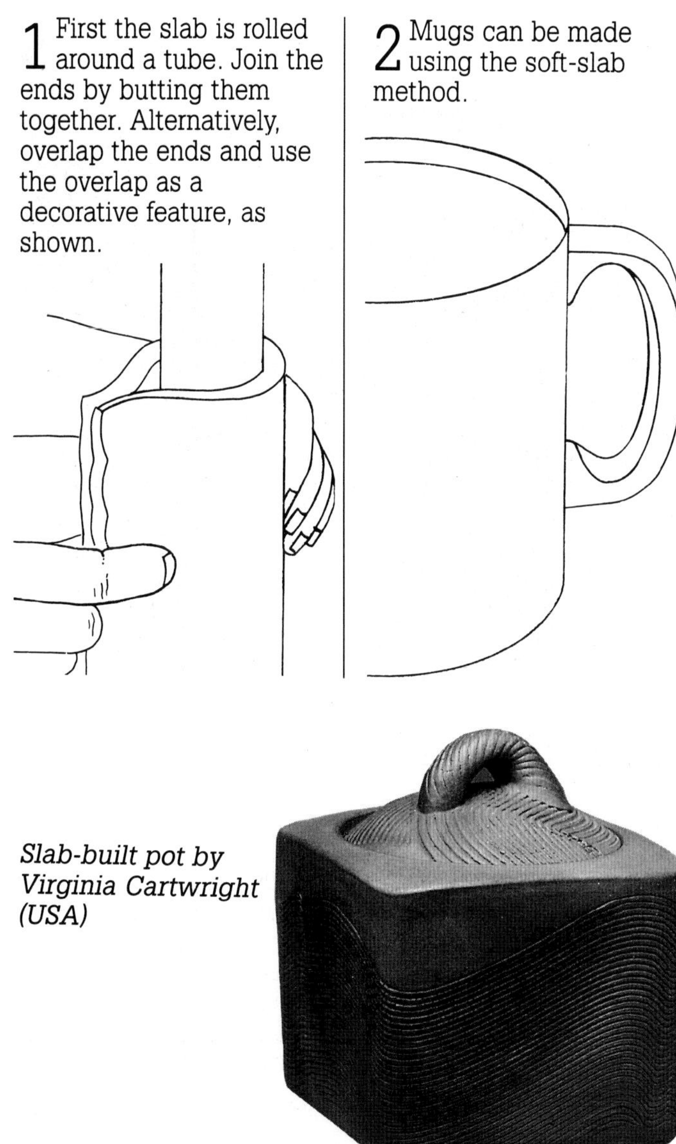

Slab-built pot by Virginia Cartwright (USA)

Handles

If the slab-built pot requires handles or has a lid that would benefit from a knob, these are usually best complemented by hand-built additions to suit the style.

A pulled handle often looks too delicate and out of place on a sturdier handbuilt form. Pierced and cut slab pieces may be attached at the required number of positions around the pot either for decoration alone or in order to lift or handle the pot.

Similarly a handle or knob to lift a lid can be a decorative construction of slabs, coils, pinched or modeled forms to reflect the design on the pot. (See also page 102)

3 An addition of free-flowing coils can make an interesting contrasting feature on a straight cylinder.

Offcuts of the main slab can be used; cut and pierced, rolled, or carved.

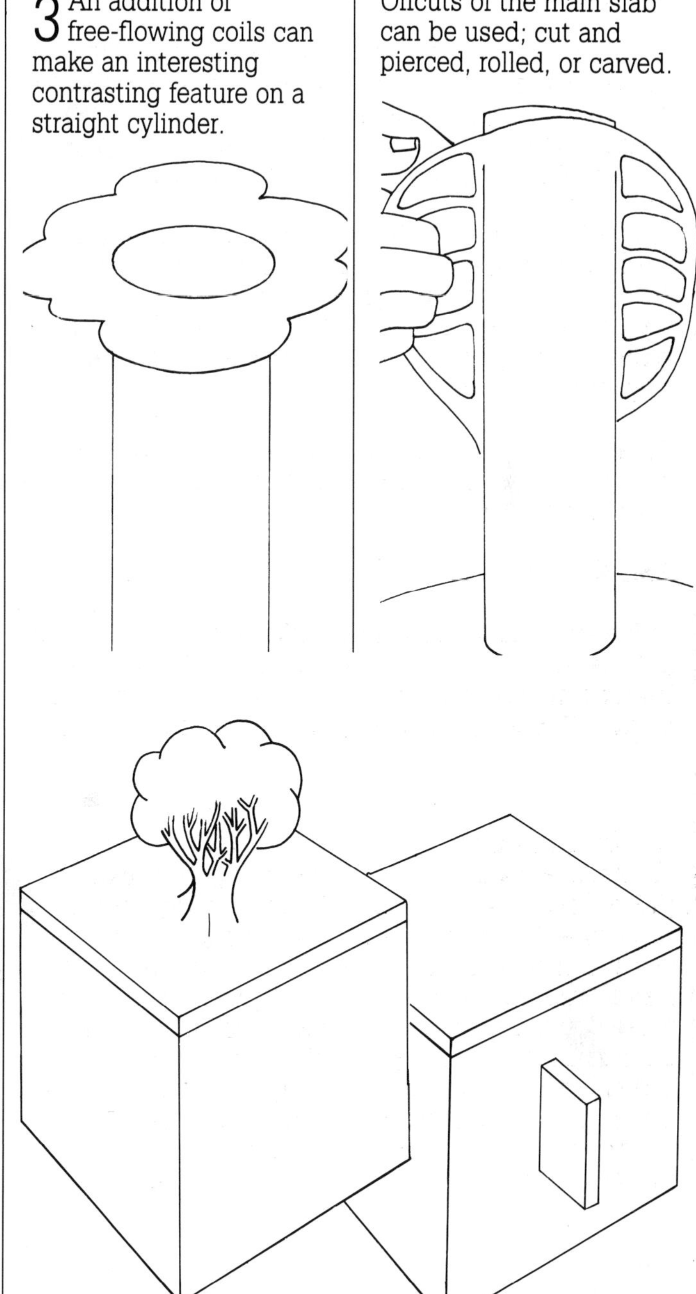

The method for attaching any additional pieces is the same as for joining any two pieces of clay. Roughen both surfaces, cover with a layer of slurry and press firmly together. To ensure handles are symmetrical, attach the first one at the chosen point. Measure its distance from the rim and place a ruler across the top of the pot from the position of the first handle, and mark the exact position required. Measure the distance from the rim.

Alternative forms

Slab forms do not necessarily need to be oblong or square. Using templates, many different styles can be created.

Henry Pim constructs his slab-built forms with cardboard templates—first to establish the eventual shape, and then as a pattern in order to replicate the shape in clay pieces. This means that the slab is not used as a flat surface with right-angled joins. Instead the sections are curved at the semi-hard stage of construction and joined carefully. The joins are smoothed with a clay plane, or open-textured blade called a 'Surform'. Care and precision is needed at this stage. Slow drying and keeping the piece damp (under plastic or in an airtight bin) will reduce the risk of these curved joins cracking or warping. Many curved or asymmetrical shapes can be created this way.

Free construction with slabs opens up many possibilities. Models of buildings, furniture, motor vehicles and so on all lend themselves to this technique. Some of the finest slab construction was created by Chinese sculptors.

Victorian washstand by Graham Flight (UK)
The straight lines of the slab-built washstand make an interesting foil for the thrown pieces standing on top. Victorian-style painted tiles complete the effect.

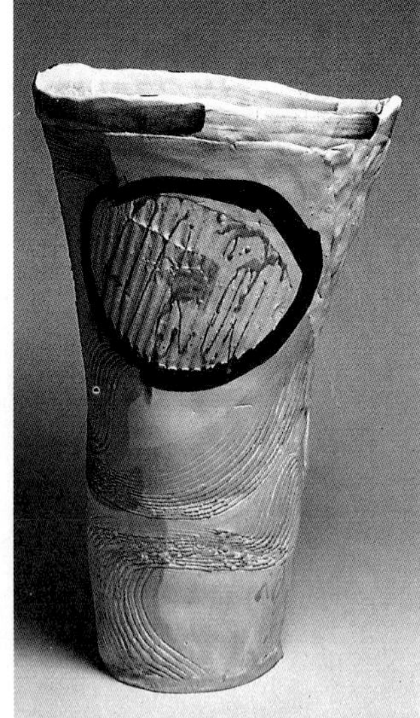

Slab pot by Sandy Brown (UK)
Textured slabs have been used to create a pot decorated with slips, engobes and oxides — added to a transparent glaze and then applied like paint.

Handbuilt stoneware peg made by Delan Cookson (USA)

Shopping cart by Paul Astbury (UK)
Slabs of tailored pieces of porcelain clay have been wired on to a wicker and wood cart. After firing they were polished with wet and dry emery cloths.

Slab form by Henry Pim (UK)
The versatility of slabs both in form and surface treatment is shown in this fascinating piece.

Vintage car by Graham Flight (UK)

Exploiting the medium

Chalice by Henry Pim (UK)
The profusion of holes in this handbuilt chalice form casts interesting patterns of light. It is $12\frac{1}{2}$ inches (32cms) high with a burnished surface.

Platters by Robert Washington (UK)
Hyplas, fireclay and T-material have been used for these two flat platters with engobes and stoneware glazes. Fired four times they show good synthesis between the clay and paint.

Teapot by
Angus Suttie (UK)
This earthenware piece $8\frac{3}{4}$ inches (22 cms) high exploits the slab form to produce an oddly-shaped multicoloured teapot.

A free-form dish

Roll out a thin slab on to a clean cloth. Suspend the cloth from each corner, allowing both the cloth and the clay slab upon it to take up a curved shape. (An upturned stool or chair will provide you with four good equally-spaced anchorage points). Try decorating the slab with free-slip patterns.

Slab-molded dishes

Shallow dishes are generally made from soft slabs. To do this, it will be necessary to use molds which are made from plaster of paris. These may be bought in a variety of shapes, or can quite easily be made to the potter's requirements. The molds are porous and dry so the clay does not stick to the dishes when it is placed in or over them. The open dish mold will shape the outside contours of the dish but will leave the inside to be shaped or decorated as required.

When the dish is leather-hard it can be removed from the mold. Place it the right way up and clean and smooth the edge with a damp sponge and your fingers. A sharp edge to a dish will chip easily when dry or fired.

Dishes can be fragile while drying, so handle them carefully, lifting them underneath with both hands. Leave them face-down on the board to dry; this will help to keep the dish level without warping. As with all slabwork, slow drying is essential.

Spherical form by Yvonne Boutell (USA)
Textured patterns have been added to a porcelain slab before using the plaster mold shown here to create a spherical shape.

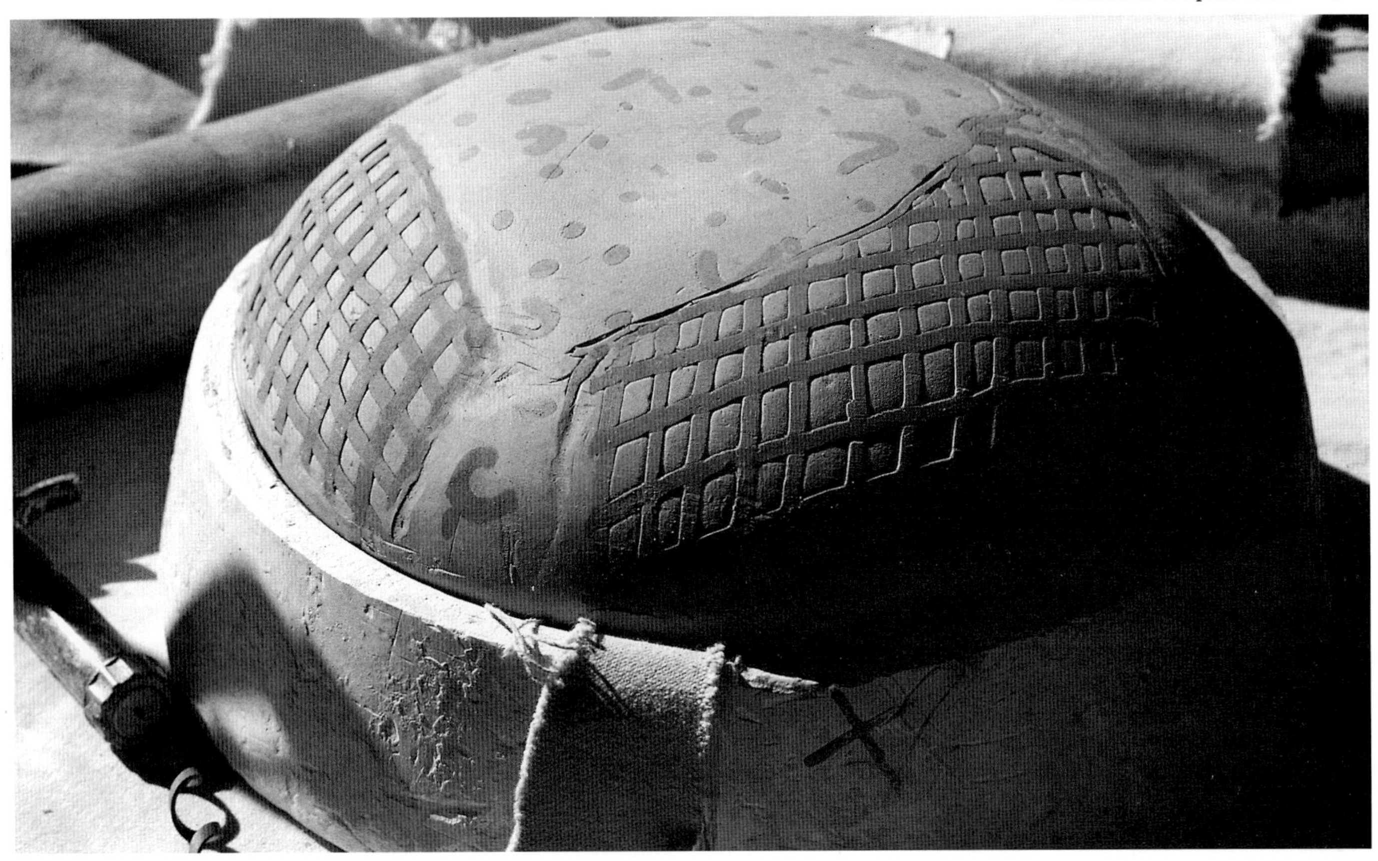

A free-form dish

1 Brush, trail or pour slip decoration on to the slab before it is suspended on the cloth.

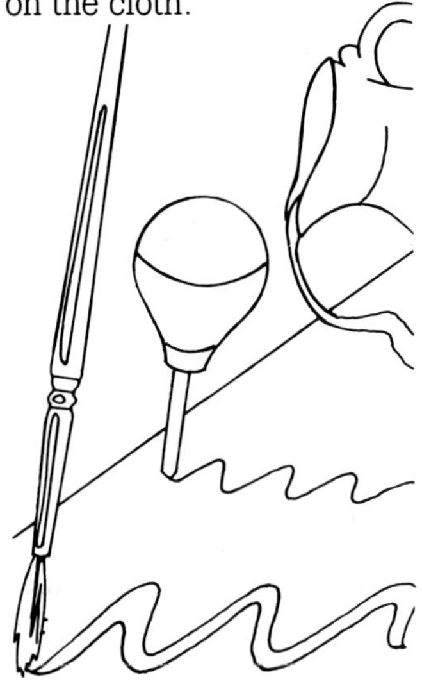

2 Using string, attach the four corners of the cloth to the chair legs and so suspend both cloth and clay in mid-air.

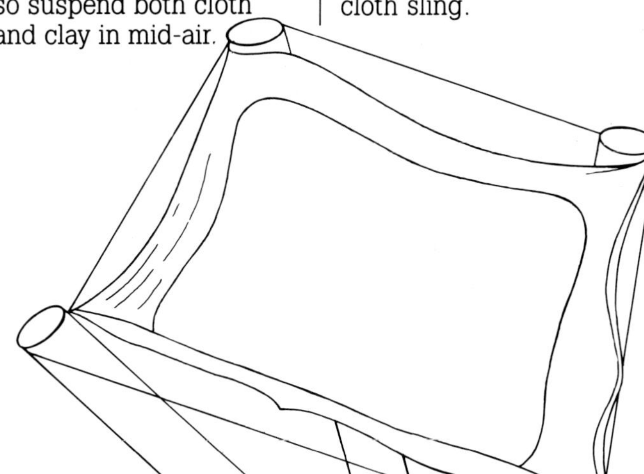

3 Leave the clay to harden, still hanging from the chair inside its cloth sling.

4 A strip cut from the slab can be linked into a loop and used as a footring; it can be made into a perfect circle by shaping it around a former, as shown.

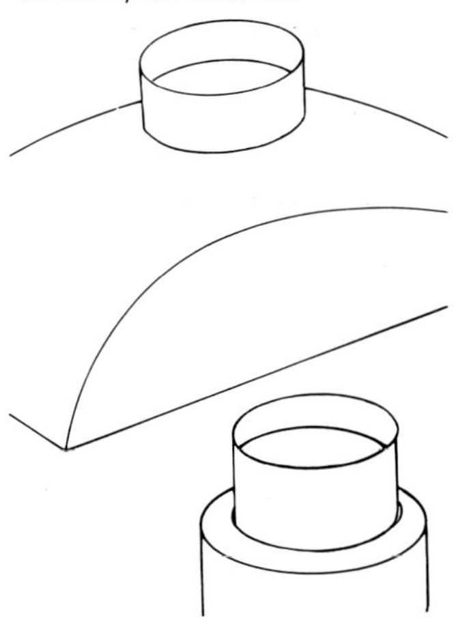

Slab-molded dishes

1 Roll the clay slab, as before on a clean piece of linen. Pick up the slab on a rolling pin or on the forearm, by sliding the hand and arm under the linen; carefully place the clay, face-down, in the mold and then peel off the linen. In this way, the smooth rolled side becomes the side which is in contact with the mold and which will eventually become the outside of the dish.

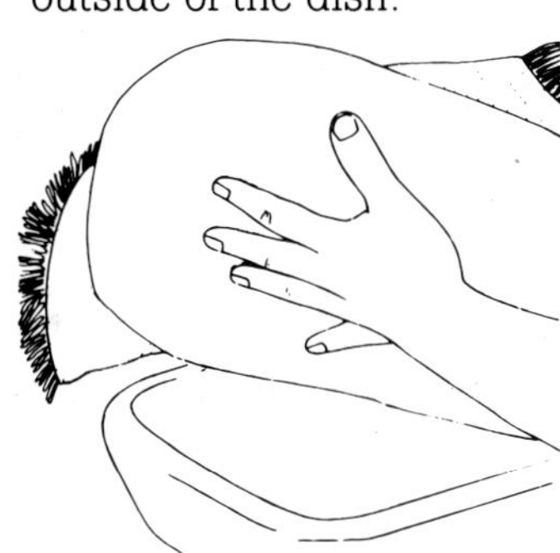

2 At this stage the slab will overlap the sides of the mold. Do not be tempted to push the clay into the mold but instead lift the sides and ease the clay down until it touches the mold all over the inside. Now smooth the inside of the dish, using a rubber kidney that has been dipped in water.

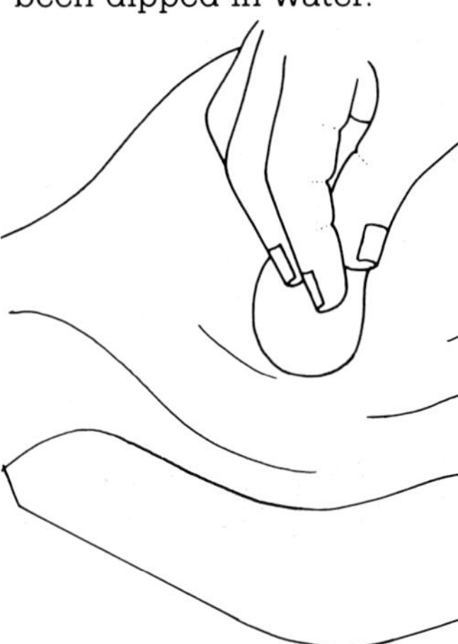

3 When this is done, stretch a thin wire over the top edge of the mold (or use a mini-harp) and cut off the spare clay so that it is level with the top of the mold.

4 It can be removed from the mold later by placing a flat board over both dish and mold and then turning out the dish on to the board to dry.

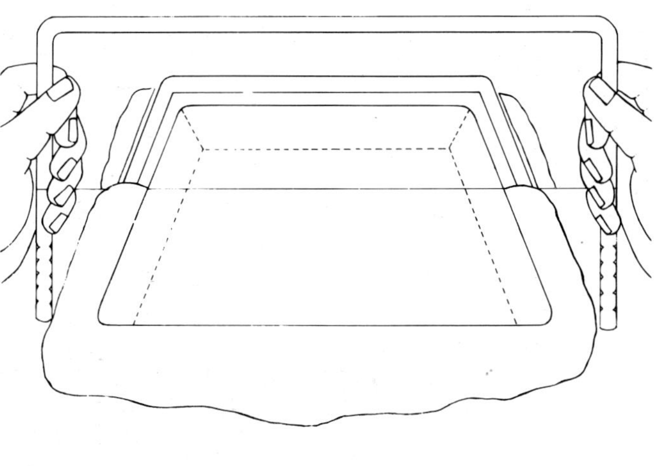

Modeling

Modeling images with clay, especially portrayals of the human figure, was one of the earliest ceramic art forms used in burial and fertility rites. It is also one of the first forms a child will attempt to make when presented with a modeling medium.

Requirements

For large pieces use a grogged clay. This will reduce shrinkage and cracking.

For smaller pieces use a smooth clay to allow for more intricate detail such as lines on faces. Smooth clay will be easier to work with. Remember that individual sections of a model should not exceed 1-2 inches (2-5 cms) in thickness or they may explode during firing; the pieces are solid and steam cannot readily escape. Wherever the clay is thicker than this it should be hollowed out—with a small hole made somewhere to allow the air to escape. ◇

To make a small figure

Individual parts of the figure are first modeled roughly and then joined with slip to create the required pose. Make sure any part which protrudes from the main body is supported (using, for example, a spare piece of clay or newspaper) until it is self-supporting. Do not rush this stage as slow drying is essential; moreover extra detail will need to be added while the clay is still leather hard.

During this process the model should be kept damp by covering with plastic material or placing in an airtight tin.

A different approach is to carve out a piece from a solid lump of clay. Removing the excess material allows the subject matter to emerge, in a similar way to stone carving. Again, ensure that any thicker parts are hollowed out or rendered safe — by piercing holes, hollowing from underneath and removing excess clay wherever possible.

C see pages 251 256 267 279 293 311

Coil sculpture

Coils can be used to create sculpture in this way. The hollow form allows larger pieces to be built while controlling the contours of the eventual shape. Remember to support any soft coils, during construction, with balls of newspaper which can be removed before firing. (If removal is impossible they will safely burn away during firing.) Do allow maximum ventilation for the escape of smoke. ◇

Heads

This coiling technique can be implemented when constructing a head. Normally a clay model of someone's head, when being sculpted, is made on a metal support called an armature. Obviously, this cannot be used when pottery is the medium—because during firing the armature would melt. The coiling technique creates a hollow form which can then be added to gradually as the coils stiffen and in this way little support is required. (If any further support becomes necessary, once again newspapers will be an adequate prop.)

Finer details can be modeled either into or on top of the shape after the coils have been smoothed together.

Sometimes heads may be made from a solid piece of clay which supports the overall shape during construction, and the adding of finer detail. To allow safe firing, the piece must be hollowed out from underneath with a looped modeling tool or sliced in half, the two separate pieces being hollowed out and subsequently joined with slip. Smooth the join and disguise it by remodeling the detail. A convenient line of cut is through the hairline, thus avoiding the facial features and allowing the final hair details to disguise the join.

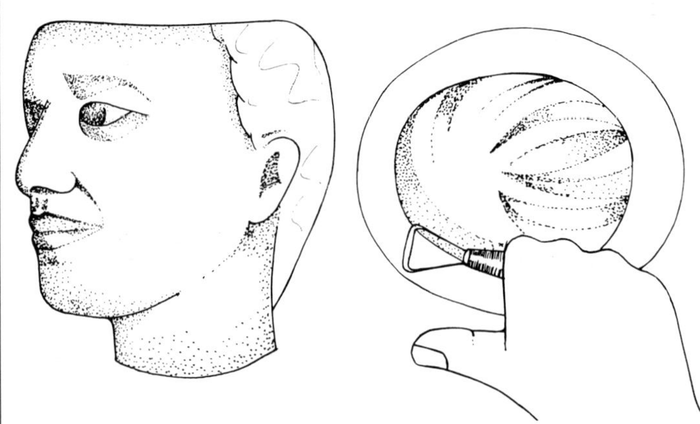

Mixed techniques

All the techniques described in this chapter on handbuilding can be combined and used together as well as being linked with the throwing techniques described in the next chapter. The possibilities are limitless and beginners should not feel restricted to using these techniques in isolation from each other. They should feel free to use any combination of techniques to achieve the hoped-for result.

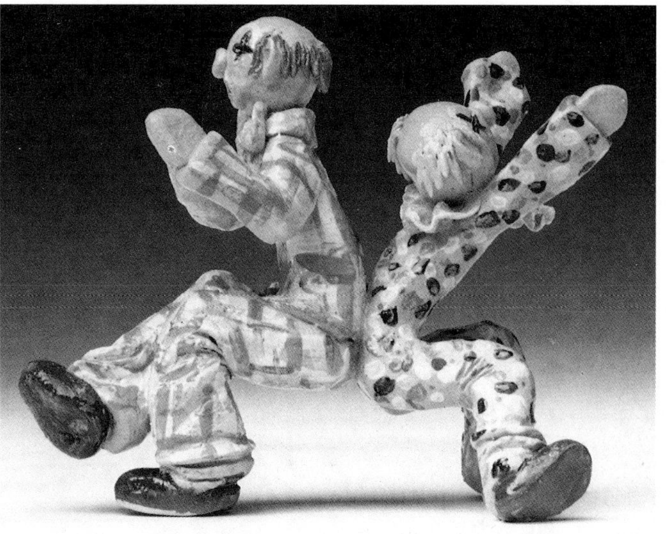

Clowns
by Sally Cocksedge (UK)
These pairs of clowns illustrate how groupings of figures often are not only more interesting to look at but also how the figures can support each other.

C see page 44

Handmade figure
by Hilary Brock (UK)
A combination of stoneware and porcelain pieces have been joined with slip and fired together to create a more intricate and sophisticated design with detailed finishing touches and decoration.

Making clay figures

1 Shape the head, remembering that no section should exceed 1-2 inches (2-5 cms) in thickness unless hollowed out. Make clay coils for the torso, arms and legs. Alternatively, create the legs by splitting the bottom section of the "torso" coil to form two legs: this in fact helps make the structure sturdier.

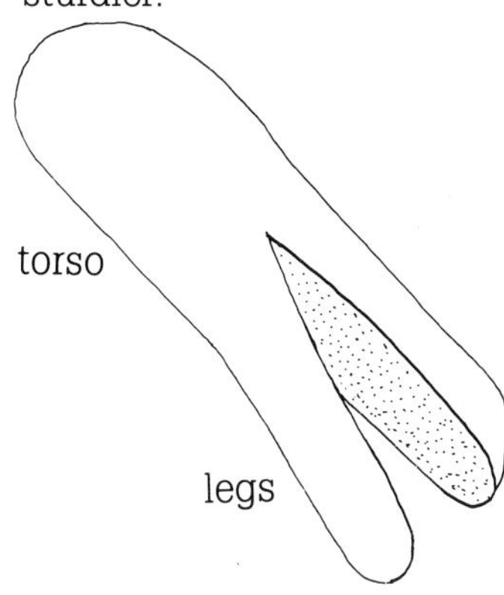

2 Attach the limbs and head. Remember that the model must be self supporting so bear this in mind in the overall design. For instance, a standing figure can be supported by a tree-trunk, bench or platform.

3 Modeled details can be added to complete the effect. This can be done by carving, using coiled additions or cutting from rolled sheets of clay to suggest clothing, features or hair.

4 Larger hollow shapes can be made by rolling clay around a tube for the body, and around a pencil for limbs (remembering to make a small hole so air can escape).

3 Throwing

The art of throwing

Throwing is a fascinating method of producing pottery that is basically symmetrical. (It may be altered or added to later and become asymmetrical.) The potter at the wheel is often the first image that springs to mind when the uninitiated visualize the potter at work. It is certainly absorbing to watch a skilled potter at the wheel, creating pieces in a seemingly relaxed and simple manner. However, a good deal of experience is required before the necessary skills are mastered.

The term 'throwing' has evolved from the way the centrifugal force of the wheel 'throws' the clay ever further from the center. For the potter holding it there it feels as if the clay could actually be thrown off. The clay must be restrained, controlled and manipulated — smoothly and rhythmically. This takes some practice in order to succeed and then to produce a pleasing shape; several factors must be co-ordinated and balanced: the state of the clay, the speed of the wheel, the bond between hand and eye, and the sensitivity of the potter's touch as he or she responds to the feel of the clay between the fingers.

Inevitably there will be many mistakes during the learning process. (See also page 320). In time, the novice learns how to control the ever-changing curve of the spinning pot and develops the techniques necessary to shape the clay as desired. Once the basic skills are mastered, throwing pots will be a source of endless fascination and satisfaction.

In general the methods described in this section are those used by the author but every potter will develop his or her individual approach so these suggestions should be regarded as useful guidelines rather than being seen as hard-and-fast rules.

Chinese wheel, being driven by assistant, showing the holes in the edge of the flywheel into which a stick was inserted.

The Indian potter's wheel was supported on a spike fixed in the ground and turned with a stick to gain momentum.

The potter's wheel

The potter's wheel was being used by the Ancient Egyptians at least four thousand years ago. Wall paintings of this age have been discovered depicting clay preparation, throwing and firing.

The use of the wheel spread gradually; by 1500 BC most of the countries of the Middle East were producing highly skilled ware with lids, spouts, handles and superb decoration. The eastern countries of China, Japan and Korea were using the wheel in 3500 BC. It is not known if the knowledge of the wheel spread here from the Middle East or if it was discovered in China and Egypt simultaneously. The Romans scattered their potters throughout their empire, spreading the development to Europe. However, the use of the wheel was not seen in America until fairly recent times — handbuilding of pots was still the accepted method.

The basic construction of the wheel differs hardly at all from today's kick wheel. A small flat disc was set on a shaft with a heavier wheel at the base of the shaft — rather like the axle of a cart with one large wheel and one small one turned on its side. The larger wheel at the bottom of the shaft was made of heavy wood or stone, the weight providing momentum for the smaller wheel. The base wheel could be spun with the foot to a reasonably high speed, so that the potter could sit comfortably to perform the centering and throwing of a pot.

A hand-operated wheel photographed at Littlethorpe Potteries, Yorkshire, England. The wheel was turned by an apprentice or assistant who operated a handle to turn the main spindle while the potter threw the pot.

An eighteenth-century English potter's wheel was driven by an apprentice who turned the large wheel at the side. This then drove the potter's wheel by a rope belt.

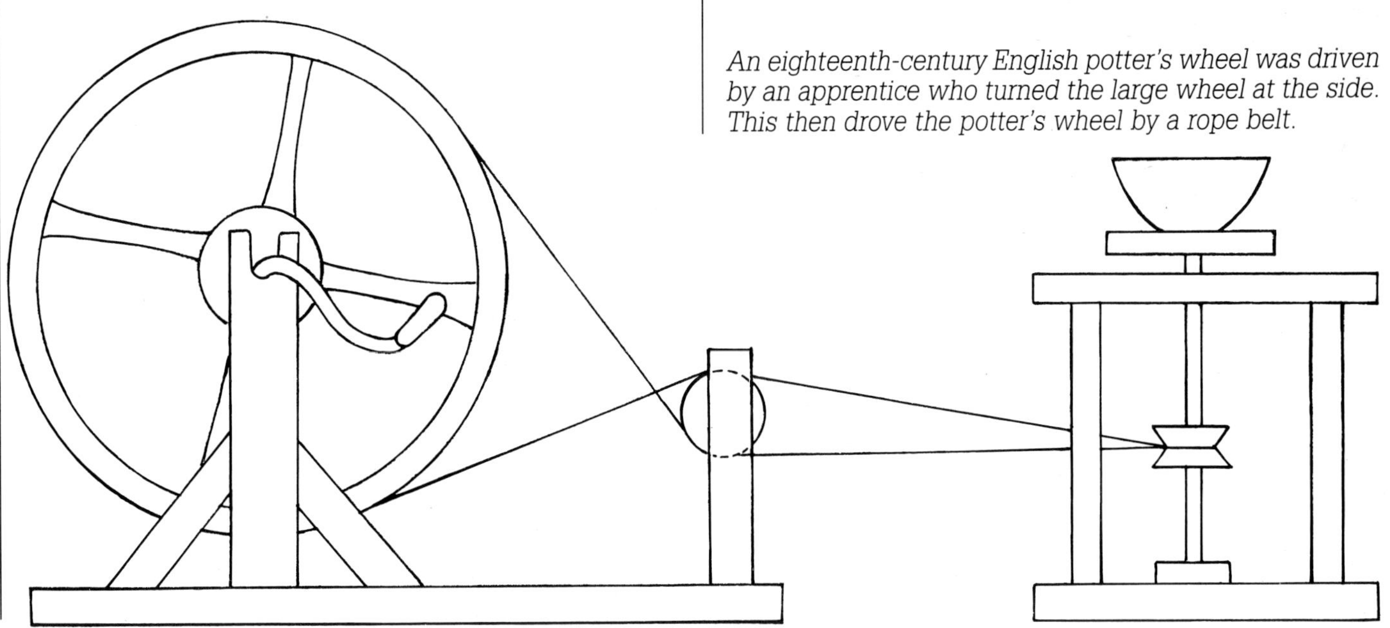

Many potters still use the foot-operated wheel, preferring the greater control of speed it gives in relation to the hands on the clay and the state of the clay being used. The foot-operated wheel is quiet in use, costs nothing to run and is fairly simple to construct. However, it can be rather difficult for the beginner to learn on this wheel; it rotates more slowly than the electric wheel and both foot and hand movements need to be controlled. An electric wheel will give the student time to concentrate on hand techniques and will enable him or her to practise the various movements required without having to worry about the wheel at the same time.

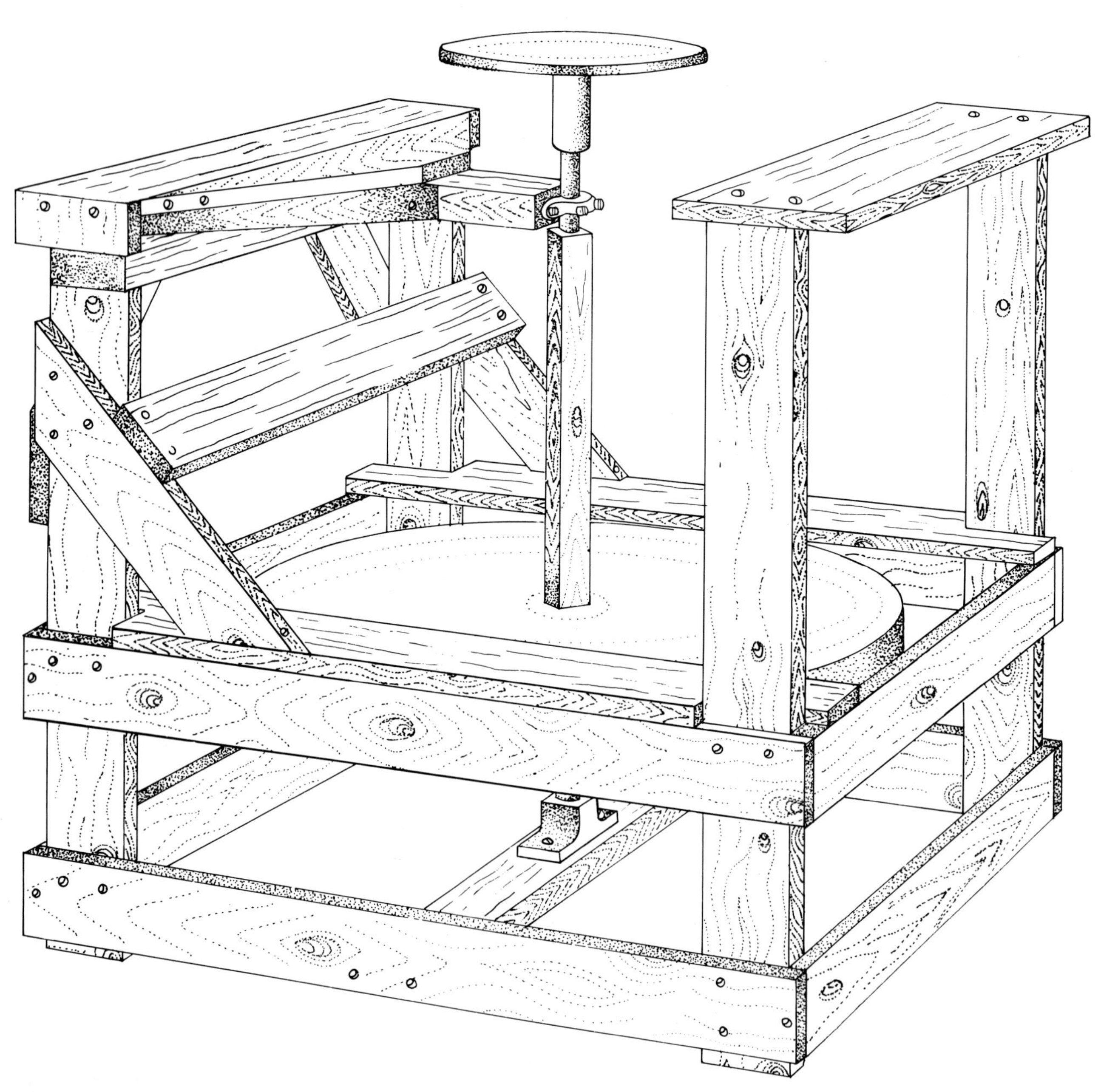

A typical homemade workbench and kickwheel

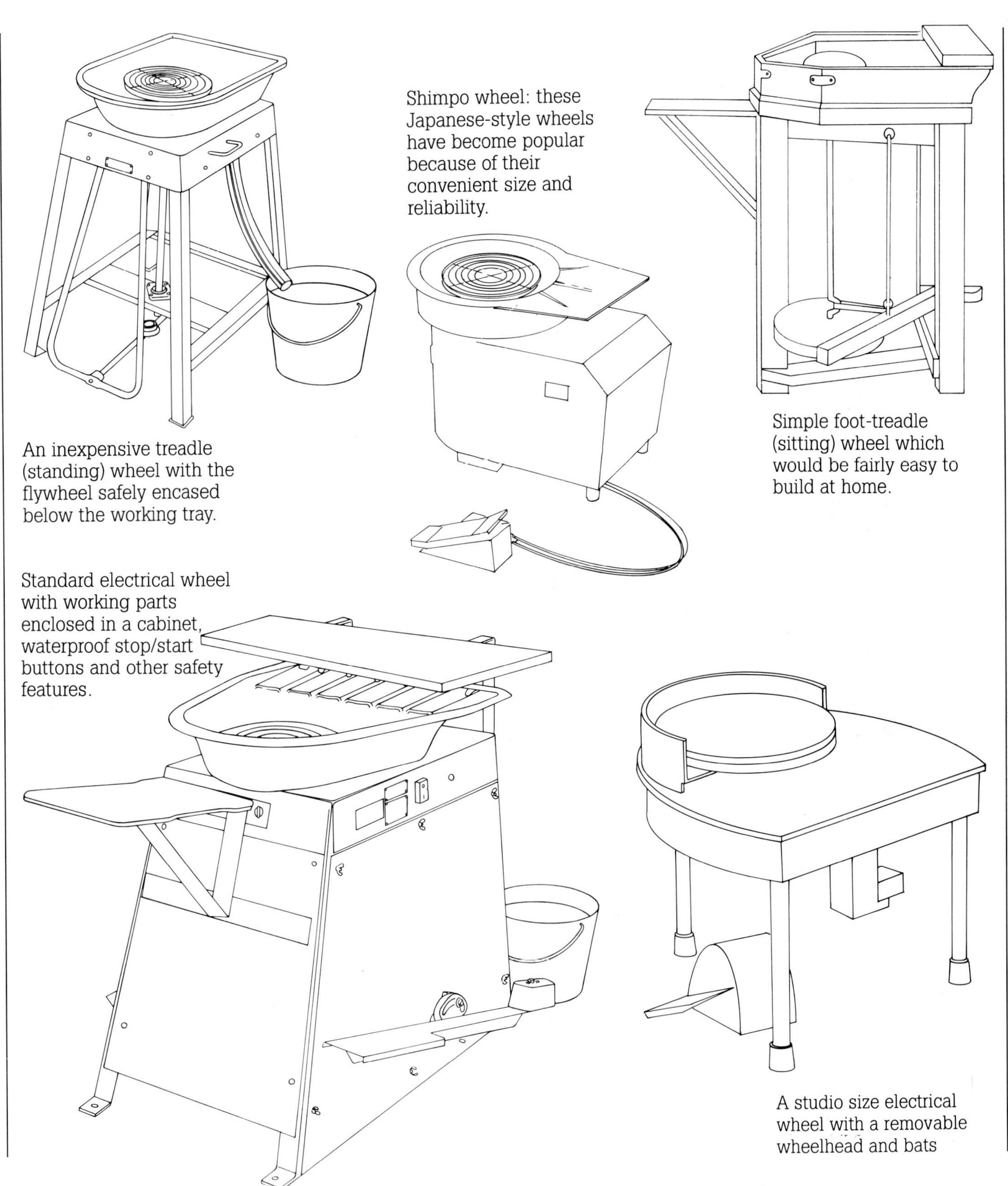

An inexpensive treadle (standing) wheel with the flywheel safely encased below the working tray.

Shimpo wheel: these Japanese-style wheels have become popular because of their convenient size and reliability.

Simple foot-treadle (sitting) wheel which would be fairly easy to build at home.

Standard electrical wheel with working parts enclosed in a cabinet, waterproof stop/start buttons and other safety features.

A studio size electrical wheel with a removable wheelhead and bats

Preparing the clay

The clay preparation for wheel work needs to be done with care until an even consistency is reached. For instance — does the clay need to be smooth or grogged — that is, does it need the introduction of some fine silica sand or some powdered once-fired clay to increase its strength or give it more texture? Does it have to be dark or light in color? This depends on the finish you want the pot to have when it is fired. Whatever the choice, the clay must be well prepared. This means a good mixing of the materials by hand or machine so that there is no variation in the overall color and texture. The clay for small batches of throwing can be done by hand rather than machine. The amount of clay you prepare will depend on your own strength (about 55 pounds or 24kg, prepared by hand will give enough for practising without leaving you exhausted). Use softer clay than you would use for handbuilding so that it is easier to center on the wheel. A beginner might think that using a harder clay would help the pot stay upright longer, but this is not the case; the harder clay is more difficult to center and uses more energy — which is better reserved for the later stages of throwing.

Beating and wedging

All air pockets must first be removed from the clay. This is done by beating or wedging the clay on the strong wedging table, slicing through with a strong wire and then beating the two halves together again. If the clay is uneven (perhaps it is a combination of soft and harder clay or contains clays of differing colors) then kneading the clays will mix them quickly. Press the heels of the hands into the clay, using the weight of your body to create a rhythm. Pull the clay forward towards you and press down with the hands again; this will make a 'dog's head' shape. More practice is needed to perfect spiral wedging; the clay must be lifted with a twisting motion from one hand while it is pressed down with the other to make a succession of spirals. The wedging action should be repeated until the clay, when cut through, shows no unevenness and no air bubbles. Although the preparation and wedging seems like hard work, after some practice you will find that the wedging and the feel of the texture and the wetness of the clay are essential parts of the making process, giving you time to think of the task ahead. I always enjoy this first touch of the clay, anticipating the throwing ahead while loosening the muscles and soothing the mind before starting. It is rather like warming up for a race.

The potter will soon be able to recognize the 'feel' of the clay and judge when it is in the right state for use. Clay badly prepared can ruin a throwing session; unevenness or air bubbles in the clay result in the pot being unwieldy and hard to center, with the air bubbles in it feeling like stones. Early attempts may end in disappointment and frustration, if the pot collapses because of poor preparation.

Storing the clay

When the clay is ready it can be cut into pieces and then made into balls of about 2 pounds (.906kg) in weight (or smaller if this size is too large to handle comfortably). While it is standing the clay should be covered with a damp cloth or a plastic sheet as it will harden on the outside surprisingly quickly while waiting to be used. If the clay is stored overnight it should be put into an airtight bin or wrapped in thin plastic sheeting (clothing stores usually throw out large quantities of this) which is ideal for clay work. A quick remix before you begin will make sure there is no longer a hard outer shell and softer middle.

1 Beat the clay thoroughly to remove air pockets.

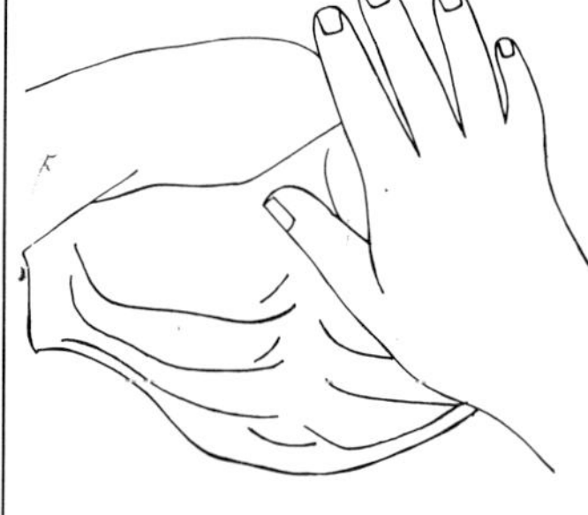

2 Slice through the clay with a strong wire.

3 Wedge the clay into a 'dog's head' shape.

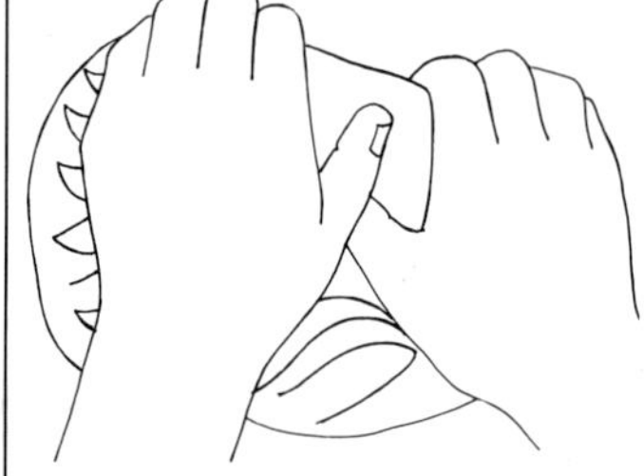

4 Lift and twist to make a succession of spirals.

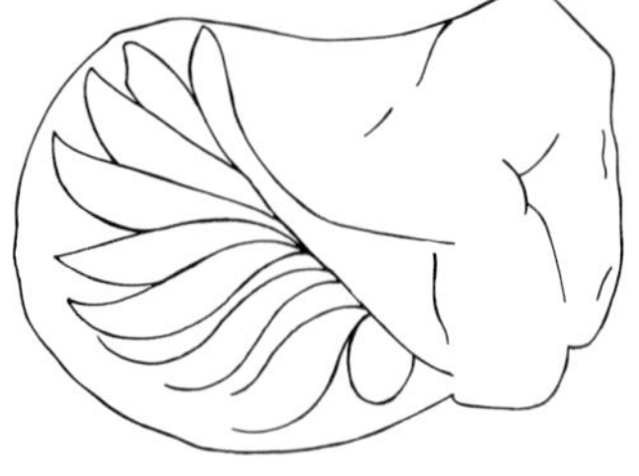

Raku lidded pot by Bryan Trueman (Australia)
Reduced color over the white crackled glaze and a blackened body add surface interest to the smooth lines of this thrown piece.

Thrown forms

Set of jugs
by Jane Hamlyn (UK)
Saltglaze stoneware jugs make an interesting group of matching thrown pieces.

Porcelain bowl
by Greg Daly (Australia)
Gold and silver leaf decoration are often best appreciated on a simple thrown form. This gently swelling bowl is only 3 inches (7.5cms) high.

Incized bowl
by Peter Lane (UK)
The delicacy of porcelain is enhanced by the surface treatment of this fine thrown bowl.

Form 1984
by Alan Peascod
(Australia)
A stoneware copper matte glaze and multiple slips help to emphasize the form of the rounded base and simple neck.

Preparing the clay

The clay for throwing needs to be well prepared so that it is even and smooth with no air bubbles. Beat it on a bench, then cut through and examine a section. See if there are any uneven parts, marbling, or air bubbles. The two halves are then beaten down on to each other.

Spiral (or dog's head) wedging will mix any number of clays together until they are one color and one texture. Further beating will then compress the piece of clay. Cut and weigh into suitable sized balls for throwing. These should be stored in a plastic bag or an airtight bin to prevent them hardening on the outside. Otherwise they need rewedging.

see page 44

Pedestal vase by Jerry Rothman (USA)

Preparing to throw

Position the wheel where there is good light to work by. Make sure you are comfortably seated at the wheel. The seat should be slightly lower than the wheelhead so that the body is over the wheel and the water thrown off is caught in the wheel tray. Place your clay on a clean board across the front of the wheel or on a bench within easy reach. You will need the following tools near to hand:

A pot of water in the well to the right (or left if you are left-handed) of the wheel — make sure it can be easily reached without stretching; a fine small sponge; a piece of wash leather for smoothing the top of the pot; a blade-ended tool of boxwood or metal for trimming spare clay from the foot and a piece of fine wire with toggles at the ends for cutting the pot from the wheel. Some potters prefer a single strand of wire or nylon or two wires twisted together.

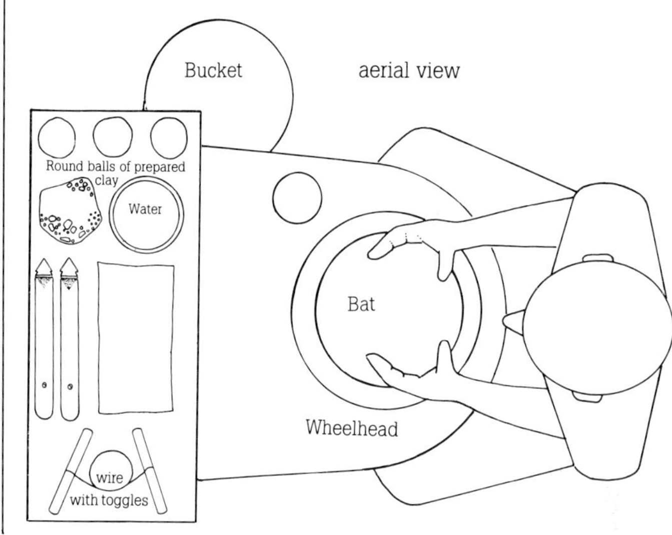

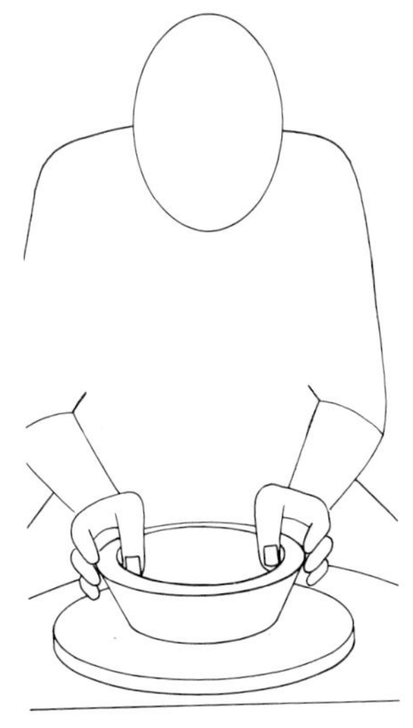

Plans showing layout of the wheel. Position the wheel so that the light comes in from the top or in front of the potter. Clay should be placed to one side, ready to hand. Boards should be placed on the bench (on the right). Make sure there is room for the finished pots to be placed there without your having to move from the wheel so that the whole operation can be carried out without too much unnecessary movement.

He or she should sit as far up to the wheelhead as possible, keeping the back straight. This will prevent strain on the back and unnecessary tiredness, especially during long throwing sessions.

This is the general layout when starting to throw. Much will depend on the studio, the light and the space available. Perhaps the left-hand side of the wheel might be available for finished pots rather than the right.

Tools for throwing can be kept on the back of the wheel or near to hand ready for use.

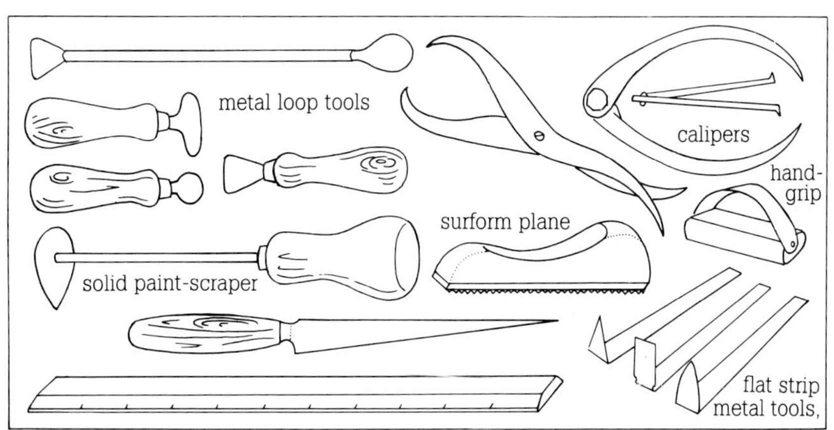

Useful tools

Different types of wheel heads

Natural sponge.

A piece of wash leather for smoothing the rim.

A sponge attached to a stick for soaking up the water residue at the bottom of tall pots.

Techniques for centering the clay

One of the main faults when learning centering is the tendency to hold the arms out from the body rather than tucking the elbows in so that they cannot be moved by the action of the clay. Another natural instinct when centering for the first time is to feel confident that the clay is fixed and centered on the wheel at last and so remove the hands in triumph only to find that the clay springs back with the sudden removal of pressure. Always remove the hands gently, letting the clay assume its natural place on the wheel. This applies to all stages of throwing when taking the hands away from the clay.

The wheel rotates in an anticlockwise direction for right-handed potters and the movements described in the text above relate to them. Kick-wheels and certain electrically powered wheels, such as the Shimpo wheel, can also be worked in a clockwise direction by people who are left-handed and the hand movements could then be reversed, left to right. As much of the time both hands are used equally, this may not be too important a factor.

Centering

1 Position yourself, sitting as close to the wheel as possible, with your back as straight as you can manage. Rest the hands firmly on the edges of the wheel tray to prevent any strain and to give maximum leverage when exerting pressure on the clay. Have the clay to hand at one side where it can be reached easily without your having to stretch too far.

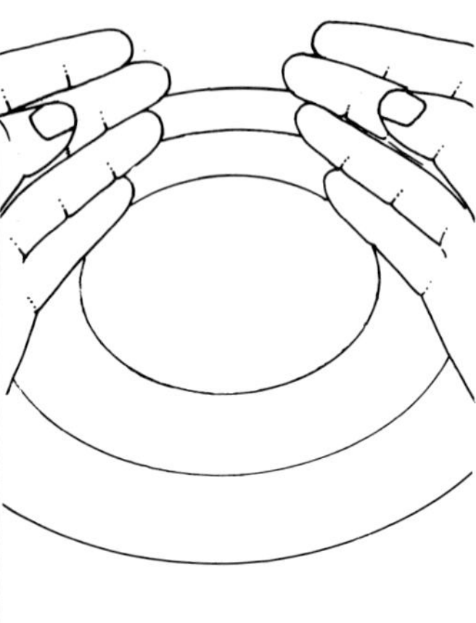

2 Dampen the wheelhead slightly but make sure there is no excess water or the clay will slide off when the wheel is started. Take the ball of clay and throw it down on to the wheelhead, aiming for the center of the wheel (Do not attempt to do this from too great a distance!)

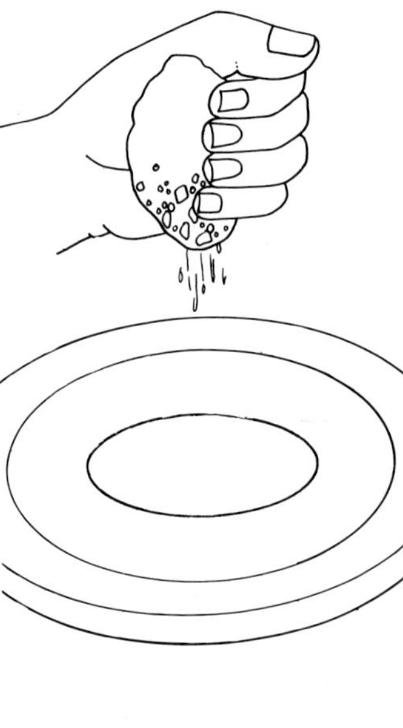

3 Beat the clay with the hands into a cone shape as near to the center of the wheel as you can. This will prevent any unevenness in the clay pushing the hands off center and will assist the centering considerably.

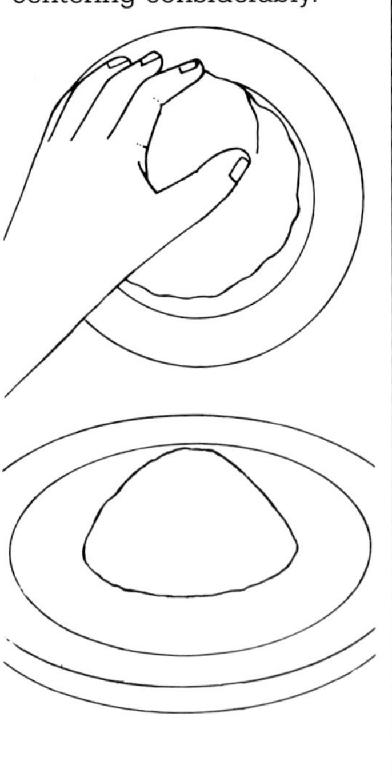

4 Wet the hands and the clay and then you will at last be ready to start the wheel.

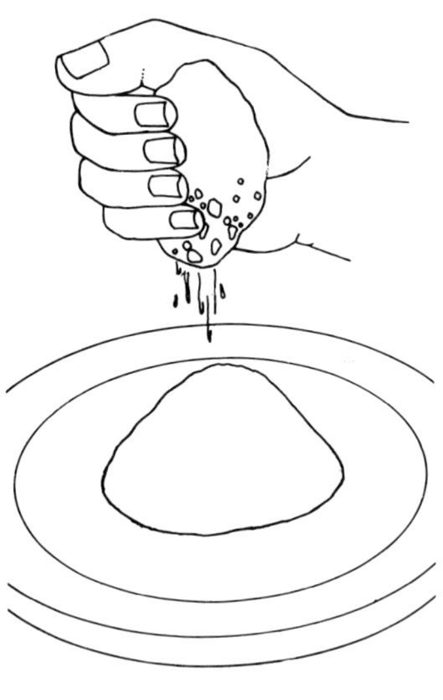

5 Set the wheel to a fairly high speed to assist centering and, with wet hands, put the left hand around the clay — with the edge of the hand touching the wheel.

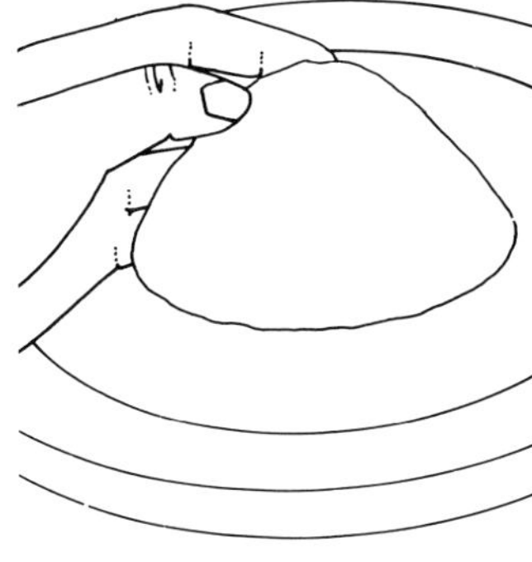

6 While the left hand is exerting pressure, the right hand is placed on top of the clay with the edge of the hand across the center. Do not cover the clay completely so that you can judge when the clay is centred. The right hand exerts pressure downwards. The clay is trapped and squeezed into the center of the revolving wheel until smooth and centred, when no movement is visible or felt.

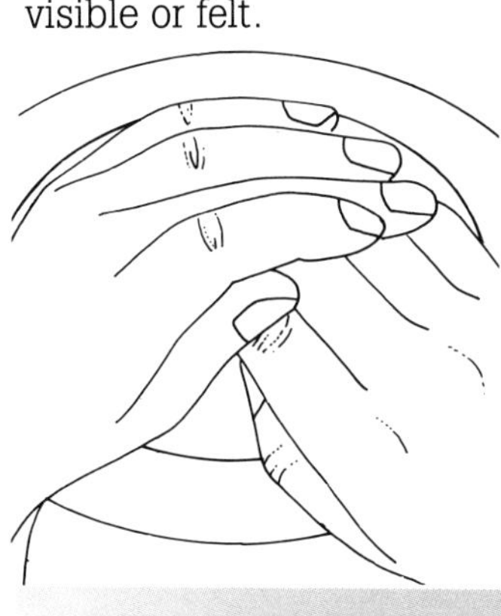

7 Do not take the hands off suddenly; the clay will spring back like rubber in the opposite direction. Instead release the hands very slowly and carefully to retain the center position. This applies to all movements while throwing. Any quick sudden movements will jolt the clay and knock the pot out of true.

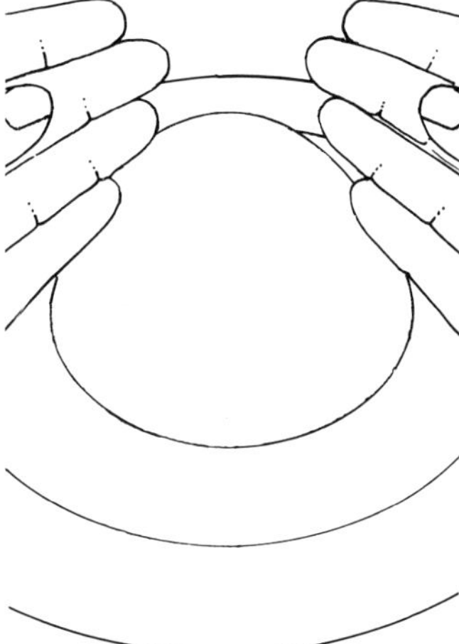

8 Some potters like to improve the centering by exerting pressure with both hands equally at each side of the clay until the clay cones into a point. This is then pushed down again with the left or right hand; repeat three or four times to mix and help center the clay, especially if the clay is slightly uneven. This method is only one way of centering; there are many variations on the theme.

Each potter develops his or her own style and way of centering. The methods will be different but the basic rules are the same.

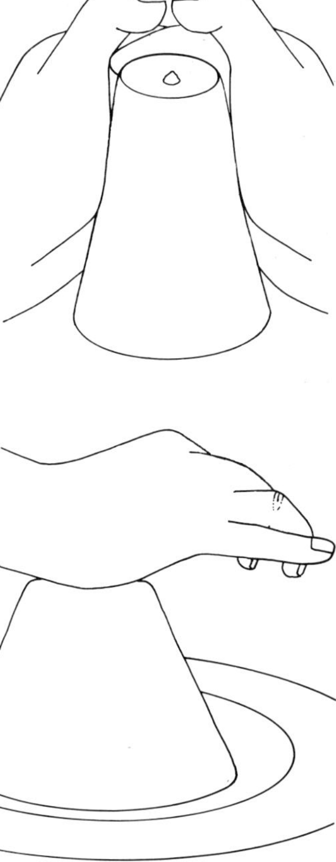

Vessel by Petrus Spronk (Australia)
This wheel-thrown vessel has been burnished and wood fired in reduction to produce a beautifully smooth gleaming surface highlighted by its contrast with the 'grittier' area near the neck.

One way of pulling a cylinder

Pulling the clay

1 Slow the wheel slightly and place the right thumb at the exact center of the clay; there will be no movement. Press down to the base without going right through to the wheelhead. The thickness for the base must be carefully judged as you press down and this judgment will improve each time you throw.

2 Keep the thumb at the bottom of the hole and pull it out towards the side of the wheel, widening the hole. Keep the thumb steady and make a flat-based opening. All these movements need steady pressure and the thumb or fingers must be removed gently so the clay does not spring back off-center.

3 Slow the wheel again slightly. The next movement is to raise the walls of the pot. Both hands and arms need to be steady, resting on the thighs or the edge of the wheel tray (or with elbows tucked in for extra support).

4 The left hand is placed inside the pot, with the fingertips at the base of the wall. The thumb of the left hand should be over the top of the wall, resting on the right hand which is held against the outside of the pot. The thumb of the right hand should be tucked into the palm, with the fingers curled into a fist.

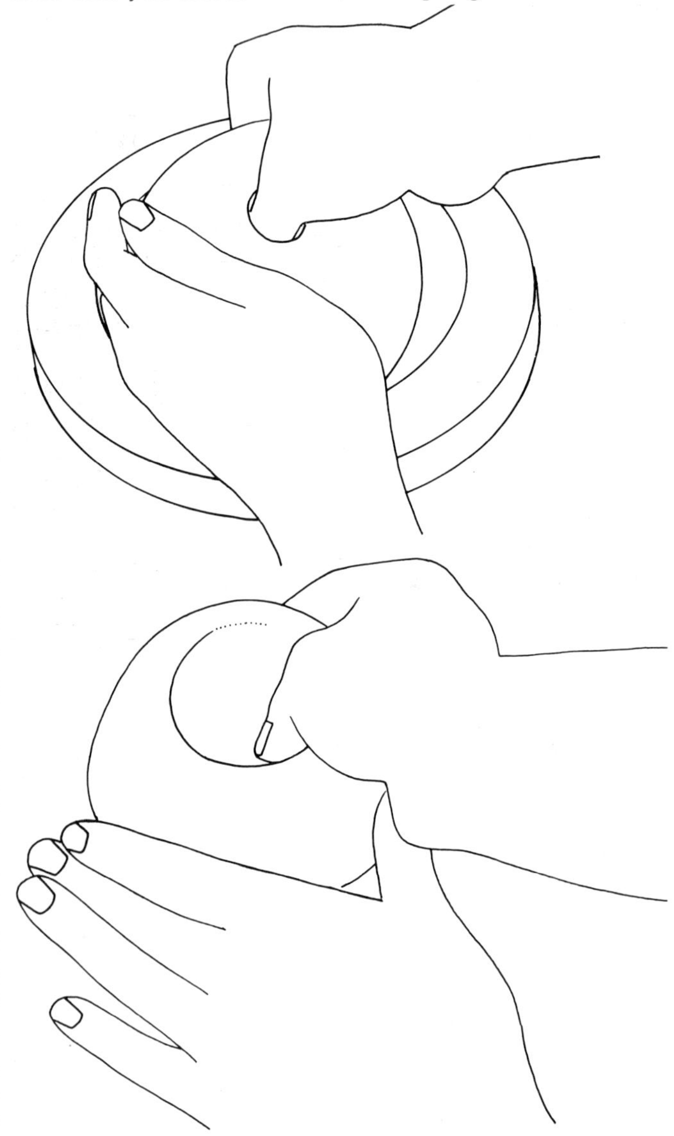

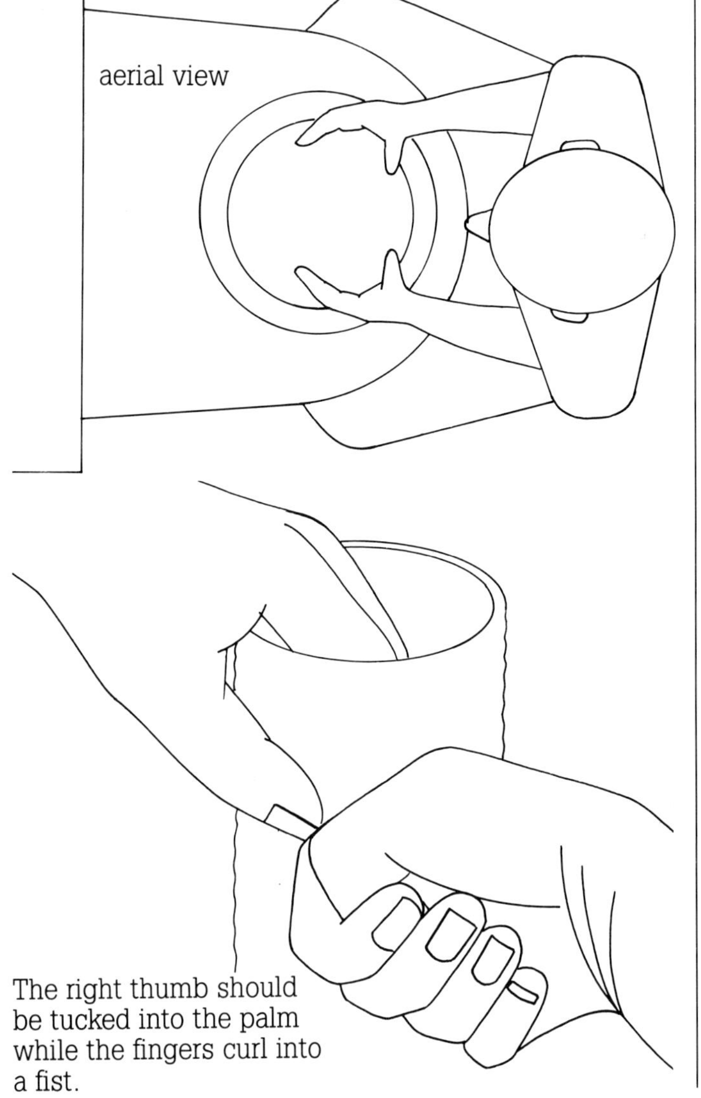

The right thumb should be tucked into the palm while the fingers curl into a fist.

5 Make sure the hands are wet and acting on the clay smoothly. Apply equally strong pressure with both hands to the wall, pressing inwards, and draw the hands slowly up the sides of the pot. Match the rate of upward pulling to the speed of the turning wheel; if the hands are pulling too quickly they will create an exaggerated spiral effect rather than smooth rings. The pressure of your hands will thin the walls as the upward movement of the hands increases the height of the cylinder by pulling up clay from the thickness of the lower walls.

6 Ease the pressure slightly as your hands near the rim of the pot until the fingers are just touching. Take the fingers away gently at the end of the pull — abrupt release of the pressure on the clay will cause the pot to spring out of shape. The steadiness of hands and arms will keep the clay still; any sudden movement now will cause the pot to wobble.

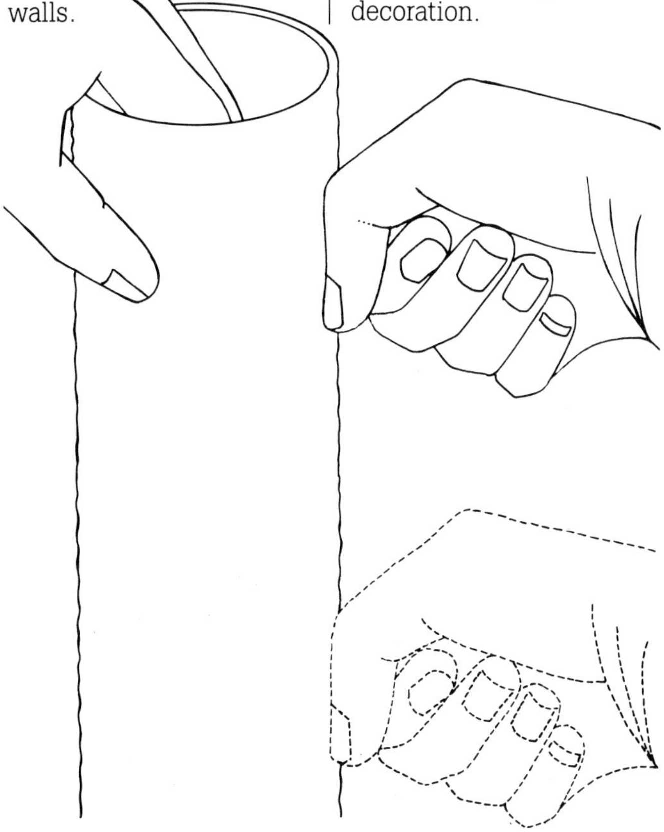

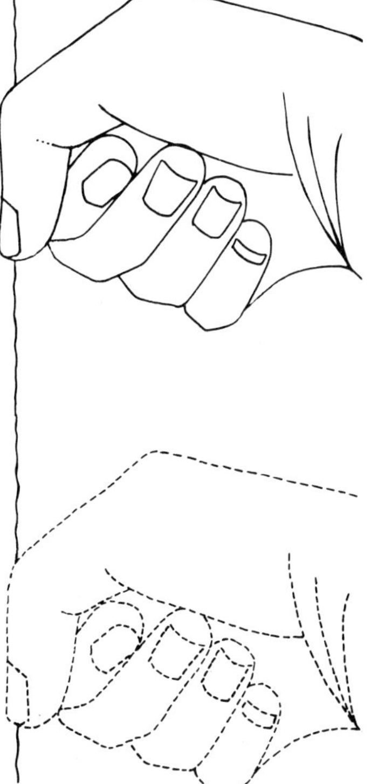

Thrown cylinders by Graham Flight (UK)
Simple but effective cylinders made of porcelain clay with a clear glaze used over painted decoration.

Possible problems

If the shape starts to widen as you are pulling up this indicates that too much pressure is being applied on the inside or that the wheel rotation is too fast. More pressure directed towards the center with the right hand will correct this imbalance and keep the cylinder straight. A tremendous amount will be learnt about the consistency of the clay and the pressure required in that first session of throwing. No amount of advice can better the physical experience of feeling the clay moving under your hands as the pot begins to take shape.

Practise making small cylinders, pulling the clay until it is as thin as possible without collapsing. Things will go wrong; the pots will often buckle and fall at first. When this happens, try to establish how and why you went wrong. Take the clay from the wheel and place it to one side; clean the wheel and start afresh with a new piece of clay. Do not be downhearted if nothing seems to go right. The pleasure felt when the first pot is made far outweighs all the failures. At all stages of making pots, there will be mistakes and the best advice is to put the attempt aside and start again. You will develop a determination to achieve a finished result which will reward you with the greatest satisfaction. It is better to pull the clay to its limits even if it collapses, than to stop when the pot is thick and heavy just for the sake of removing a finished pot from the wheel. Beginners find it tempting to remove their first attempts without considering their quality. This is understandable and these earliest creations are often discarded once the pots improve with practice! However, the thrill of removing those first pots from the wheel makes everything worthwhile. To have taken a lump of clay and formed it into a glistening wet pot which stands on a board in front of you gives both satisfaction and a desire to improve. These are the main driving forces behind most potters and will never diminish no matter how much experience is gained.

When you have managed to pull a reasonable cylinder, stop the wheel and cut the pot down the middle with a thin wire. This sounds drastic after all the effort that has gone into making it, but the two halves of the pot in section will show the sides clearly. They should be an even thickness, widening slightly at the base.

1 If the speed at which you raise the side of the pot is too fast compared with the speed of the wheel, deep grooves will rise up the side of the pot in a spiral effect. This will make the pot uneven. Moreover, the top will be impossible to cut straight. This can be corrected by pulling again from the bottom slowly, smoothing out the spiral. Then trim the top straight.

2 The pot widens too much. This can be caused by an over-fast wheel throwing the pot outwards. It may also happen if the pressure of the inside hand is too great and is forcing the clay outwards. All these factors must be well balanced if the sides are to be straight.

3 A very common fault in first throwing is to exert too much pressure at the base. This results in the top half of the pot being torn from the base completely. If this happens, clean the wheel, select a new piece of clay and start again.

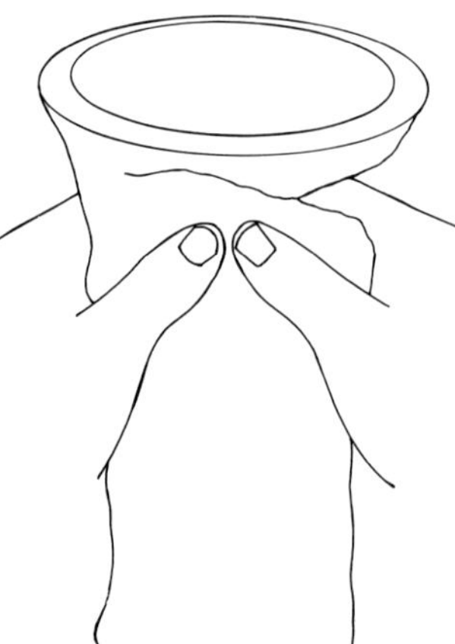

4 If the hands and arms are not supported properly while pulling, the hands will wobble from side to side and create uneven thickness up the sides of the pot. The thin walls will gradually collapse as the pot grows. Make sure your arms and hands are supported and cannot wobble.

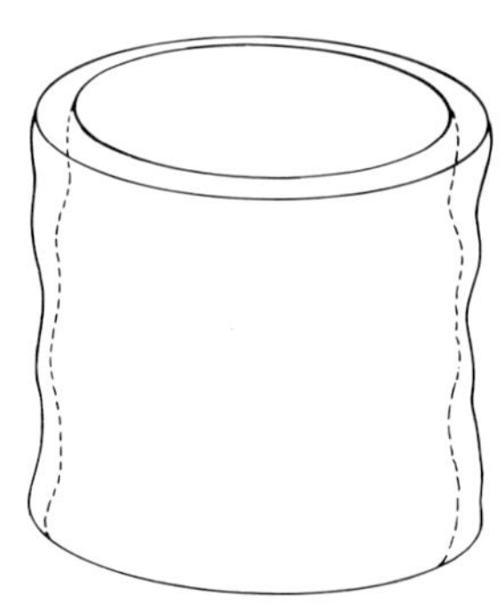

Throwing different shapes

1 After the first achievement of making a simple cylinder, you must next attempt to control the shape — making the clay go exactly where you want it to go and so changing the profile of the pot. Aim to have a particular shape in mind at the outset. If you are throwing with no idea of the finished pot in mind then the chances are that the result will be a mixture of design and accident. You might like to progress from a cylinder to a simple pitcher shape — that is a pot with a swelling belly that narrows in again to the original cylinder neck.

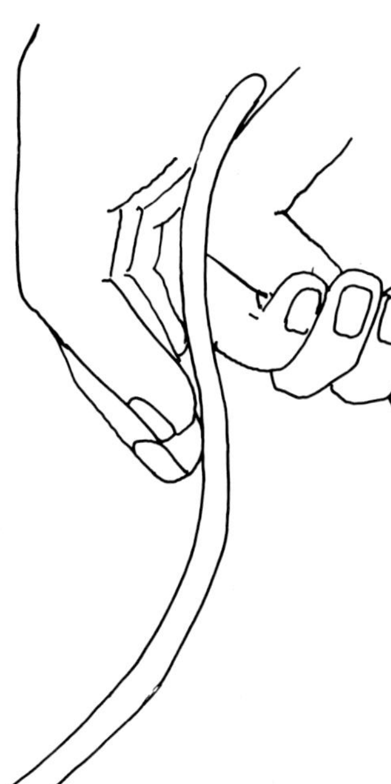

Placing the fingers that are outside *below* those inside helps to draw the shape outwards.

2 When the slightly tapering cylinder is nearing completion, make another pull upwards from the base, but this time press outwards slightly with the inside against the outside hand which is supporting the pot. Continue to widen the shape until you have a shape which pleases you. A mirror placed at the back of the wheel will help you to see the profile as it grows.

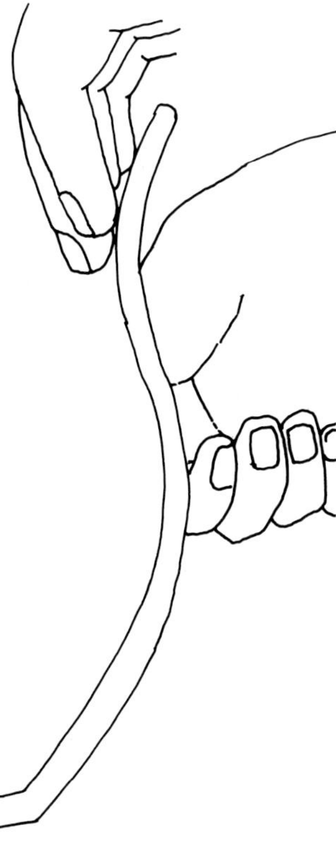

Placing the fingers that are outside *above* those inside helps to draw the shape inwards.

3 The pressure for shaping the pot should be reduced as the walls are becoming thin and weak; too much pressure at this stage will collapse the pot. Pulling up at the same time as you widen the pot will achieve a growing shape, one which springs from the wheel rather than one which looks fat and sagging — as it can do if the widening is done too near to the base of the pot, if you press the shape out without pulling up at the same time. A gradual widening of the belly, rather than a sudden movement, will be less likely to cause a disaster.

Widening the shape out too soon will result in a heavy sagging pot.

4 The top of the pitcher may be changed after the belly of the pot is finished, perhaps by widening the neck or flaring out the rim slightly, creating the ideal shape for a jug.

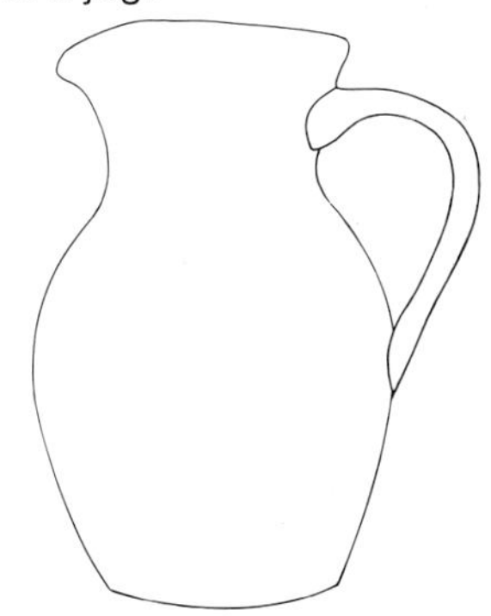

One pull may not be enough to finish the pot so the process must be repeated with increased sensitivity, until the pot walls gradually become thinner. The basic shape will dictate the final result. A straight cylinder can produce a swollen jug shape; a bowl can be made from a lower, flatter shape which is opened wider to create a curved base.

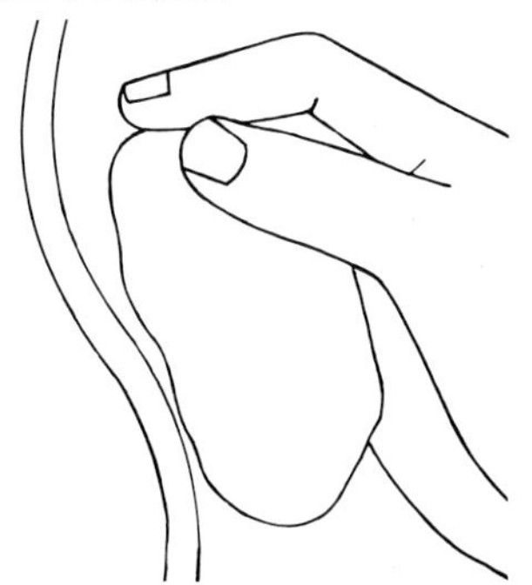

To finish off, smooth the edge with a piece of wash leather.

Removing a pot from the wheel

Preparing to remove the pot from the wheel

1 Smoothing the rim as the pot revolves gives a pleasing roundness to the edge. This is done either by holding a piece of wash leather over the rim, or holding the rim between the thumb and finger of the left hand and smoothing the rim with the forefinger of the right.

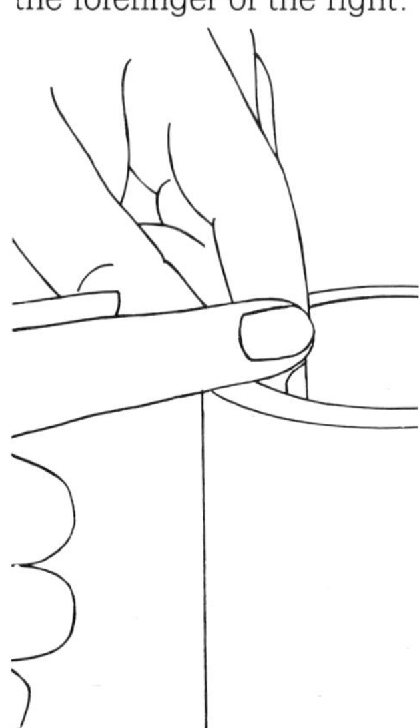

2 Remove the excess water from inside the pot with a fine sponge attached to a stick. The sponge is wound on with wire.

3 Using a pointed tool held against the foot of the pot, remove any clay which has spread on to the wheel. This makes a neater foot and ensures that the pot can be cut from the wheel cleanly.

Releasing the pot

The pot can be cut from the wheel with a fine twisted wire held taut between the hands. The thumbs should press the wire down on to the wheel so the wire is pulled through the base of the pot. A pot may be lifted straight from the wheel to a tile or board. Wet the wheelhead before pulling the wire through so that the water is dragged under the pot by the wire, repeating the action to make sure the pot is loose from the wheel. Push the pot with a twisting movement at the base, where it is thickest, towards the edge of the wheel and from there on to the hand before carefully placing the pot on to a tile or board. The pot may become slightly oval as it is lifted but it will regain its circular shape when it is placed on a flat surface. To stop distortion some potters place a piece of paper, cut to size, across the top of the wet rim before moving the pot. Others lift the pot directly upwards after cutting off to release the pot cleanly. Another way of lifting a very large pot, if the hands are not big enough to surround it completely, is to place a thin slat of wood along the back of the pot. By holding the two ends, the arms and the slat of wood make a triangular frame with which to lift the pot from the wheel.

Many potters use a wooden wheelhead bat which has two holes in it, one at the center and one about 2 or 3 inches (5 to 7cm) from the center. These correspond to raised nuts which have been fixed into the head of the wheel. This simple adaptation means that at the end of throwing, the bat, together with the finished pot, can be lifted readily from the wheel and any distortion of the clay is avoided. This is particularly helpful when making larger pots or wide bowl shapes which are all too easily distorted when being cut from the wheel. Some wheelheads have devices that can be pressed upwards to release a bat from the wheel or have a rubber wheelhead into which a tile can be inserted for throwing purposes.

Methods of twisting the wire.

Attach a piece of clay to the center of the wheel and stick two pieces of wire into the clay. Holding one piece in each hand, revolve the wheel slowly. Hold the ends firmly and twist together from the bottom to the top. Another method is to use a loop of brass wire with the two loose ends clamped tightly together in a vice or clamp. Keeping a toggle in the loop end, turn the toggle to twist the second two strands together. Finally attach a toggle to the other end.

Methods of removing pots from the wheel

1 Tightly grip the wire at each end. Hold the thumbs firmly down on the wire and press it against the wheel head.

There will be less risk of distorting the pot if the wheel rotates slowly while cutting with the wire.

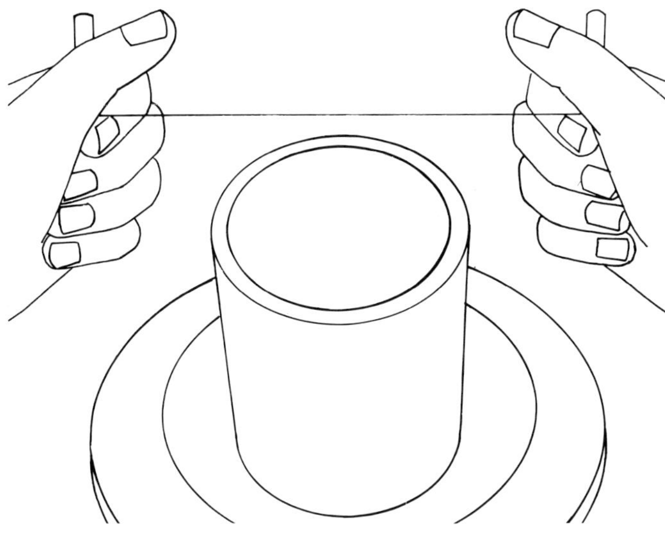

2 There are several ways of actually lifting the pot off the wheel. In this first method the hands must be dried after the cutting and then simply lift the pot cleanly off the wheel in an upwards movement.

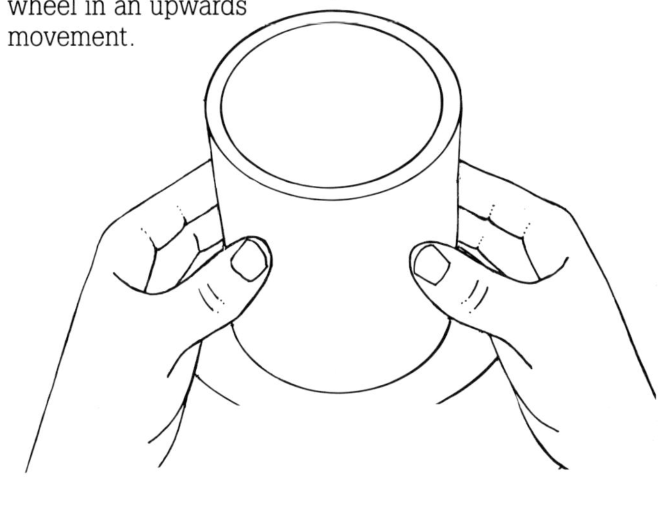

3 Alternatively pour a little clean water on to the wheel head and then cut through the base of the pot; drag water under the pot as you do so.

The pot can be removed with a twisted, double-stranded brass wire which has toggles attached to each end.

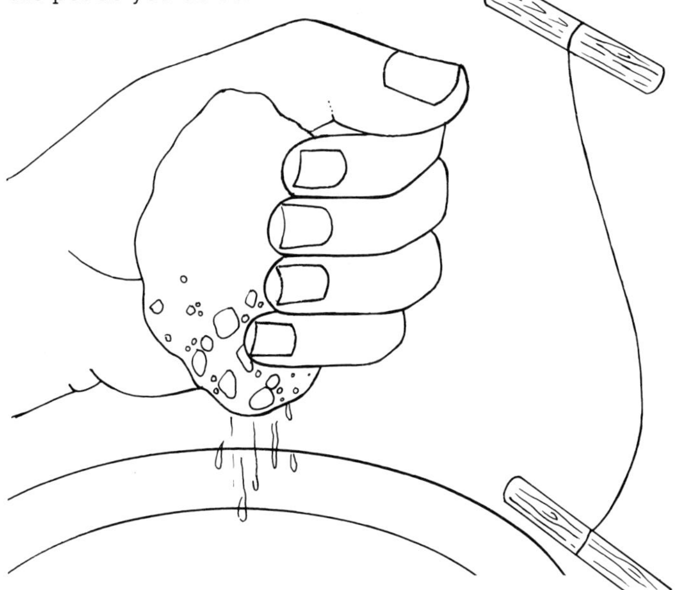

4 Next press the fingers firmly at the thickest part of the base and slide the pot off the wheel on to the bat or your hand. If you have wheel heads with removeable bats then it is simply a matter of lifting the pot and the bat off together; this helps prevent distortion.

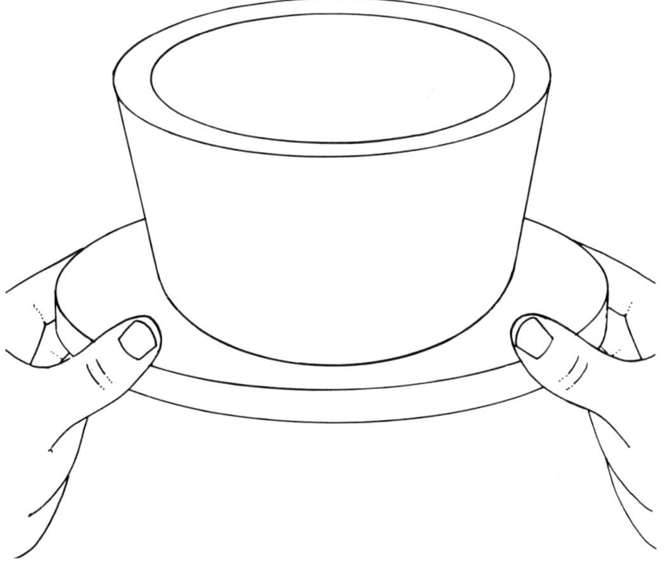

Throwing a high bowl

Another basic pottery form to attempt is the bowl, an open curved surface which relies on the proportions and finish of the rim and foot, and the relationship between them for its tactile and visual appeal. The inside of a bowl is important because it is far more open to view than that of a vase or jug; the inside and outside are viewed in relation to each other which is not the case with a narrow-topped shape, so you should aim initially to achieve a smooth unbroken line.

Experiment with the type of rim; flatten it, round it off with the wash leather, turn it outwards, inwards or downwards. The rim is an important element which should enhance and complement the bowl.

High bowl by Peter Lane (UK)
Air-brushed ceramic stains have been built up into a fascinating design.

Making bowls

1 Center the clay in the usual way but open it up with a slightly curved base, pulling the thumb smoothly upwards as well as outwards.

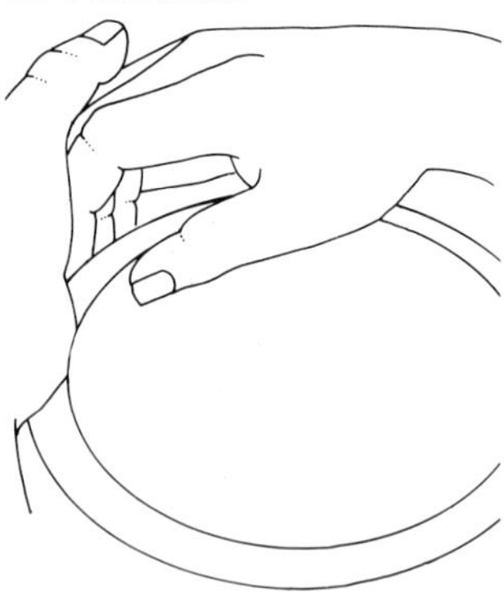

2 Pull the cylinder in the usual way, keeping the walls upright until the final stages of throwing.

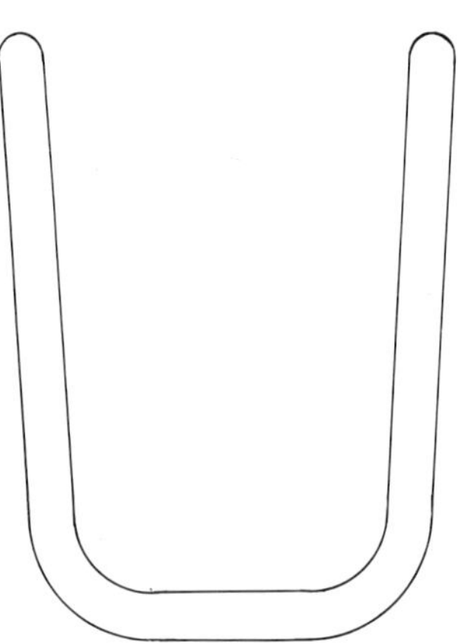

3 Now create the bowl's shape by flaring out the clay gradually. Slow the wheel as the bowl widens, and pull the inside hand from the center of the bowl to complete a steady curve in the pull.

4 Pulling from the center reduces the risk of a ledge forming where the inside fingers meet the base; otherwise you will create a wide-topped pot rather than a well rounded bowl without a break in the curve of its inside surface. Of course, the central well may be a design feature and commence at the edge of the base.

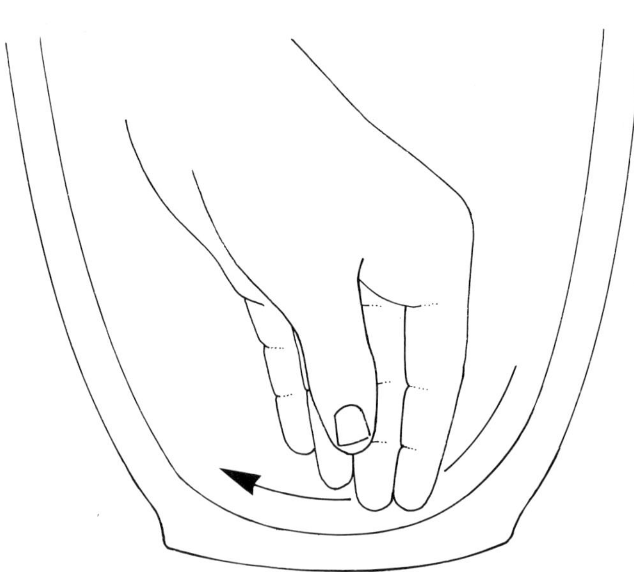

There is a wide variety of rims which can be used to finish off the bowl. Here are just a few shown in cross section.

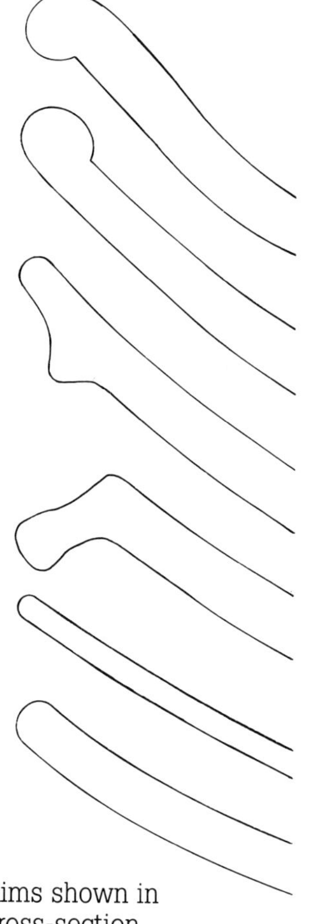

Rims shown in cross-section.

Throwing a wide bowl

1 The clay needs to be a wide low shape, almost as wide as the wheel itself. Center the clay in the usual way but flatten the piece with the heel of the hand.

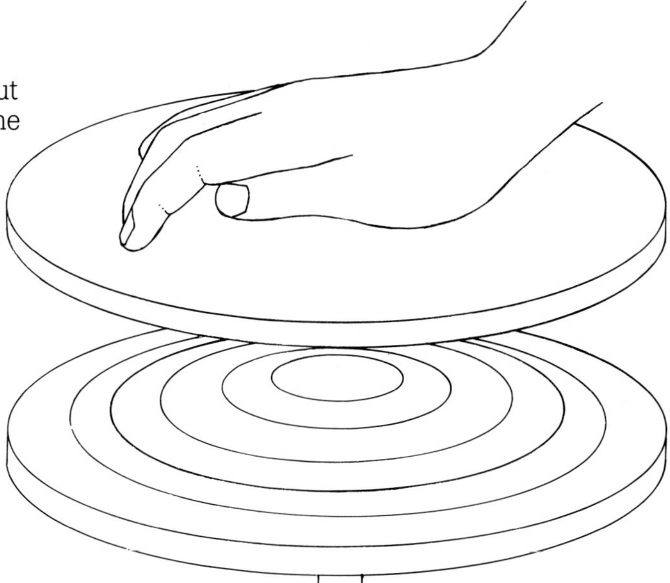

2 When opening the clay use *all* the fingers of the hand. Working from the center, pull outwards and upwards to create a curved round base.

3 Open up the clay with the fingers of the left hand, pressing down in the center and drawing the clay open. Keep the base curved. Meanwhile the right fingers are following the action of the left hand but are drawing up the clay on the *outside* of the bowl.

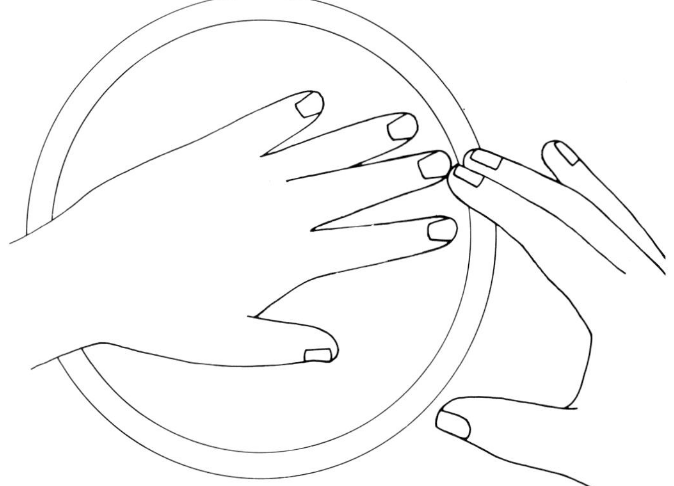

4 Repeat the pull, taking it from the center of the bowl and spreading its walls a little. Gradually increase the pressure on the inside as you draw out the bowl but do not attempt to make it too wide or the sides will collapse. It will soon become obvious when the clay has reached its limits.

5 A selection of rim edges and shapes enhance the finished bowl: the possibilities are endless.

Wide bowls are easier to remove from the wheel if they are thrown on a wooden bat and lifted from the wheel on the bat to avoid distortion. Fit the bat to the wheel with coils of clay or use a standard bat which fits on to ready-made studs on the wheelhead. The pot can be cut and carefully slid on to a tile or bat but some distortion might occur. You may even lose the bowl if it is too soft.

Throwing a wide dish

A large dish is best thrown on a wooden bat fixed to the wheel to minimize the chance of distortion when removing it from the wheel.

Fix the clay to the bat and gently beat it into a wide flat shape before centering. A wide plate needs to be supported from as far out as the wheel will allow. Center the clay in the usual way but this time press the clay down into a wide flat shape with pressure from the heel of the hand. Then make a shallow indentation with the fingers to form the base of the plate and open the clay to the width of the wheel. Pull wall of dish upwards and slightly outwards to form the unsupported edge of the plate. Support the clay with the outside hand while pressing out with the fingers of the other hand. Pull the side of the plate again and gradually flatten it, making sure that the wheel is slowed down as the plate grows wider.

A point will be reached when the width and flatness of the plate will stand no more pressure without flopping. At this stage the wheel will be turning very slowly, with the fingers hardly touching the edge; the slightest sudden movement will cause the rim to wobble and flop. Draw a wire under the base of the dish before removing it from the wheel (still attached to the wooden bat) and put it aside to shrink and dry without sticking.

Fruit plate by Ray Finch (UK)
The wide sweep of this thrown plate is emphasized by finger-control decoration. Great care is needed when removing wide dishes from the wheel.

Throwing a flat dish

1 The flat dish is best thrown on a bat fixed to the wheel. Otherwise the wide flat shape will buckle when trying to remove it. Throw on a purpose-built bat which fits on to studs or fix a bat to the wheel with coils of clay.

2 Flatten the clay by gently beating. Then, with the wheel revolving, press downwards and spread the clay with the heel of the hand to the width of the wheelhead.

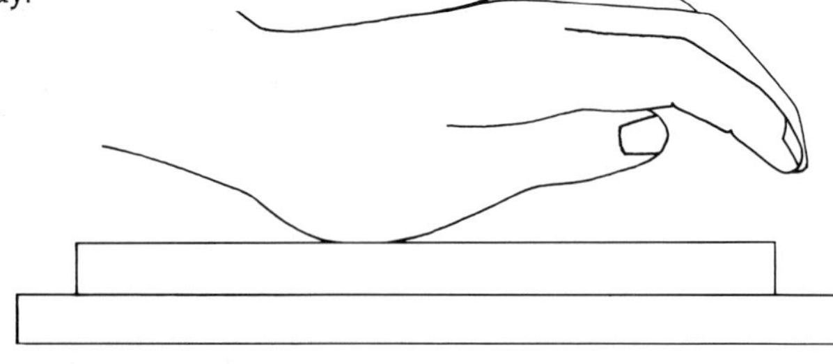

3 Pull the edge of the flat disc of clay slightly upwards and outwards, pulling from the center of the base.

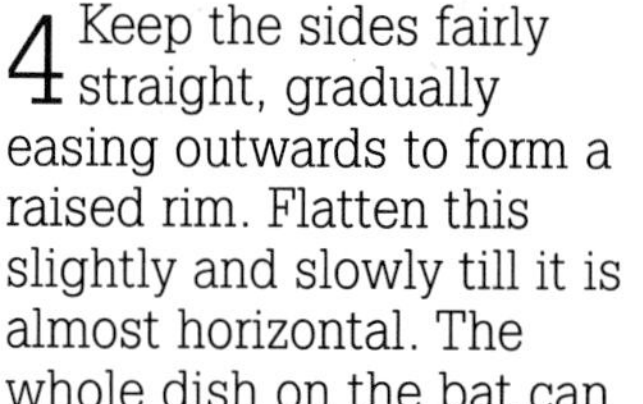

4 Keep the sides fairly straight, gradually easing outwards to form a raised rim. Flatten this slightly and slowly till it is almost horizontal. The whole dish on the bat can then be removed. Slice through the base before leaving it to dry. This will enable the dish to be removed easily when you are ready to turn the foot.

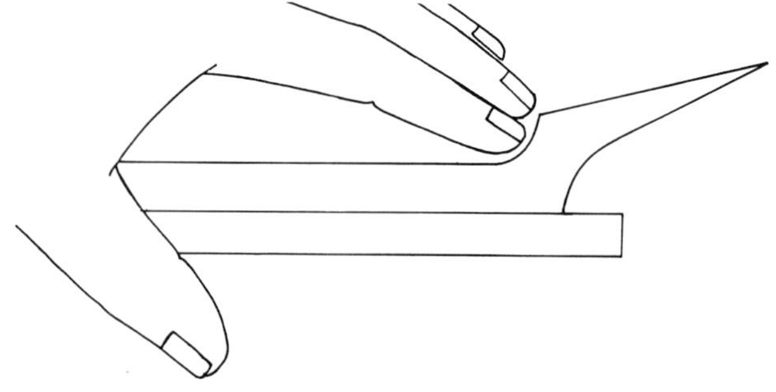

Throwing a bottle shape

After centering and opening apply slightly more pressure on the outside to keep the top tapered in. After each pull from the base to the rim, collar the top. Keep the top narrow. When you need to put your hand back inside the pot ease it in gently. This will widen the top a little but it will be easy to narrow when doing the final collaring. It is the only way to keep the top narrow while throwing. Gradually exert more pressure with the inside hand as you pull up so as to fatten the belly of the pot outwards. To bring the top half of the bottle back in, the outside fingers must be kept above the fingers on the inside. This will create a shelf, the clay being flattened over the inside hand and pressed into a narrow neck. Leave enough clay at the top of the pot to finish with a bottle neck. If the clay does run out, let the pot dry, add a coil of clay and throw again. However, if the bottle can be formed in one piece, so much the better.

Bottle shape by Ian Sprague (Australia)
This flagon achieves a satisfying full roundness before narrowing in to a tiny neck.

1 Collaring your hands around the pot, keep your thumbs at the front and gently squeeze the neck inwards.

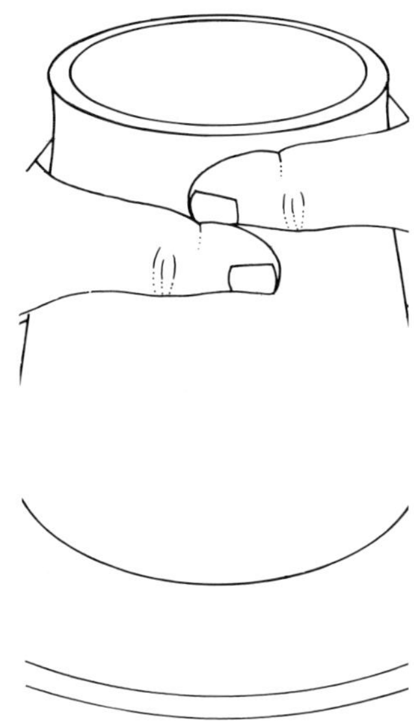

2 Pull from the base again, applying more pressure from the inside and swelling the belly of the pot outwards. Reverse the pressure and press from the outside to narrow the neck. The clay is pressed over the inside hand slightly downwards.

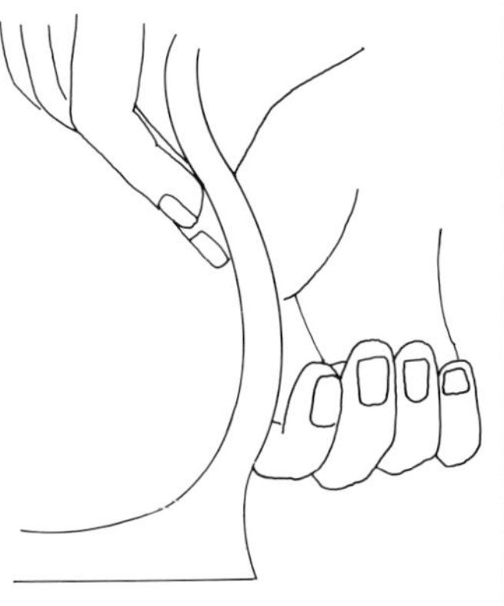

3 Collar the clay into a very narrow neck, using finger and thumb.

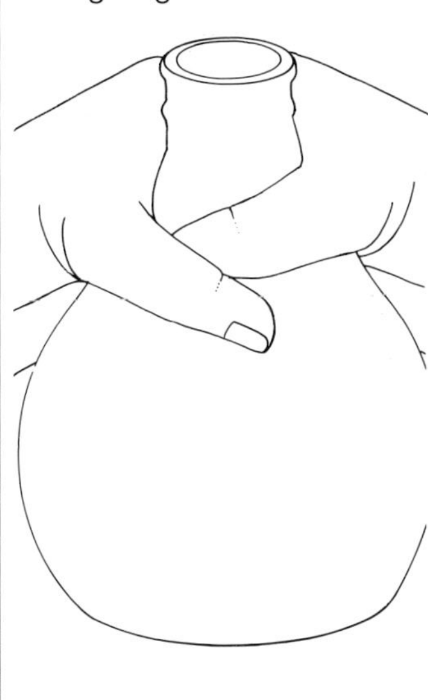

4 Ease just the tips of your fingers into the narrow neck and pull the clay very gently upwards into a bottle neck.

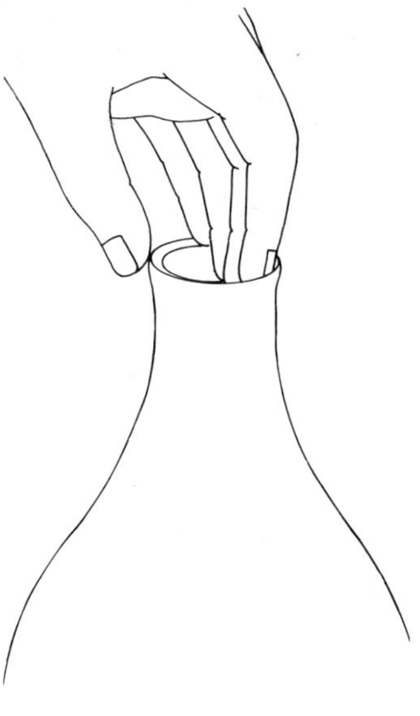

Porcelain form by Delan Cookson (UK)
This bird-like form has been made by assembling together separately thrown pieces. The oxidized barium-nickel glaze has produced a lovely speckled finish.

Joining shapes

At some stage in the course of throwing and building pots you may decide on a form which will require more than the basic thrown shape. The design lines may change direction suddenly, the size may be beyond the capability of the thrower, or the amount of clay too heavy for the wheel. To overcome these technical problems you will need to throw and join separate sections. To make a raised bowl, for instance, a combination of cylinder and bowl may be required.

Selecting a bowl and base which will suit your needs, cut the bowl from the bat and turn the base on a prepared chuck (see page 94). If the bowl is a wide one the chuck needs to be a wide, low hollow ring thrown on a bat and dried to a leather-hard state. Feel the thickness of the base between the fingers to gauge how much is to be removed. All the bowls can be turned on this, so that the thrown edge is always protected.

Recentering on the wheel

A more direct approach is to trim shapes by recentering them upside down directly on the wheelhead, securing them with dots of clay pressed on to the wheel around the rim of the pot. Turn the underside of the bowl in a curve following the contours of the inside so that there is no foot and no spare clay. Measure the top of the cylinder to be used as a base and mark a circle of the same width on the base of the bowl while it is centered on the chuck. Three or four close lines will serve as a guide for the joining and allow for adjustments in placing — a pair of calipers will help you to judge the width required.

Remove the bowl and chuck from the wheel and replace the cylinder, still fixed to its bat and still centered, back on the wheel. Turn away any unwanted clay from the outside of the cylinder and decide on the height needed. If it is too tall for the proportions required, cut the top down with a sharp thin knife or a needle slowly eased into the rim while the wheel turns. Roughen the top edge with a knife and smear it thickly with the slurry, coating the base of the bowl at the same time.

Position the bowl level on the top of the cylinder, turning the wheel slowly to make any adjustments to the level. (Alternatively, the stem can be placed on the inverted bowl which is more stable but the two joined pieces will have to be finished and finely adjusted in the upright position.)

When both pieces are running as true as possible (there may have been some warping of the bowl which cannot be corrected), press the two halves together by tapping with the fingers. Roll a thin coil of clay and press it around the outside of the join, smoothing it into the pot. With the wheel turning slowly, trim and scrape the area where the two halves join, making a smooth flowing line. Finish with a fine sponge, a rubber kidney and fingers.

Place your design close to the wheel so that you can refer to measurements easily. To make a series of tall-footed bowls, cylinders and bowls are thrown separately on the wheel, the tops of the cylinders and the bases of the bowls being carefully measured to make sure they will fit together.

Throw these shapes on separate bats and leave them to harden slightly. They need to be dry enough to handle without being damaged but still damp enough to join. The edges of the bowls and cylinders will tend to dry out where they are thinnest; to prevent this happening keep them damp with a cloth or an occasional fine spray of water.

Joining a bowl to a pedestal

Throw a selection of cylinders and bowls. According to the designs you have in mind; these can be tall, thin, short or broad while the bowls may be open and wide or tall and narrow. There are many interesting combinations.

Allow the thrown shapes to stiffen and then select the pairs of bowls and stands that you wish to join together.

A tall container

The next stage depends on your design. If the pot is to be a tall container, then the bottom of the bowl has to be cut out with a needle or knife to the width of the base cylinder; use the calipers to measure again, then lift the cut portion out and smooth the inside of the join. When this is done the combined pot is cut from the wheel bat and turned upside down on to the chuck so that the base of the cylinder may be turned.

If the pot is to be a bowl supported on a tall foot, then the bowl is left intact and the base of the cylinder is turned or trimmed right out, leaving the base as a tall foot.

A combination of shapes to be fixed together will benefit from being sprayed with water and wrapped in plastic. This will give the pieces time to dry slowly, reducing the possibility of cracking at the join as the pots dry out.

Large spherical pot with a narrow neck

Another use of the joining technique is to combine pots to

increase the size. Again the design needs to be thought out beforehand. Perhaps the shape you require is a large rounded one with a fairly narrow neck. In that case, throw two wide bowls with thick edges to form the two halves of the sphere. Throw them again on wheel bats, measuring the top edge of each until they achieve the same diameter. When the bowls have hardened slightly place one back on the wheel, roughen the edge and coat it with slurry. Repeat this process on the second half (which can stay fixed to its bat) and then place it rim to rim with the pot on the wheel, turning the wheel to center both halves. Press them together to make a firm join.

The bat must now be cut from the top half of the pot. Place a thin coil around the join of the two halves and smooth it over the joint. Then cut a hole with a knife through the thick clay at the neck where the board has been removed. The clay here will be thick and soft and can be thrown and shaped into a finished neck of the style required, or a separately thrown neck can be added at this point. The outside of the join is smoothed as the wheel turns and the inside of the join may also be finished when the top is opened. When the outside shape is complete leave the pot to harden slightly and then turn the base in a prepared chuck.

Additions

A third variation in making a large pot is to add clay to the rim of the thrown base to increase the size, following the method potters have used throughout history.

Add thick coils of clay to the top edge of the base pot and smooth them in as the wheel turns. Rolled coils may be used or flat strips of thick clay cut from a slab and fixed to the rim of the thrown shape; or add a freshly made cylinder with a thick wall to be thrown into shape directly. All these methods work well if you allow the joined piece to harden slightly before adding another coil. The size of the pot is limited only by the size of the kiln. (see also page 105)

Cutting and reassembling

Making pots to be joined provides an opportunity to conduct further experiments into how cutting can alter the basic shape of the thrown piece. By cutting parts out of a circular form and compressing the shape, oval shapes can be made. Or remove the base altogether, press the hollow form into the shape required (oval, square, oblong, and so on) and then fix it to a slab or a base of a different shape. Try slicing some of the shapes and reassembling the pieces in a different form. Animal or bird shapes may grow from constructions of cut pieces. Many of the experiments will not be worth keeping and firing but they will help you to understand the material you are using and the best ways of joining clay without its cracking. They may even provide inspiration for a further series of thrown shapes.

H page 309-10 **C** page 76

Tools

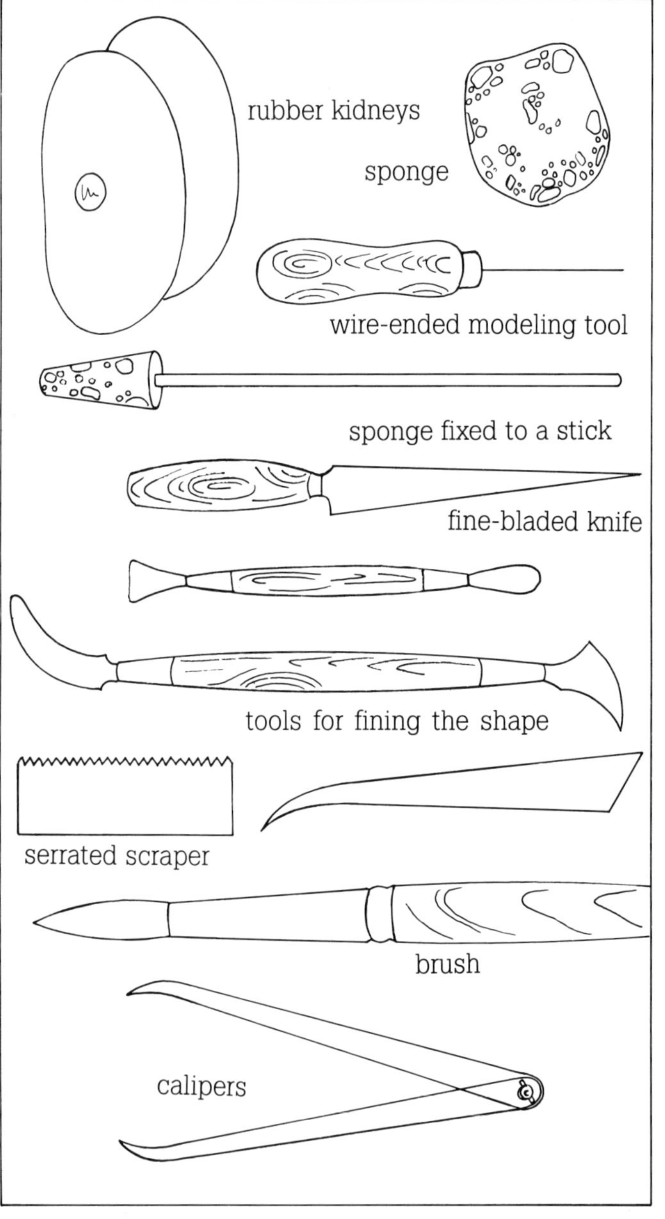

Measuring and fitting

1 Replace the cylinder on the wheel. Trim the outside with a flat scraper and then cut the top flat and straight.

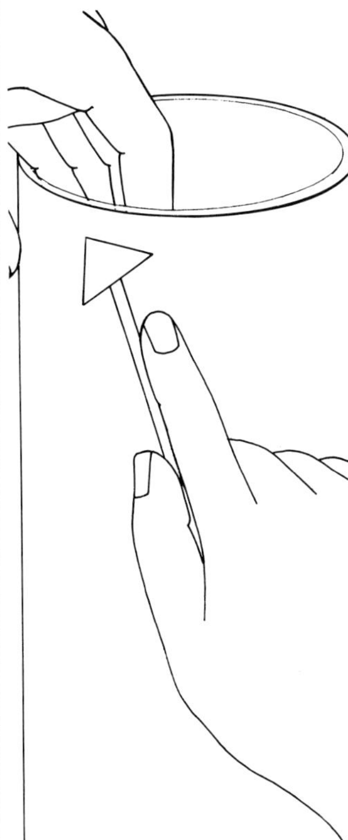

Cut the cylinder off with a needle or sharp knife at a slight angle in order to accommodate the curve of the bowl.

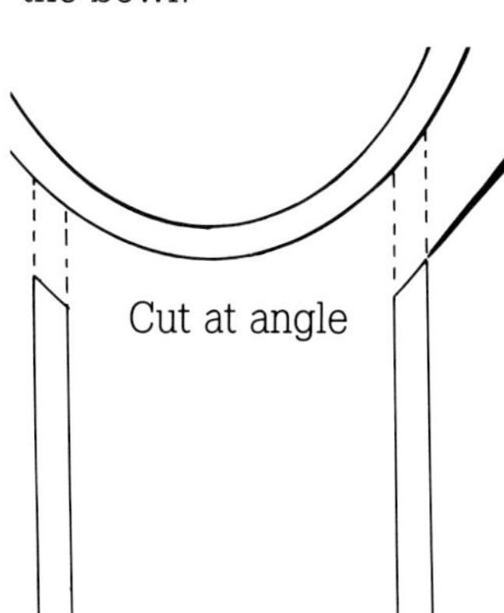

2 Put the cylinder to one side for the moment. Cut the bowl from its bat and center upside down on the wheel. Trim away any spare clay to leave the bowl outline an uninterrupted curve.

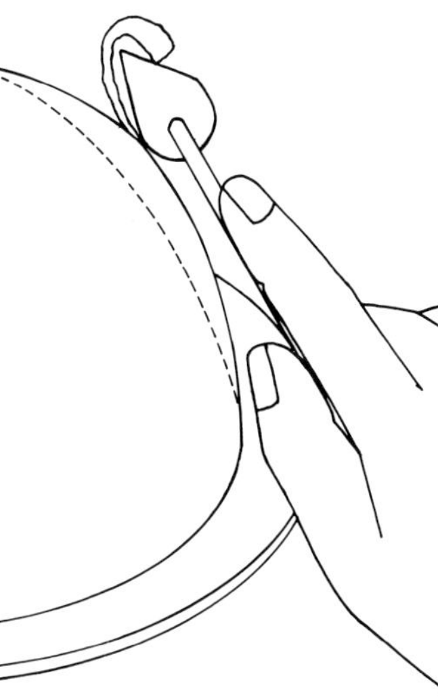

Now mark a circle on the bowl. Make sure this is the same diameter as the top of the cylinder. Roughen it and paint the area with slurry.

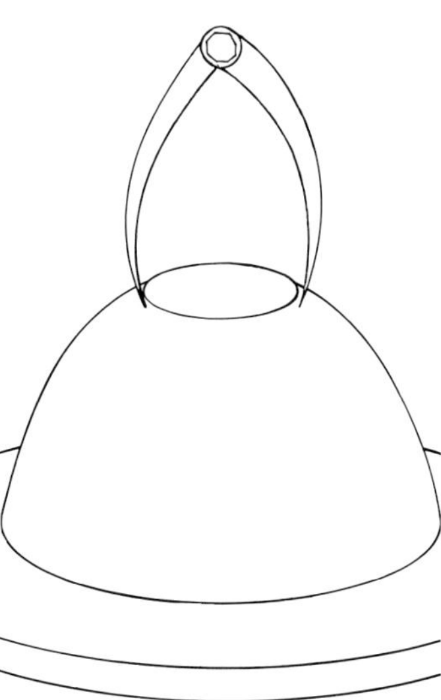

3 Still fixed to its bat, replace the cylinder on the wheel. Roughen and slurry the rim. Position the bowl on the base, press it on firmly and then clean and smooth the join.

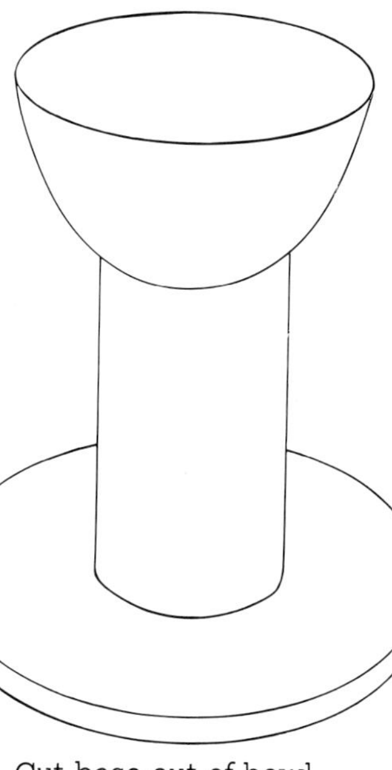

Cut base out of bowl and make one hollow center.

Cut here with knife as the wheel revolves slowly.

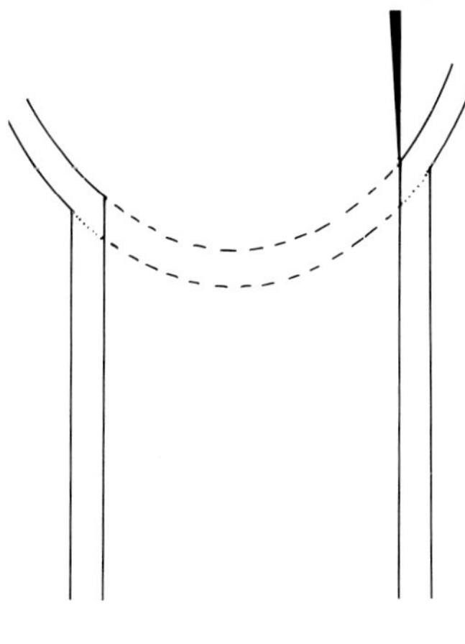

4 Either drill a hole in the base to allow air to escape or, when removing the pot from the bat, cut slightly above the base so as to leave the base open.

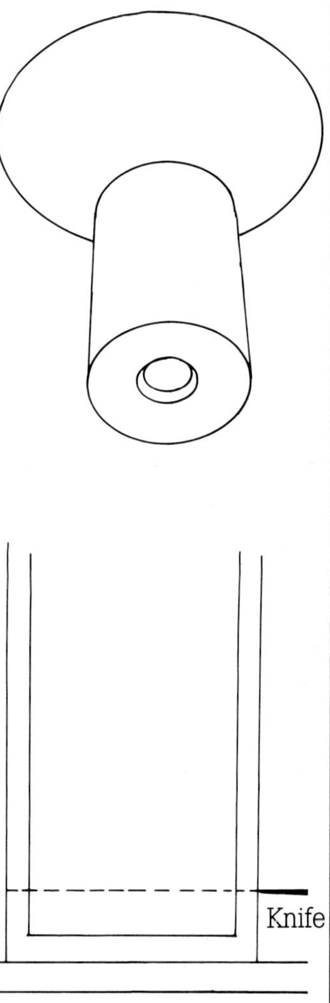

Trim base of the pot at the wheel head and cut it away from the bat.

Turning the foot

Most pots will need some work on the foot before they are considered finished. This stage is called turning or trimming and is done by placing the pot upside down on the wheel when it is 'leather hard' and cutting away the excess clay at the foot.

Providing a steady anchorage

When the pot is dry enough to handle without suffering any damage, cut it from the wheel bat and place it rim down on a clean board. This will help the base (the thickest part of the pot) to dry evenly and also stops the rim of the pot warping as it dries. When the base has a soft leather-hard texture place the pot upside down on the wheel, centering it by the rings on the wheel or by eye. When it is correctly positioned hold the pot in place with one hand and press three or four dots of clay around the rim to anchor it — pressing the dots on to the wheelhead, not on to the pot.

Using a chuck

Another way of fixing the pot to the wheel is to use a chuck, that is a turned shape of clay made on the wheel so that it is centered and accommodates the shape of the pot to be turned. The pot rests on the chuck rather than the wheel. This is particularly useful when the pot to be turned has a fine rim which could easily be damaged. The pot may become unstable when balanced upside down, in which case the chuck is hollow to allow the pot to be held inside safely while it is being turned.

Tall pots

A tall pot may be turned after it has been recentered the right way up. Fix it to the wheel with a little water and suction and press the excess clay at the foot on to the wheel to secure it. The final cut of the trimming is made through the spare clay to the wheelhead in order to release the pot.

Trimming the foot

Whichever way is chosen, the pot is held steady while the wheel is turned so that the spare clay at the base can be formed into a foot. The proportion of the foot is an important part of the overall shape and should be considered with care as it can affect the whole character of the piece. When the pot is firmly in place, turn the wheel at a slow speed. Decide first how wide the foot is to be and cut away the spare clay on the outside with a sharp-edged tool. Some trimming tools can be made by bending a strip of metal at a ninety-degree angle and sharpening the edges to a blade. The sharpest and most versatile of trimming tools can be made from packing-case strip steel. Shaped in a loop and bound to a handle, the loop can be bent to any curve required and as the metal wears it becomes thinner and sharper.

Turning the foot

When the width of the foot is established, carve out the clay from the center outwards as the wheel rotates, leaving a foot ring as thick as the sides of the pot. Do not be tempted to carve the clay right down the sides of the pot as this will take away the freshness of the throwing and make the shape mechanical. Just take off the bare minimum to define the foot and the pot. Later, as experience grows, the pots you make may require trimming all over the inside and out, or even rubbing down after the biscuit firing.

Tools for turning

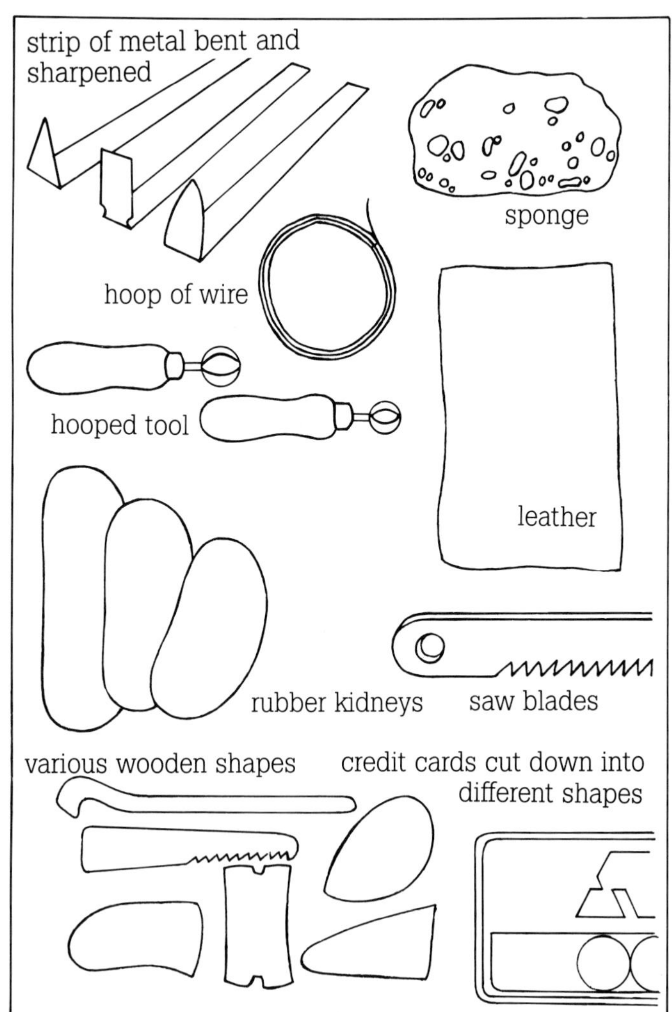

Turning or trimming Turning or trimming a foot usually takes place when the pot is leather hard.

1 First the pot must be inverted and fixed on to the wheel. The pot is centered upside down on the wheel and held in place by small coils of clay thumbed on to the wheel head.

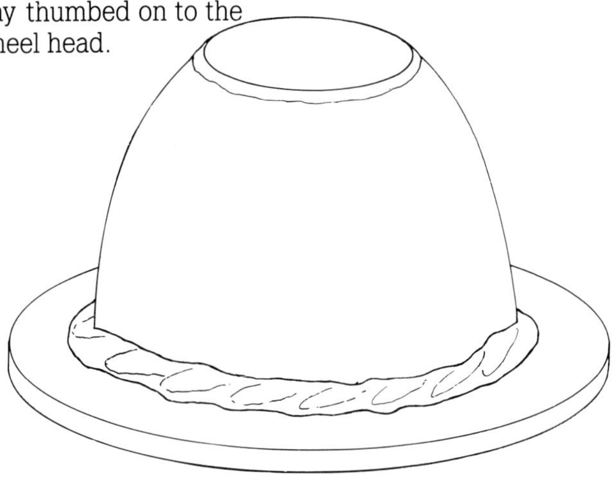

2 You will need to remove the excess clay on the outside before you can decide how wide the foot will actually be.

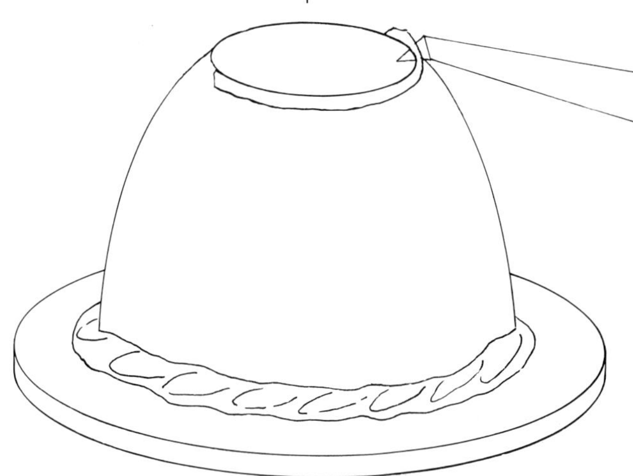

3 Carve out the clay from the center of the foot.

4 For shapes which will not stand upside down on their own, a thrown hollow form is used for support. This is called a chuck.

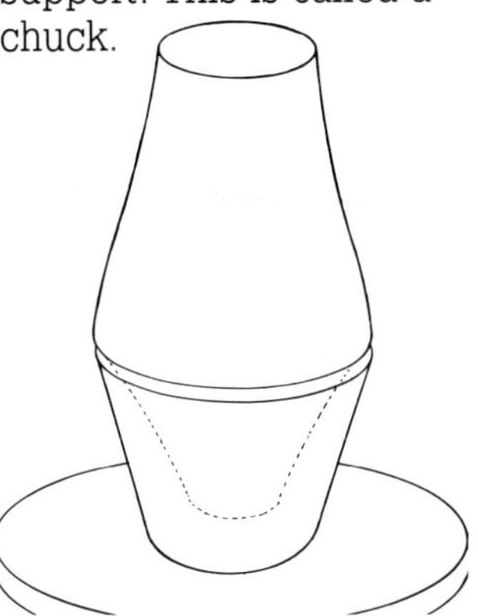

A wide chuck is often used to prevent damage to a delicate rim.

A cone-shaped chuck will support a goblet type of pot.

A hollow chuck is used for bottle-necked pots.

5 You can also trim excess clay from the lower walls of upright forms such as bottles by recentering them when almost leather hard and attaching the foot to the wheelhead by smearing clay downwards. This 'spare' clay joining the pot to the wheelhead is cut away at the very last moment when all the trimming has been completed.

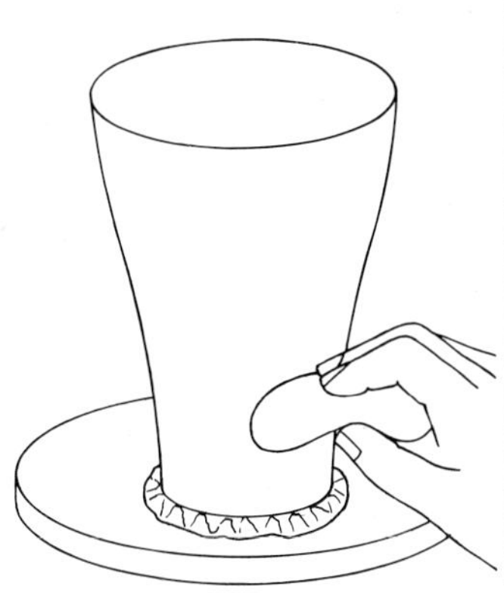

Lids

Type of lid

A lid is usually a cover for a pot designed to keep the interior contents warm, cool, dry or clean. A basic lid may be as simple as a flat slab of clay placed on the rim of the pot. More functional lids may contain some form of fixing which will stop the lid sliding off when the pot is moved. Lids generally fall into two groups; those that fit into the pot and those that fit on to or over the pot. Some are thrown the right way up, the knob and the outside being finished during the throwing; the underside will then be turned and trimmed, or simply cut off directly as a complete form. Others are thrown upside down to ensure a measured fit with the top of the pot.

Lids thrown the right way up may consist simply of a shallow bowl with the opening of the clay started from outside the center, the center piece being left to form the knob. This lid rests on the rim of the pot or on a ledge built in the inside of the pot while this is being made. Throwing lids the right way up allows the potter to create a greater variety of knob shapes; they may be small and solid, large and hollow, or carved and modeled.

Lids made upside down are usually best for pots that need a tight-fitting lid or a long flange on the lid itself to stop the lid falling off as the pot is tipped (on a teapot for instance). The inside of the rim of the pot and the width of the lid have to be measured with calipers; slight adjustments to the size and fit must be made when the pot and lid are in the soft leather state. Try the lid for size before it is turned; if the lid is too small it may be drawn out slightly while the clay is still damp; if it is too large it may be trimmed on the wheel to fit the opening.

Upside-down lids may be turned in place on the pot so that they are shaped to fit the pot and complement its shape. The final shape of the lid can be finished on the pot, taking into account its relationship to the spout or handle. There are other types of lid; some fit right over the top of the pot and are made like small or shallow cylinders, the tops being turned. A lid can be formed from the top half of a sealed shape, exactly matching the bottom half so the join scarcely shows; lid and pot are separated at the leather-hard stage by cutting the two apart with a sharp knife.

Lid components

neck (rim)
knob
lip of lid
mouth
flange of lid
shoulder of pot

flange of pot
knob
lip of lid
mouth
gallery
shoulder of pot

Rim-resting lids

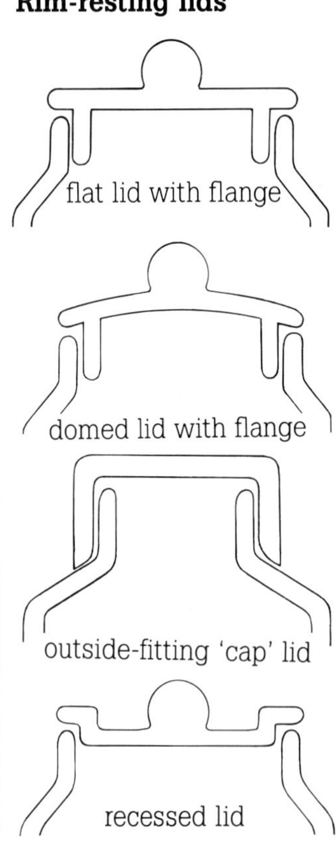

Lids on a gallery

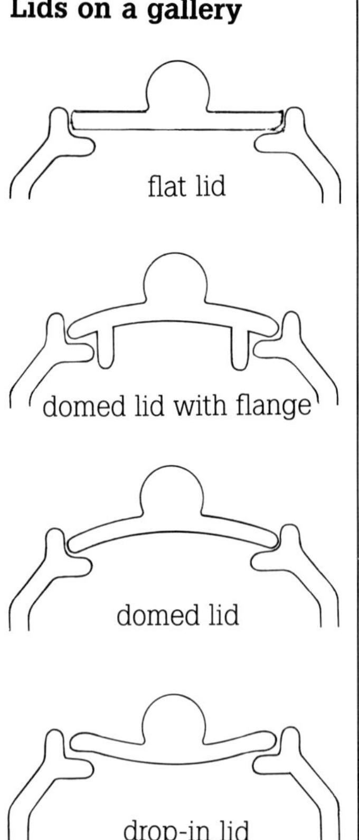

Sealed-pot lids

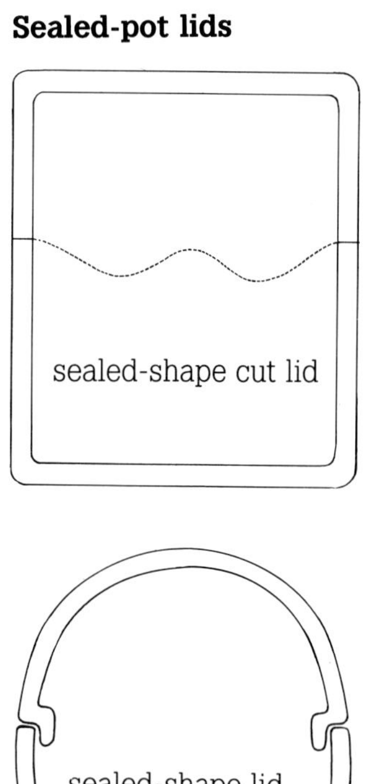

Throwing a pot with a gallery and lid

A gallery (or flange) is a small ledge which is formed on the inside of a rim. The lid simply rests on this. A casserole is a suitable object to make with a flange and lid.

Pots and lids should be made at the same time, measuring them carefully; shrinkage of both will then take place at the same rate.

1 Throw the pot in the usual way — while the walls are thick and short, split the rim into two by pushing the inside half downwards to make a step.

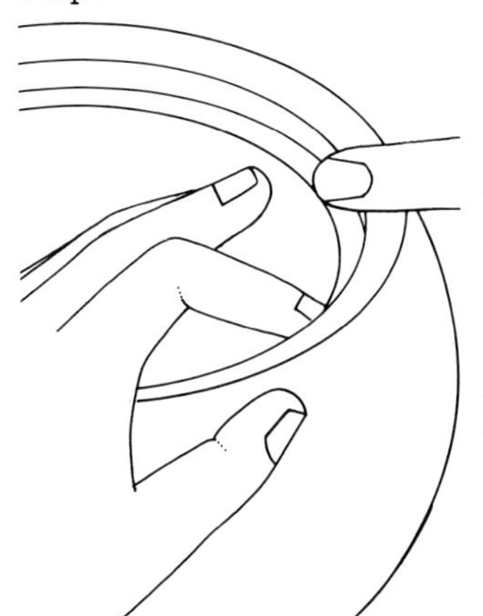

2 Carry on throwing the casserole and complete the shape. You will find that the step is still on the rim, although not quite so distinct now.

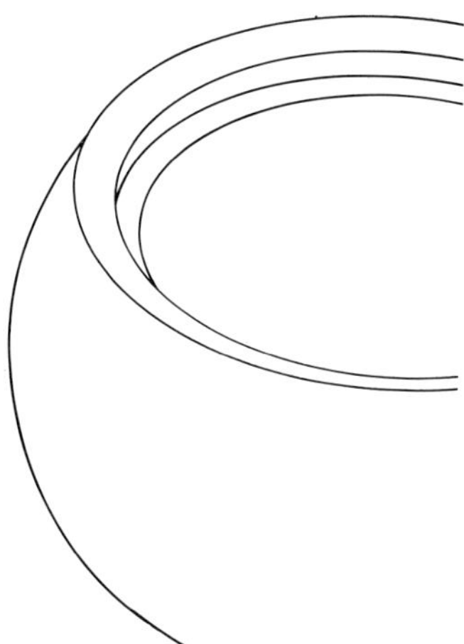

3 Remake the step, placing a finger on top and a hand underneath the ledge for support. Finish and smooth the gallery with a square-ended tool.

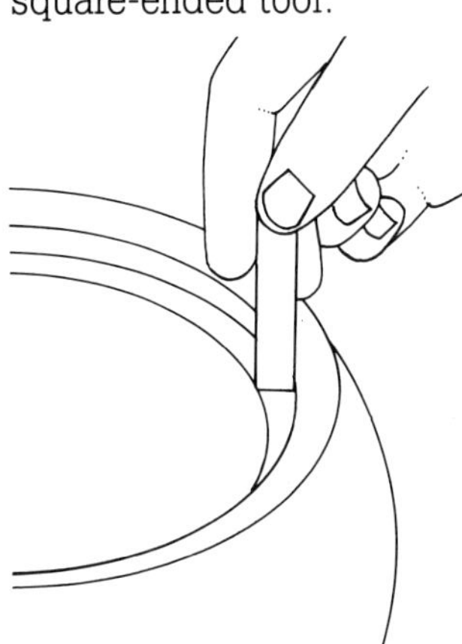

4 You must measure the inside width of the gallery before it is removed from the wheel in order to be able to make a well-fitting lid.

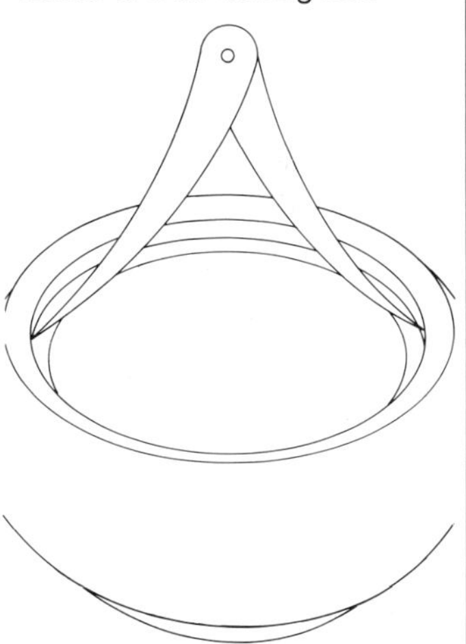

Casserole lid: flat lid with added knob

First flatten a piece of clay on the wheel and cut it to the width of the casserole rim. Leave it to stiffen a little.

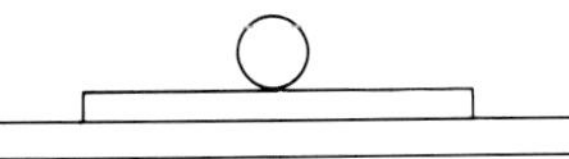

Attach a ball of clay to the center of the lid and throw the ball into any shape required.

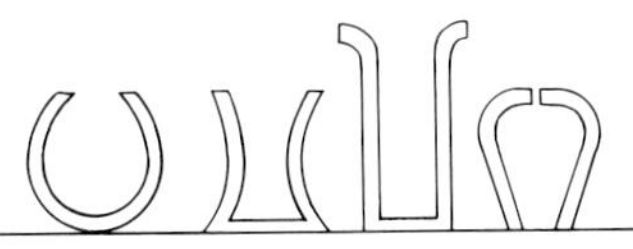

Flat lid with integral knob

When flattening a piece of clay, leave a lump in the center which can then be thrown into a knob of the required shape.

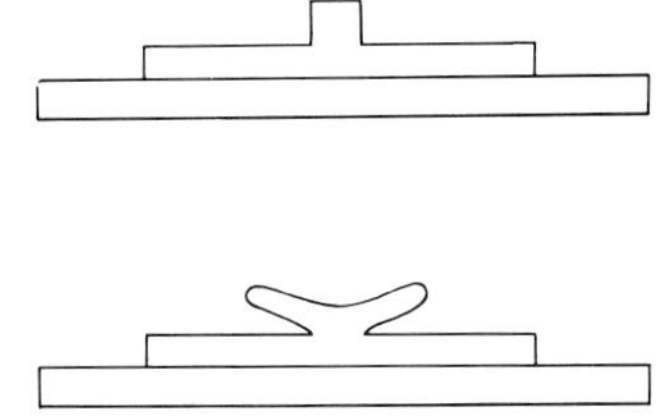

Turned domed lid

Center a piece of clay. Shape it into a mushroom with a stalk and head.

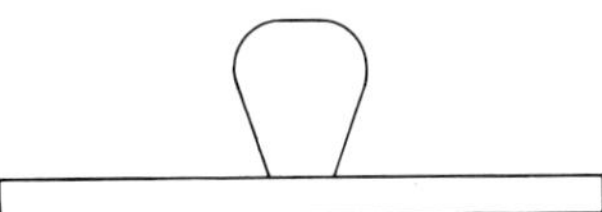

Open up the piece into a shallow bowl with a flattened rim.

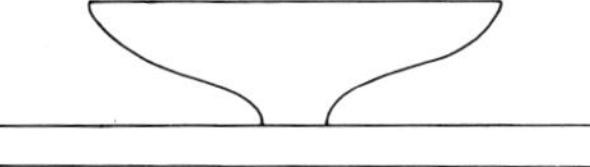

Use the calipers to ensure the lid is the same width as the casserole.

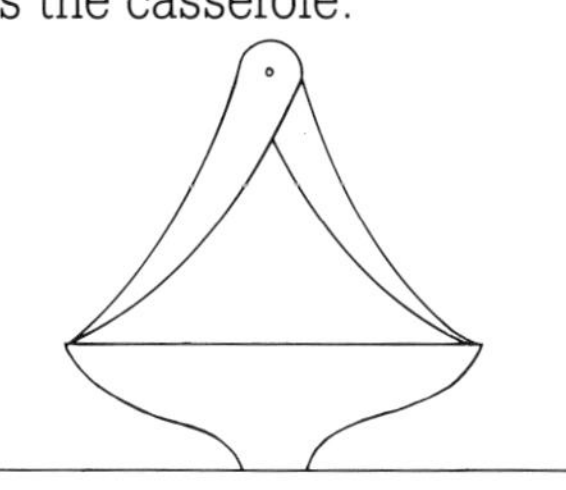

When the lid is leather hard it can be turned over and the knob trimmed.

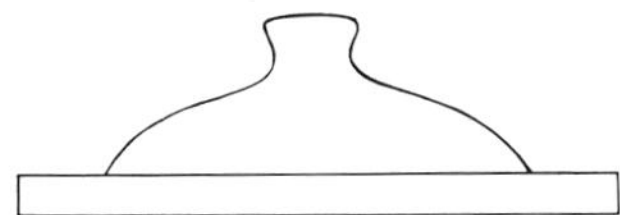

Lid thrown right way up

1 Measure diameter of neck (or rim) with calipers placed on exterior surface of rim.

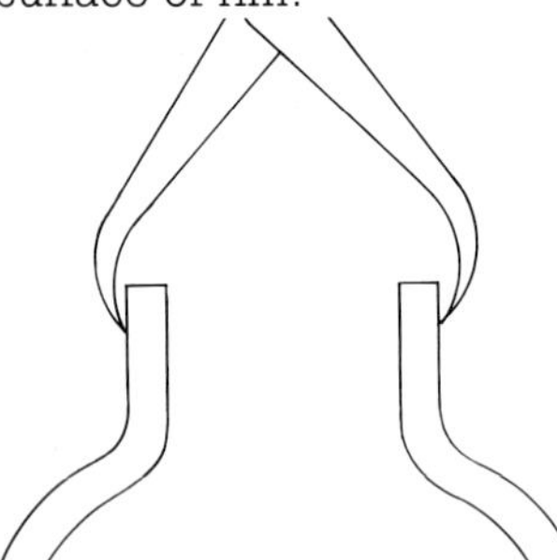

2 Throw right way up, measuring the width of the lid so that it will fit the neck exactly.

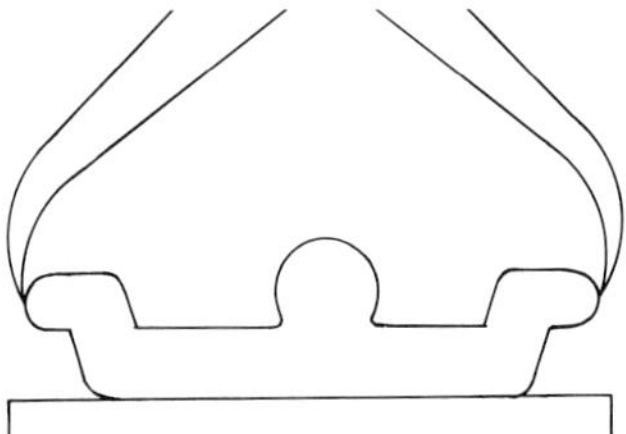

3 Measure mouth (inner diameter of neck).

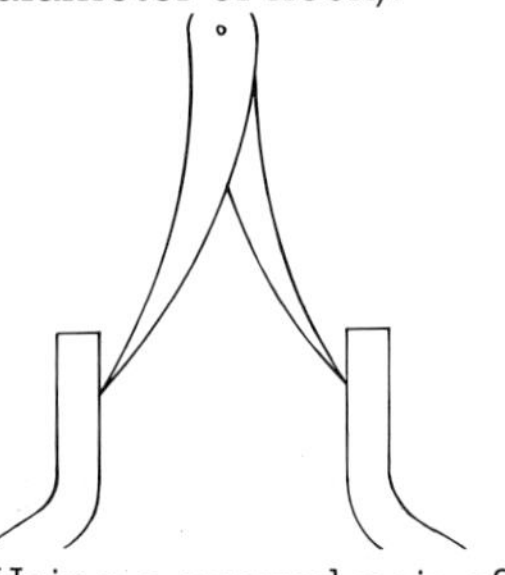

4 Using a second pair of calipers, open these to same width as calipers in 3. You can now use these for external measurements.

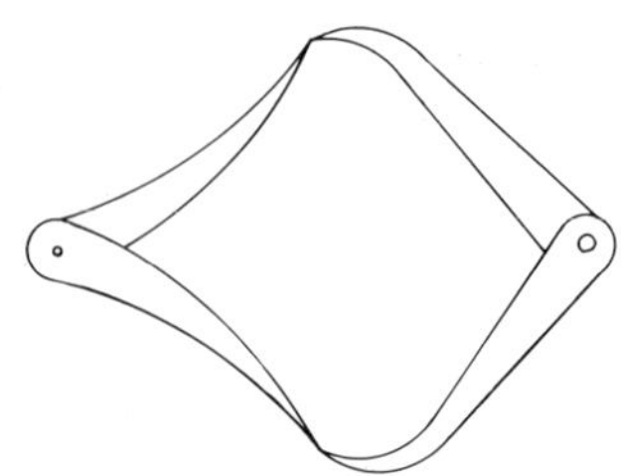

5 Turn and trim the spare clay on the lid until it is the correct size for the mouth, constantly checking diameter with the second pair of calipers.

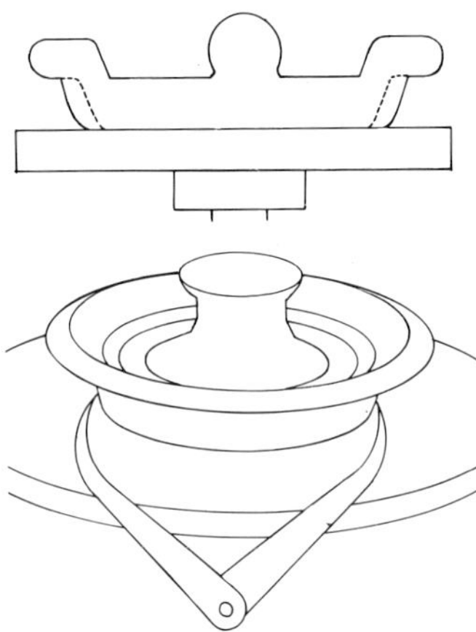

6 The lid will then fit both the outer edge of the neck and the mouth.

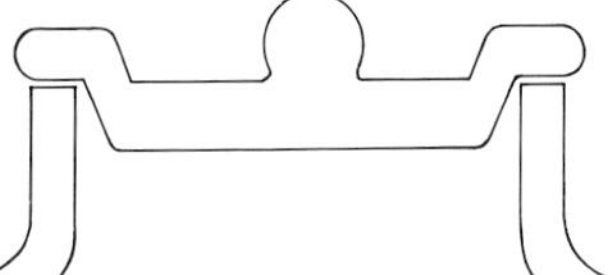

7 Turn a small pot on the wheel and use this as a support for the lid so the knob is not damaged. Now trim or hollow out the underside as required.

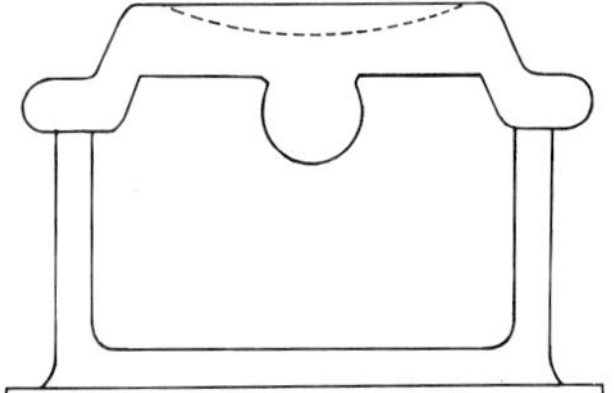

8 The same sequence can be used for making a lid that fits a gallery, but measure the mouth of the neck first to check the diameter of the lid and then measure the diameter of the gallery's mouth. (See also *A lid for a teapot* page 99.)

Outside fitting 'cap' lid

1 Measure diameter of the external neck (or rim) of the pot.

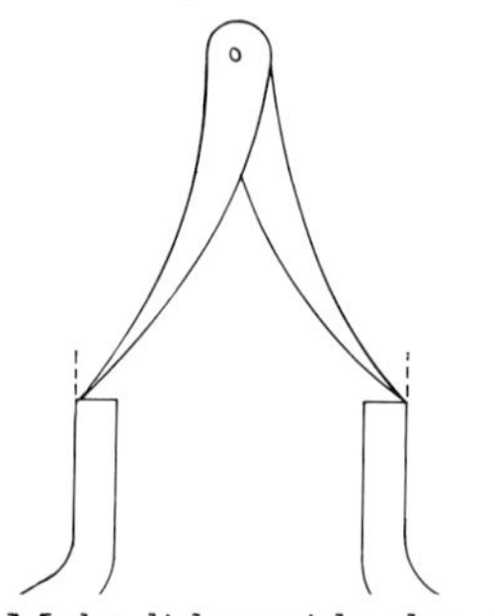

2 Make lid upside down. Keeping calipers in the same position measure interior of flange, making sure it is the same size all the way down.

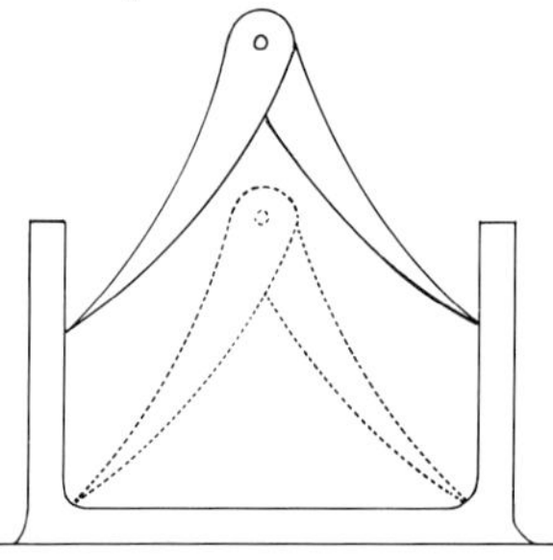

3 When you are sure clay is firm enough, cut the lid flange at an angle to fit shoulder of pot.

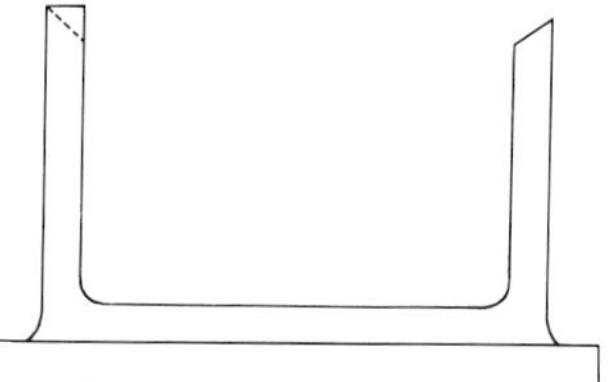

4 The lid will then fit snugly on to the pot; the inside of the lid cannot touch or damage rim.

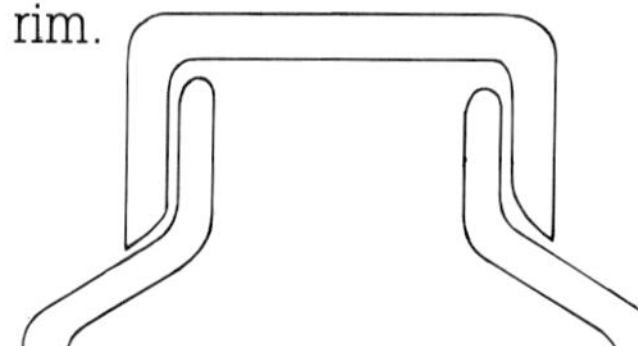

Sealed shape cut lids

1 Sealed pots can be made from a thrown or slab-built pot. The top is cut off completely, using a sharp knife or needle. If the cut is made in a wavy line, both halves will lock together and prevent lid sliding off.

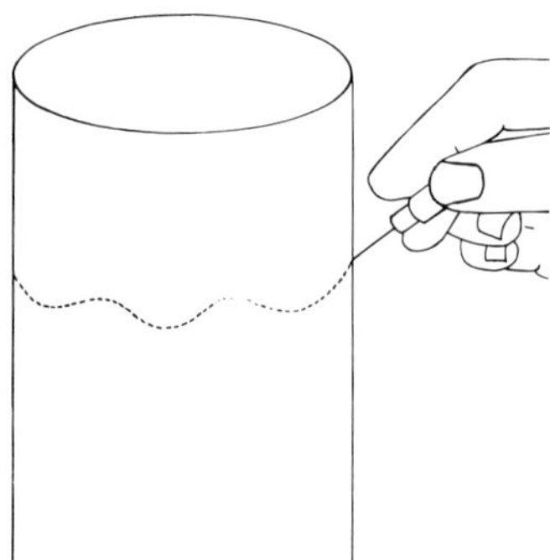

2 If using a double thrown or pinched round pot, the cut line may be almost straight except for one or two interlocking shapes.

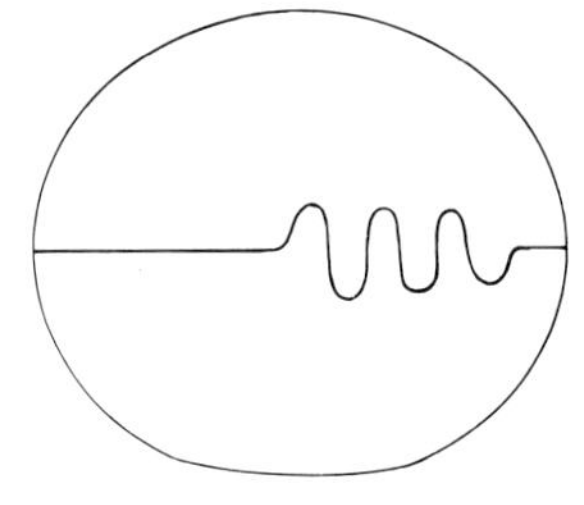

See also *Slab pots* page 55
Pinched pots page 39

Further information on measuring accurately is given on page 314

Sealed pot lids

1 Use a thrown, cast or handmade sphere or rounded shape.

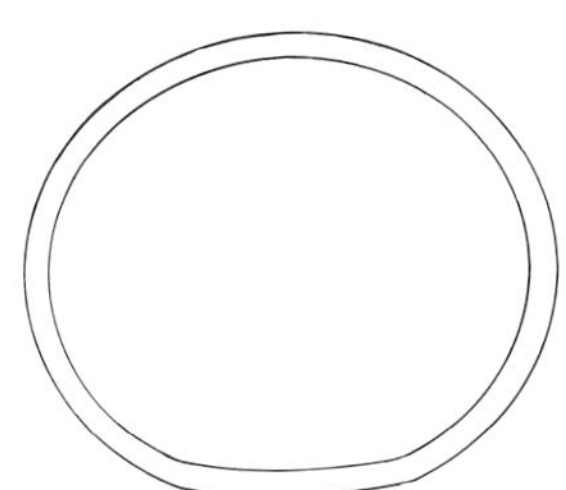

2 With a square-ended tool, indent a groove all around the sphere.

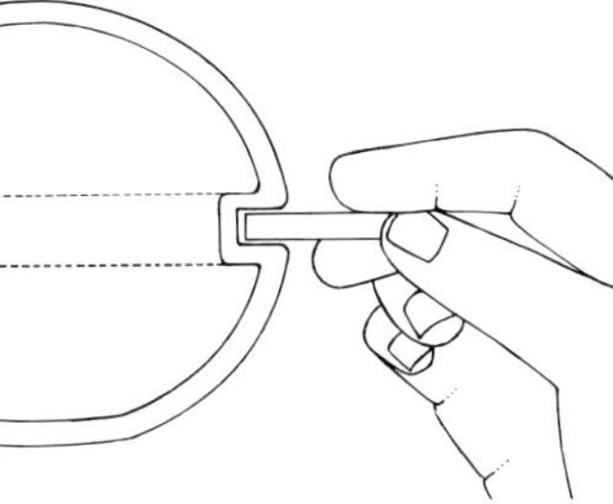

3 Cut carefully with a needle at the lower end of the indentation. Trim off slight spare clay on bottom section.

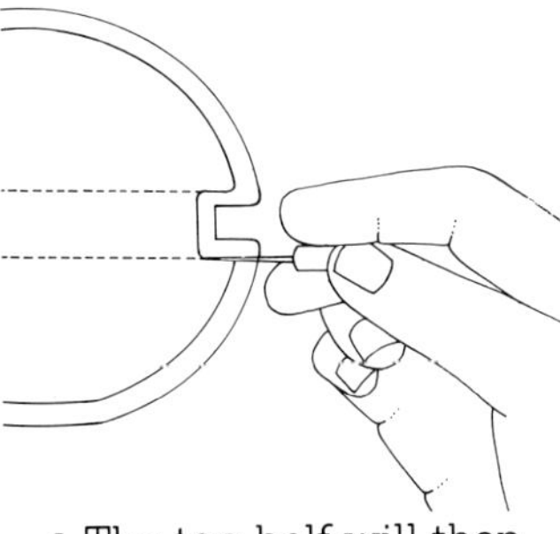

4 The top half will then drop neatly into bottom half.

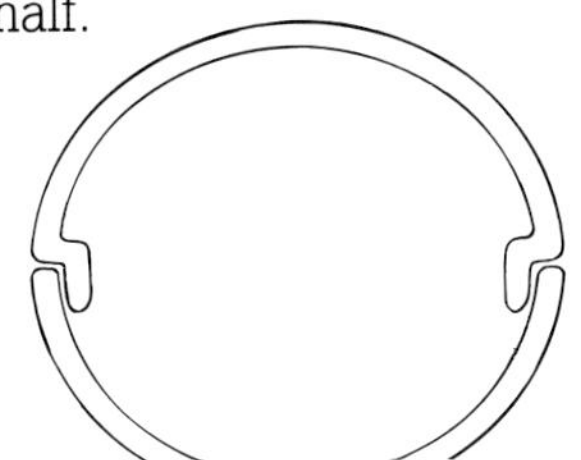

Drop-in lid for a teapot—made upside down

1 Measure diameter of mouth and gallery.

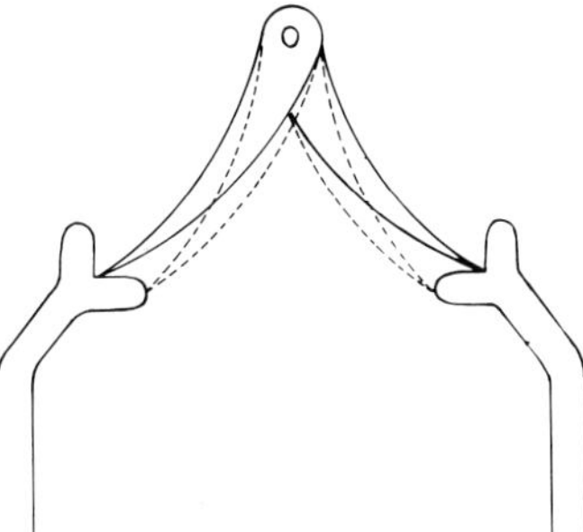

2 Throw a shallow thick-rimmed bowl. With fingers, split the rim into two halves, pushing the outside half downwards.

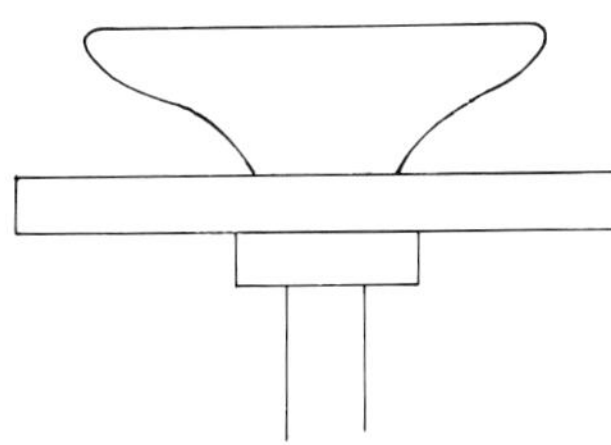

Note: A teapot lid has a long flange to prevent the lid dropping off when the pot is tipped.

3 Measure and adjust the width of the flange (that is, the piece that will fit into the teapot). Pull flange upwards.

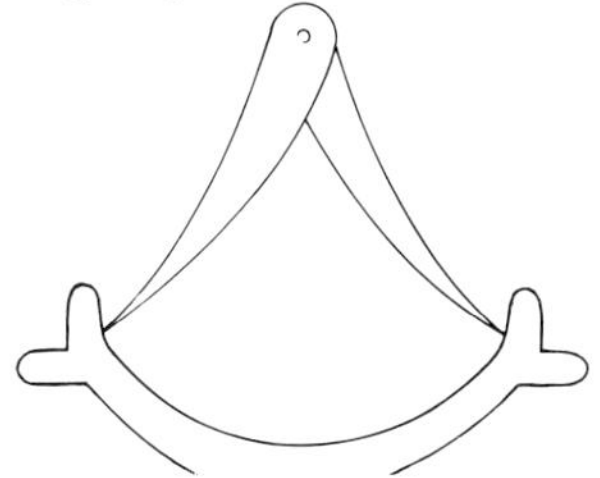

4 Square the flange and the rim with a square-ended tool.

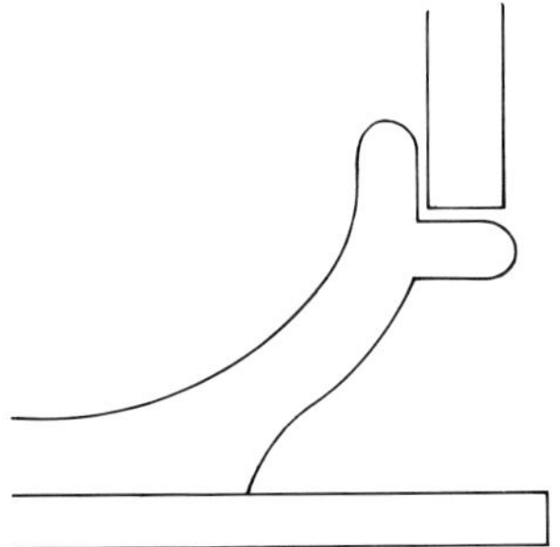

5 Cut off the lid from the wheel and allow it to dry until leather hard. Then turn the knob, with lid well supported on rolls of clav and a chuck.

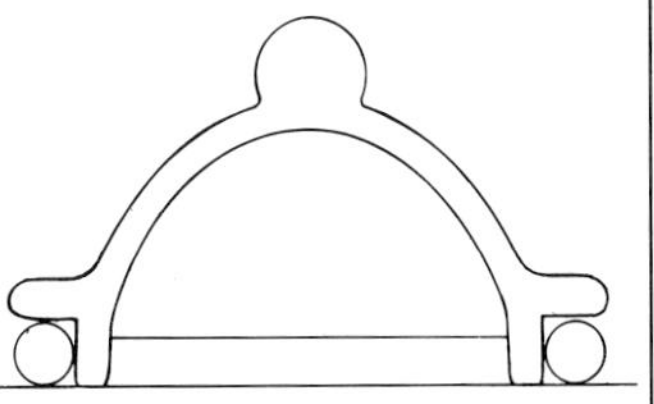

6 A teapot lid has a long flange to prevent the lid dropping off when the pot is tipped.

Teapot with stand by Jenny Beavan (UK)
The porcelain teapot has been surmounted by a neatly fitting lid with an unusual 'winged' knob.

Lips

Forming lips

Lips can be formed on wet clay by holding the dampened finger and thumb of one hand on the outside of the jug and the forefinger of the other hand on the inside of the jug, the inside finger being placed in between the outside fingers. Draw both hands up the jug, pressing the inside finger between the outside fingers so as to form an indentation in the side of the pot that will lead to the edge of the lip; this is pulled outwards and slightly downwards. Some vessels will require only a slight shaping of the top of the lips; others will need a lip which develops from the top third of the jug and sweeps up to a wide flowing lip — to ensure a good pouring action. In some cases the top can be cut away to leave a spout or lip.

A cut lip is a cross between a spout and a lip and is made from a separate sheet of clay or a section of pot cut in half-moon shape; this is then fixed around a similar shape cut in the rim of the pot.

Lips are pulled on rims before the clay becomes too hard.

Stem jug by Frank Hamer (UK)
The lip is an integral part of the form and balance of this thrown piece.

Lips

1 One finger goes inside the jug. Place the forefinger and thumb of the other hand on the outside. Now draw up both hands to the rim.

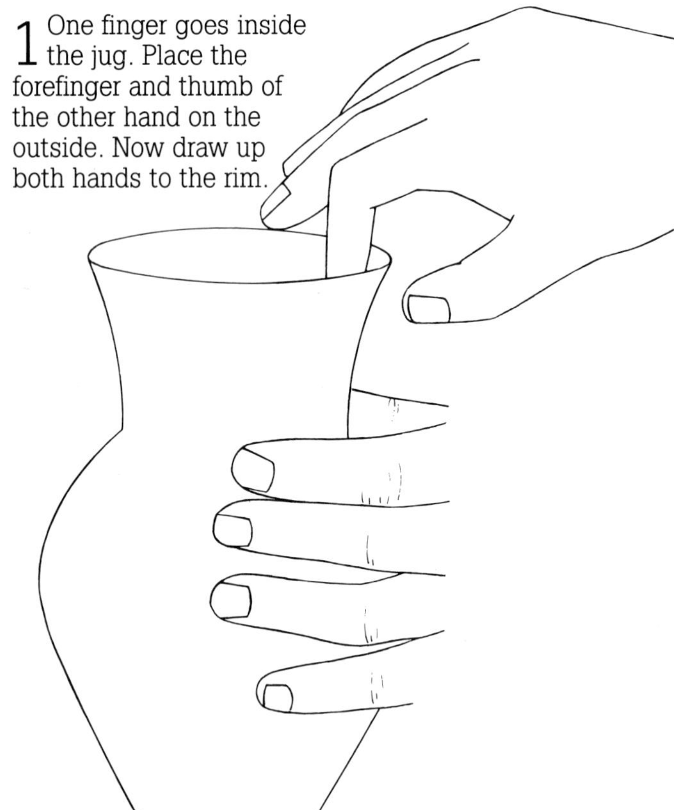

2 Finish the lip by pulling the inside finger over and outwards over the rim.

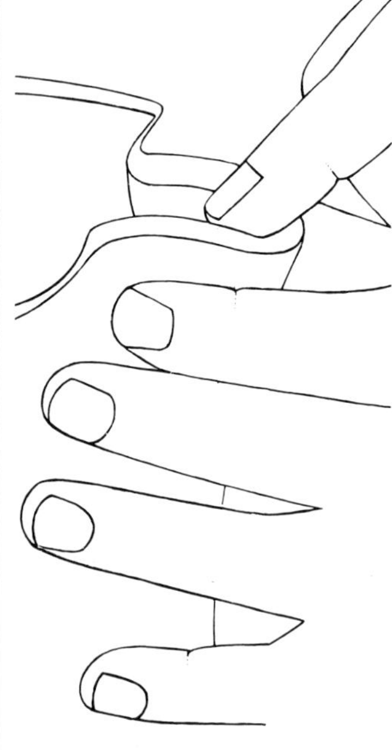

3 Another method is to cut the rim into a lip shape by using a needle or wire to remove the unwanted clay.

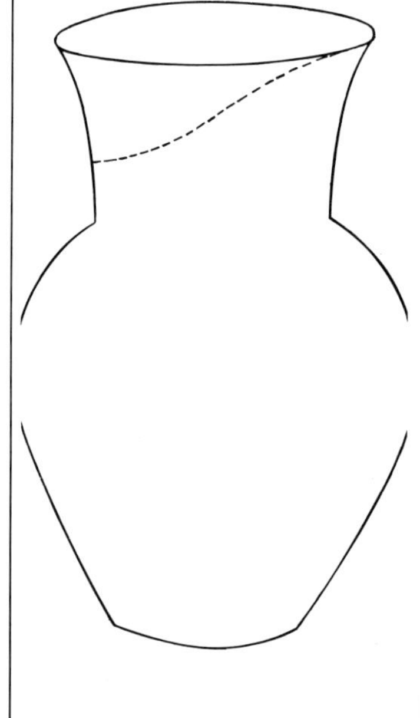

Spouts

Additions and subtractions to or from pots of all kinds are important features both as elements of visual design and for functional purposes.

The most common use of spouts is in the making of tea-pots, the spout being the most efficient way of directing hot tea into the cup. A spout should be made separately but at the same time (and of the same clay) as the pot and the lid so that they can all dry and shrink together. However, the spout (being usually thinner and smaller than the pot and the lid) may have to be wrapped up to stop it drying too quickly.

The spout is thrown as a small narrowing bottle shape with a tall, narrow neck to allow for trimming later. It is attached at the leather-hard stage when the pot and the lid have been turned and the lid has been placed in position to gauge the proportions and the position of the spout; the lid is removed for the fixing of the spout. Once the position of the spout has been determined the base is cut at an angle to make the spout fit on the teapot's side so that the top of the spout is above or level with the open top of the teapot. If it is below that height the tea will obviously spill from the spout as the liquid fills the pot. Hold the cut spout against the pot and carefully mark the point of attachment.

Pierce a regular pattern of holes in the circle that will be covered by the spout. Cut them cleanly with a hollow hole cutter or a twist drill bit to avoid burr. Do not press the holes through with a spike or the clay around the holes will split. Clean off any spare clay around the holes and smooth with fingers and a damp sponge. If the holes are too small they will block with glaze during the firing and if left rough they will gather sediment and become blocked when the pot is in use.

It is sometimes useful to thin the wall by shaving clay from the outer surface at the point where the holes are to be made. This reduces the risk of blockage which can occur in a thick-walled piece. When the spout is attached to the pot, cut the end of the spout at a slight downward angle and round off the cut edges with a damp finger, widening the end slightly to create a good pouring action.

Throwing and joining a spout for a teapot

1 Throw a narrow-necked cylinder, alternately pulling and collaring with the fingertips until it has a wide base and a narrow neck.

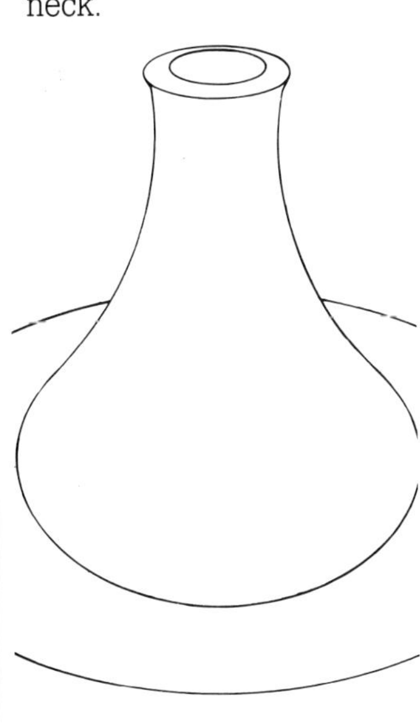

2 Allow it to dry until leather hard. Then hold it behind the teapot to test the angle of placement.

The spout will be cut to fit on to the curved side of the pot.

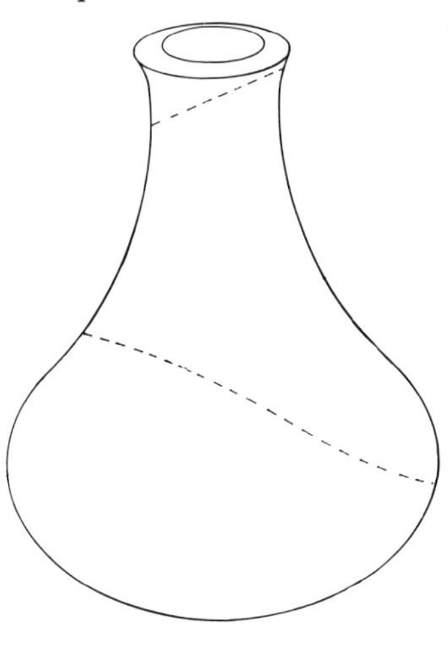

3 Before attaching the spout, drainage holes must be pierced through the body of the teapot with a hole cutter. Mark where the spout will go and then cut the holes.

Score and slurry the edge of the spout which will be joined to the pot.

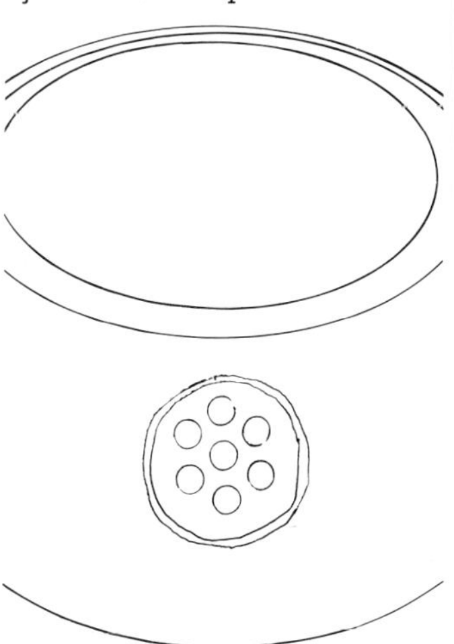

4 Make sure the spout is cut either level with or above the rim. Otherwise the tea will overflow when the pot is filled.

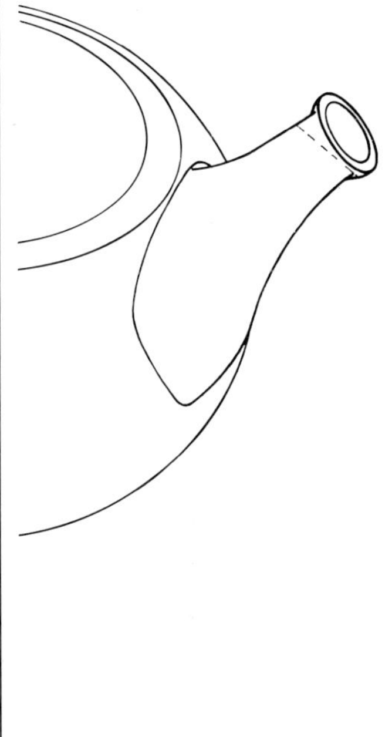

Handles

Some convenient method of picking up and holding is necessary on many pots. Mugs, jugs, teapots, urns, casseroles and some lids need a handle, knob or raised section on the outside to grip the pot and make it easier to lift, especially if hot or wet. Handles are usually pulled from a pear-shaped piece of fairly firm clay. Use plenty of water, squeeze clay with one hand and adopt a downwards pulling action which gradually thins the clay to a flat strap; its size depends on the size of the host pot.

The thin straps are laid on a clean board to stiffen before use and set to dry in a curve. They may be left attached to the lump of clay and allowed to dry in a natural curve. The handles will have to be watched to make sure they don't harden too much; a combination of stiffness and flexibility is needed to make the attachment to the pot successful. The pot and the handle should be at the same state of dampness when they are joined.

Some potters prefer to attach the clay to the pot first and then pull the handle on the pot itself, but this requires considerable practice (as does all handle making and pulling). Handles can also be fashioned from a flat slab sliced into strips; from a tapering coil slightly flattened into a D-shaped section by lightly running a rolling pin over it; or from a loop of wire being pulled through a piece of clay — the shape and size of the loop dictating the shape of the handle made.

If you own a pugmill or similar extruding machine then endless handles may be squeezed through a template fixed to the mouth of the machine; this makes a continuous strip which is then cut to the required lengths. Lids do not always have knobs of course; a strap handle fixed to a wide lid will give a pleasing effect and will also make the lid easier to pick up with a cloth when hot.

Attaching handles

Hold the handle against the pot to judge where it will suit best, mark the position and roughen the surface at both joining places. Paint some thick slurry on the two points and on the top edge of the handle. Press the handle on quite hard, curving it down to the bottom fixing point and pressing firmly into place.

A small bench wheel would be useful here to turn the pot and so allow you to study the profile with the handle attached; adjust the curve with the finger if needed, cleaning off any spare slurry and generally tidying. A handle thus fixed is almost impossible to pull off or remove without cutting the clay.

Strap handles

A roll of clay, slightly tapered to one end, can be flattened with the hand or with light pressure from a rolling pin to form a strap for a handbuilt jug. This will be more appropriate than a pulled handle. Also, it allows for various widths or strengths and additions of pattern before the handle is applied.

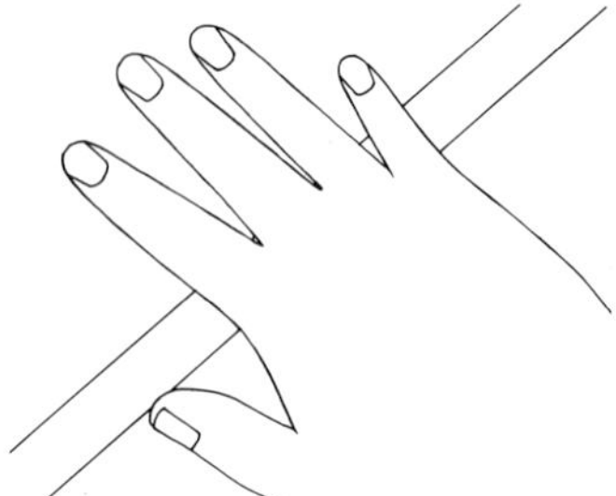

1 The handle is attached to a scored area at the top of the pot.

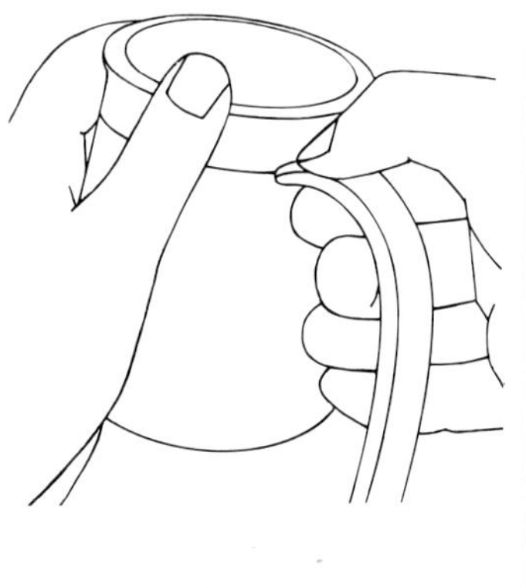

2 The base of the handle is pressed carefully and firmly in position, bearing in mind that the length of the handle will determine the grace and comfort of the inner curve.

3 Finally, smooth the ends of the handle into the surface of the pot.

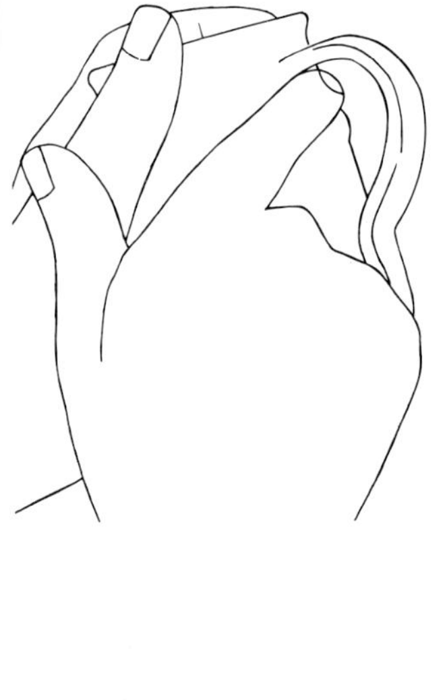

Pulling handles (using one hand)

1 Handles can be pulled from a well-wedged, pear-shaped piece of clay. Use plenty of water and hold the piece of clay in one hand.

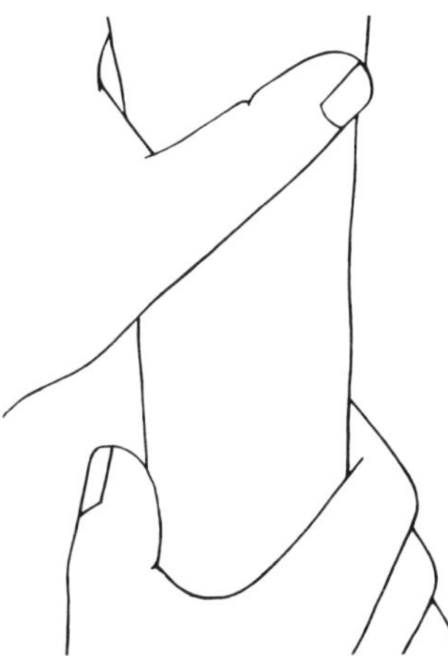

2 Squeeze the clay with the thumb and fingers around it, gradually running the hand downwards to increase length and turn the handle.

3 Finish off with the thumb flat and running down the center.

4 Break or cut off the handle and place this carefully to dry — either laid flat on a board or set in a curve to harden.

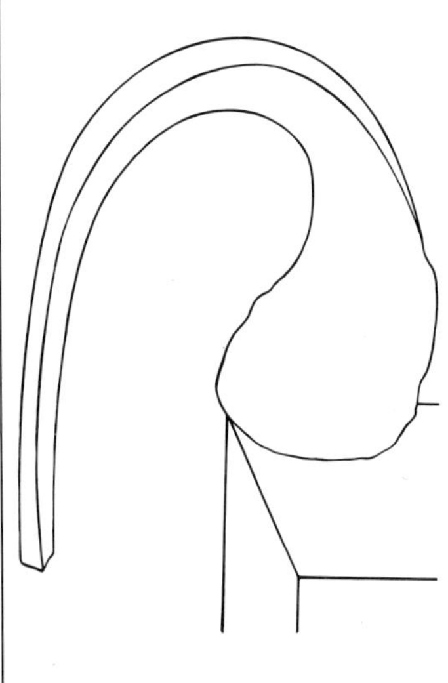

Drying in a natural curve

Pulling handles with both hands

Here is another method which is basically the same but which allows you to use both hands.

1 Fix the clay on the edge of a bench.

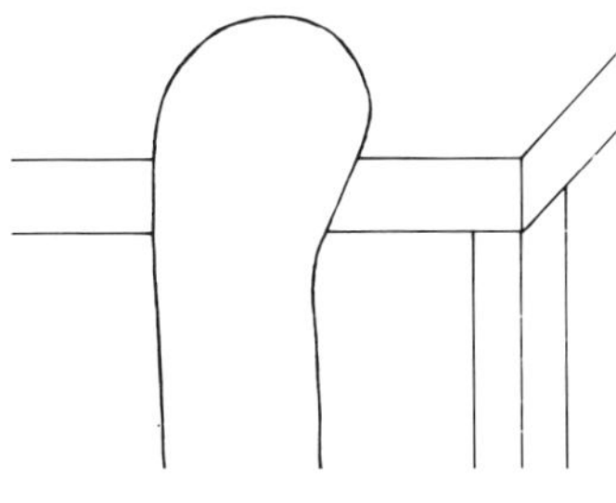

2 Using both hands, alternately smooth and turn the handle out, running both thumbs down side by side to form a central rib.

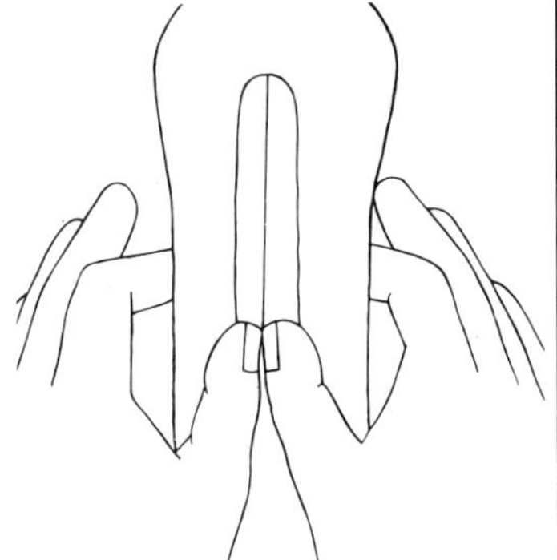

3 Fix the soft handle by its wide end to the top of the pot, not too near the rim. Hold it in position and decide where the bottom join should be.

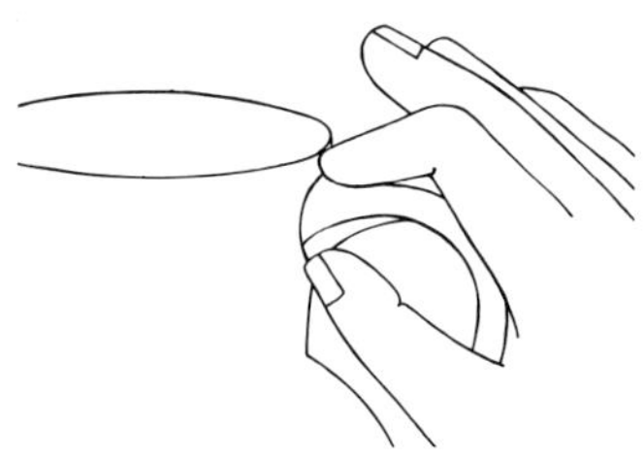

4 Carefully join both ends to the pot, using slurry.

Tools for making decorative or repetitive handles

The wire cutter is a development from the simple wire loop trimming tool. A variety of stiff wire shapes can be made at home or bought ready made. The loops are drawn through a block of clay and the resulting handle becomes an exact copy of the loop. Additions of patterns, twisted into the loop before the handle is made, will result in a raised pattern of ridges continuing right down the length of the handle and creating an unusual effect.

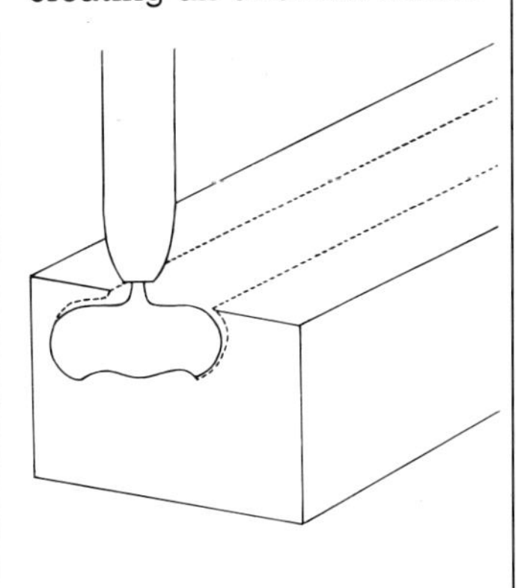

Lugs

Lugs

Lugs or handles are often fixed on the sides of pots which need to be lifted with both hands. A short section of a pulled handle can be used, attached along its edge in a curve. Make sure the two lugs are symmetrically placed at the same height on each side of the pot. Lugs are made by throwing solid shapes, circles or cylinders of clay which are cut in half or divided equally to ensure an exact match for the lugs.

Pairs of matching curved pieces can be luted on in order to provide an easy comfortable grip on a pot. Cooking pots, especially, need lugs to grip when hot.

Forming lugs

1 Cut two pieces to the same size and shape. Make sure you mark the sides of the pot at the same height for each lug. Score the surfaces and apply the lug with slurry.

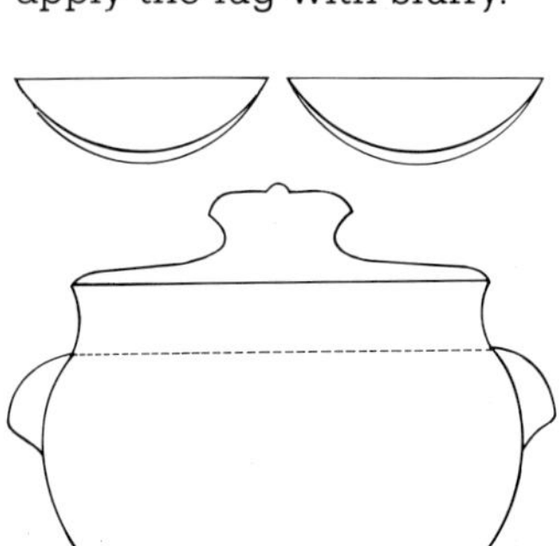

2 You can make sure the two lugs are set opposite each other by placing a piece of wood across the center of the pot.

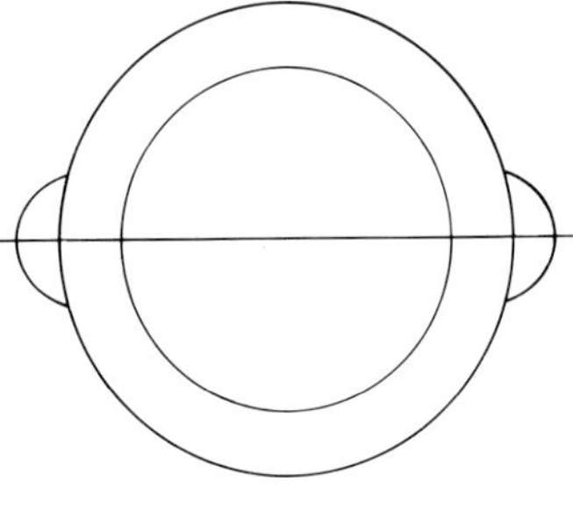

A flat strip of clay can be cut into two lengths and joined on edge, as shown.

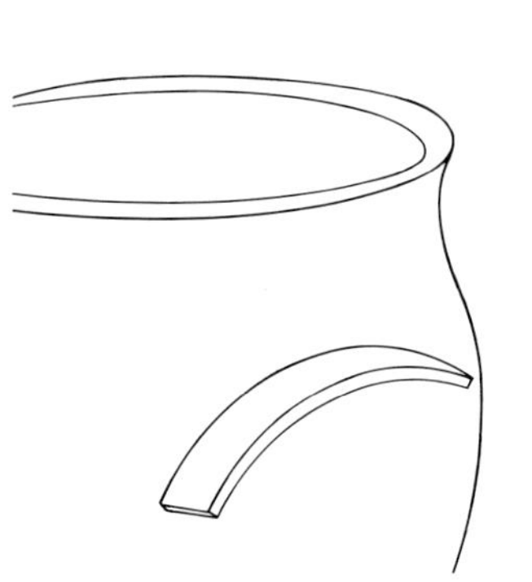

3 As an alternative, small thrown knobs may be attached.

Pairs of matching lugs can be luted on to a pot. →

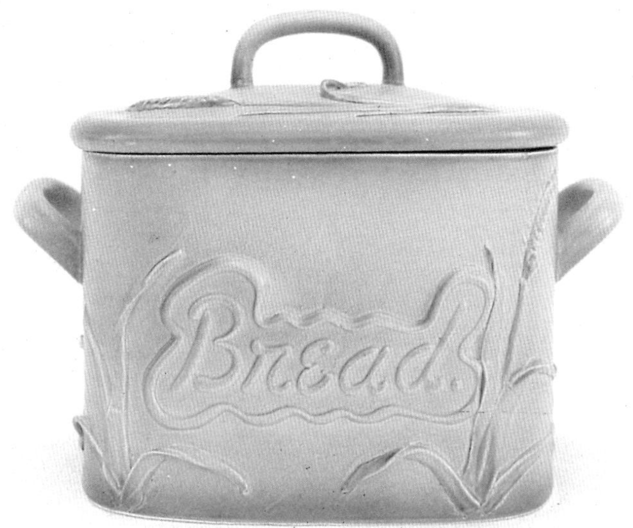

Breadbin by Lisa Katzenstein (UK)
Slipcast in a mold, this breadbin has been finished by the addition of lugs and a handle; these were cast in models that have been hand carved in plaster.

Adding coils to a thrown pot
(see also pages 48 - 51).

This method is used in many countries as a way of producing large pots without having to center large pieces of clay. It also allows the bottom section of a large pot to harden before completing the top half.

The base section of the pot is thrown in the normal way with a fairly thick section of wall. It is allowed to harden slightly until the rim of the pot is dry enough to handle.

A large rope or coil of clay is rolled or squeezed out and (making sure the rim has been roughened) attached to the rim of the pot with slurry. Place it evenly on the rim and pin it to the thrown pot section by scraping the edges of the clay down with the thumbs to form a firm seal.

Turn the wheel slowly; wet the coil with a sponge while it is turning. Apply a little downward pressure to ensure coil is firmly attached to the base. This will help smooth the coil into an even thick rim.

Continue to pull the walls of the pot upwards to increase the height of the pot, using the coil to give additional height. Then smooth in the old and the new parts of the pot by pulling right up from the base.

This method of increasing the size of a pot is ideal for tall narrow-necked bottles.

Repetitive throwing

Method 1

This method aims to produce sets of matching pots made by throwing. First prepare the clay in the usual way.

1 Decide on the design and shape of the pot you wish to make.

Weigh out sample amounts of clay to determine the size of the pot. Throw these sizes to the required shape and decide what quantity of clay suits the size of pot required. (Every potter will throw differently, and less experienced potters, in particular, will not necessarily throw evenly each time and so cannot rely on the suggested measurements stated in books.)

2 Weigh out the clay carefully so that a suitable number of articles of equal size can be made, bearing in mind the measurements reached in the 'prototype' throw.

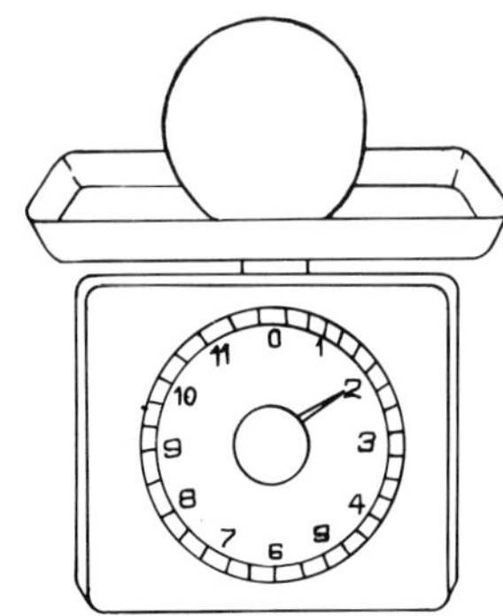

3 Throw the first pot with one measure of clay. When thrown, this pot must be measured for height and width so that its size and shape can be repeated. A measuring stand, or some form of reference point, will be required.

4 If each piece of clay is thrown in the same way, opened to the same width of base and pulled up using the identical technique each time — with practice the pot shape will be repeated. There will obviously be some variation with handmade pots, but when these are glazed and decorated such slight differences will not be obvious.

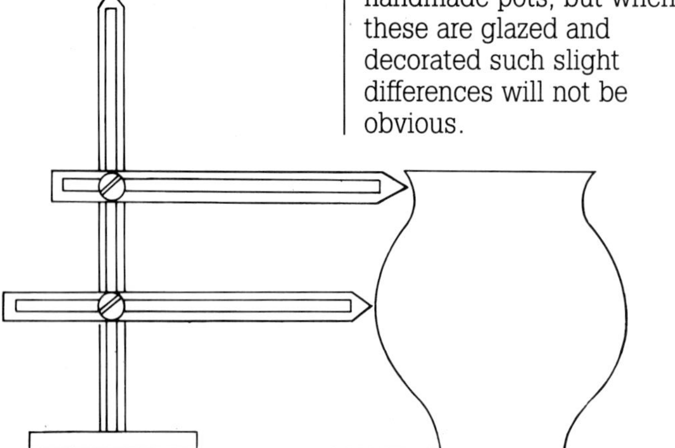

Method 2

Another technique is 'throwing off a lump'. This involves centering only one large piece of clay. Squeeze out a small piece from the top of the lump and throw this to the required shape. This is then removed from the large lump of clay with a flat thin slat of wood which is eased under the base of the small pot while the wheel rotates slowly. Alternatively a thin wire can be used to slice the pot cleanly away from the 'parent' lump of clay. Repeat this operation until the required number of pots have been thrown.

This technique does need quite a lot of practice as control of the small pot on top of the lump can all too easily be lost. For small numbers of matching pots it is simpler to use individual balls of clay.

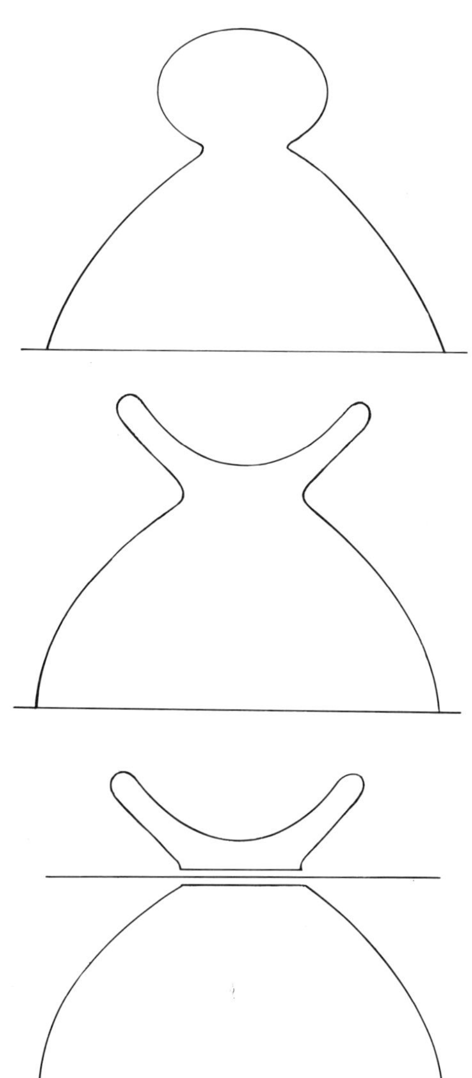

Throwing problems

Uneven walls

When looking down on to the pot, if the sides appear thick and thin as the wheel revolves, the pot is off centre.

Correction:

Place both hands around the pot, with the thumbs nearest to you, and collar in the pot by even squeezing. At the same time recenter the whole, and so stop the pot moving unevenly. Otherwise the top rim will become uneven as the pot is collared.

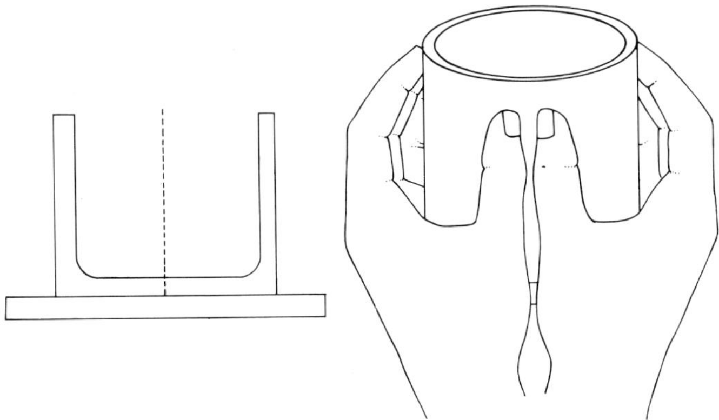

Air bubbles

If there are air bubbles in the clay it will be difficult to throw evenly or successfully. Moreover, if the bubbles are left in the clay and the pot is fired, they will expand into warts or bumps on the surface. Alternatively, they may burst and create volcanic-like holes!

The air bubble will feel like a hard stone in the clay, and will strike the fingers each time the wheel turns, jolting the fingers out of position as the pulling continues. The air bubbles may move up into the walls of the pot, eventually throwing the pot off center or making the side tear away at that point. So it is vital to make sure the clay is wedged, mixed and beaten properly in order to remove all the air bubbles before starting.

If an air bubble does appear and the pot is to be kept, puncture the bubble with a needle and release the pressure. Smooth the mark over and continue throwing. If air bubbles continue to appear in the clay, discard the pot and rewedge the clay. Then start again.

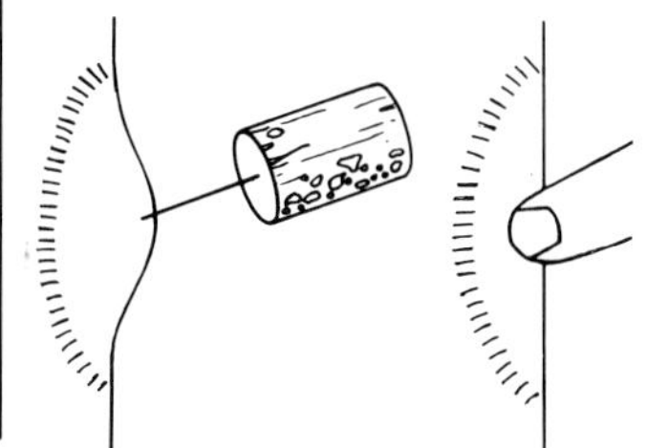

Uneven rims

Take a sharp pointed instrument (a needle in a cork is ideal). While the wheel revolves slowly place the fingers of the left hand inside the rim on the righthand side. Gently place the needle against the outside of the rim and increase the pressure until the needle is pushed right through the pot and, as the wheel revolves, cuts a strip of clay from the rim. Any excess should then have been removed.

The loop of clay is removed on the needle, leaving the rim level.

Renew the shape of the pot by pulling up again from the base to the rim with a steady even pull.

When the walls are uneven up the sides of the pot (if the walls feel thick and thin as the fingers are pulled up the sides) this is a result of uneven pressure or movement of the fingers whilst pulling up.

Make sure the arms, wrists and hands are fully supported on the legs and knees or on the wheel tray of the wheel, so there can be no side-to-side movement of the fingers as they are pulled up the sides of the pot. Or — if the pot is a tall one and you cannot rest your arms on the knees or wheel — you may have to stand up; tuck the elbows and arms into your sides and stand as close to the pot as possible.

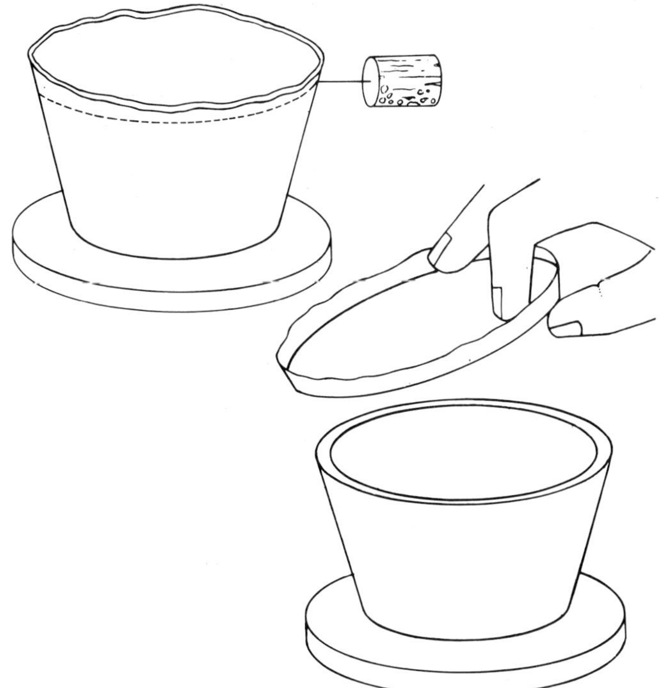

4 Alternative Methods

Detail of sculptural form by James Robison (UK)

Mold making

Many potters are not content to use just the obvious means of production as discussed in the preceding chapters but prefer to experiment with different ways of making a piece. They may combine many methods to produce the result they seek. The work may be very innovative and can produce interesting new results.

Molded clay with additions of thrown or hand-modeled pieces: slip-cast items which are cut and joined again in a different format or attached to thrown sections of pot — these are all ceramics, meaning objects made in clay, but are not classed as pots in the true sense of the word. The artist has chosen to extend the range of clay, working beyond the confines of throwing, coiling or slabbing.

No one book could discuss all the possibilities. Trial and experimentation with all manner of techniques may lead to many diverse avenues of exploitation into one's personal creativity.

The following examples of work show this experimental exploration and the many aspects of ceramics. The field is enormous, and one chapter can never encompass all the possible methods, stylistic development or versatility of the medium. However, by seeing how just a few potters have explored the creativity of clay may inspire fresh ideas and remind the potter that there are many ways to achieve an end result.

Plaster

This is dehydrated gypsum which, when combined with water will form a porous stone if mixed with water before it sets hard. It may be poured in liquid form over or into objects and will adopt the shape concerned. It was used by the Greeks and the Egyptians but its modern name comes from deposits which are to be found on the outskirts of Paris, France.

The absorbency of plaster makes it superb for the casting of clay or slip or for drying out very wet clay. Although it is invaluable in ceramic creation, due care must be taken when using plaster of paris. Small amounts left in a pot will explode in the kiln, blowing craters or cracking the surface. If it is dried too fiercely (for instance, if left on top of a hot kiln) it will become crumbly and soft.

Mixing plaster

As a rough guide to the proportions required when mixing plaster, allow 2½ to 3 lbs. of plaster mixed with every 2 pints of water (1.25 kg to 1 litre).

Always add the plaster to the water. Use a good-sized bowl or bucket and half fill it with clean cold water to allow for expansion when the plaster is added. Sprinkle large handfuls of plaster into the water (having already removed any lumps).

DO NOT STIR.

Keep adding plaster until the water is saturated and the plaster heaps up above the surface.

Only then may you stir. Do so gently, keeping your hand beneath the water. You must avoid creating air bubbles in the mix.

As soon as you start to mix the plaster and water, it will begin to thicken and is soon ready for pouring. Pour when thin rather than too thick, so that the plaster will readily fill all the hollows and forms of the mold.

Clean around the top edge of the mold when plaster has set and remove the clay model. When the mold is dry, a negative convex mold can be made from the first one. This will be a hump or mushroom mold. Soft soap must be applied to the inside before attempting to cast a hump mold from the convex mold to stop the new plaster from sticking (see page 113).

Fleeing horse by Tony Bennett (UK)
The horse (2¼ inches or 5 cms high) is made in slipcast earthenware with glazes and stains built up over many firings. Enamels and lusters have completed the effect.

Forming a shape to be cast

1 The shape of the dish is made upside down. Using a piece of thick glass or marble (or plastic-coated board), cut out a template. This template should be exactly the same shape as the rim of the mold you wish to create.

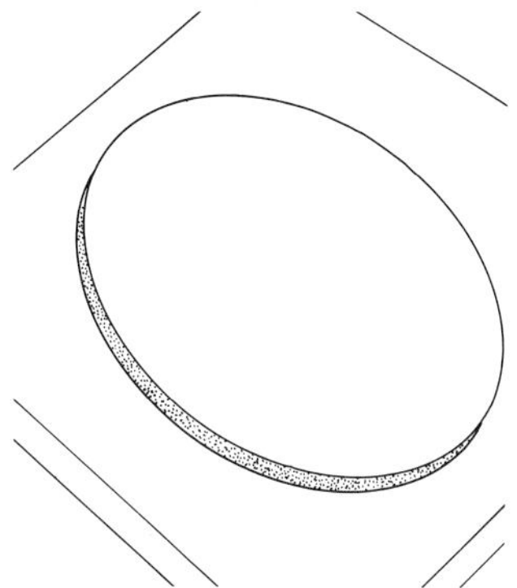

2 Draw around the template and then remove this. Within the drawn outline start to add pieces of clay to make the shape three dimensional. (A coil placed around the outside of the shape will give a useful guideline.)

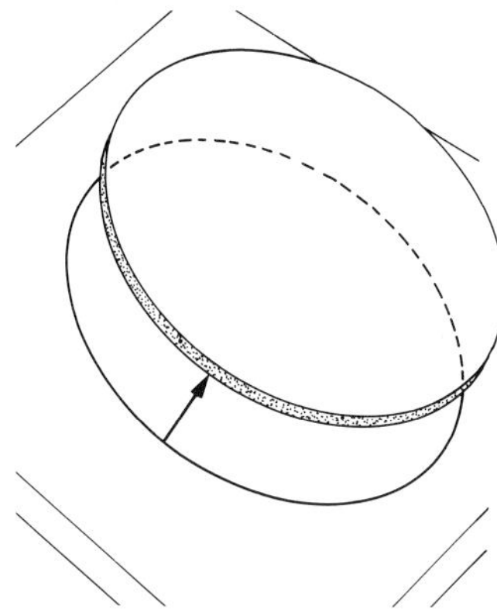

3 With a wooden paddle, beat the pieces of clay into a mound to form the rough outline of the shape planned.

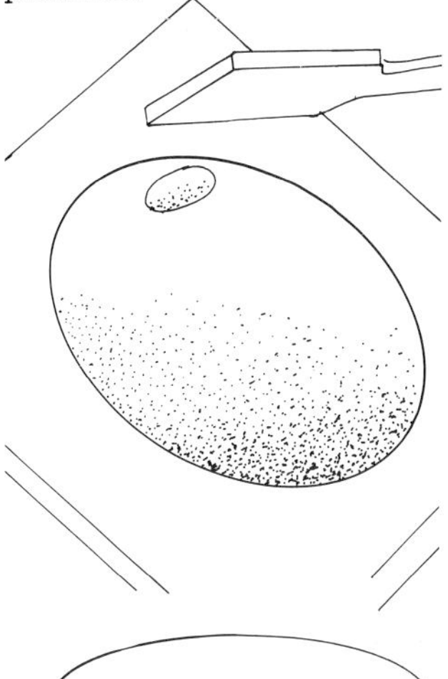

4 Smooth off with a scraper and rubber kidney.
Check the sides with a template to make sure they are even all the way round.

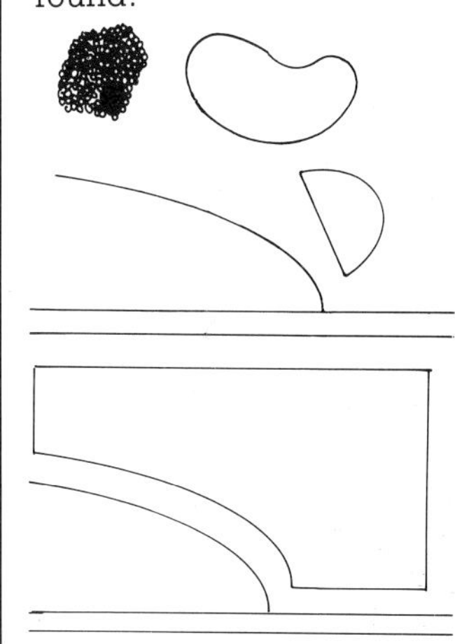

5 When the form is finished surround it with a casting box. This consists of a frame made from adjustable pieces of wood held together with angle iron and sealed at the bottom with clay.

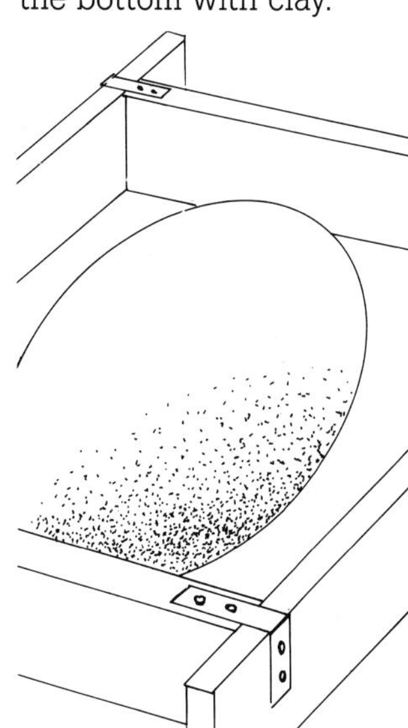

6 The casting box should be adjusted to within $1\frac{1}{2}$ inches (4 cms) of the edge of the model. Leave about 1 inch (2.5 cms) clearance above.

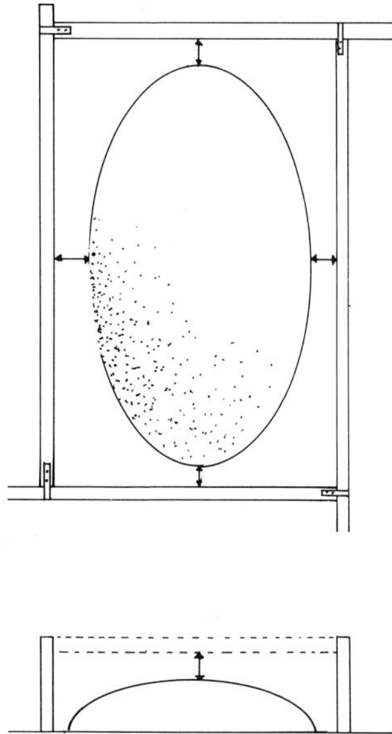

7 Pour the mixed plaster slowly over the model from the side. Raise the level very gradually so as to avoid trapping air bubbles.

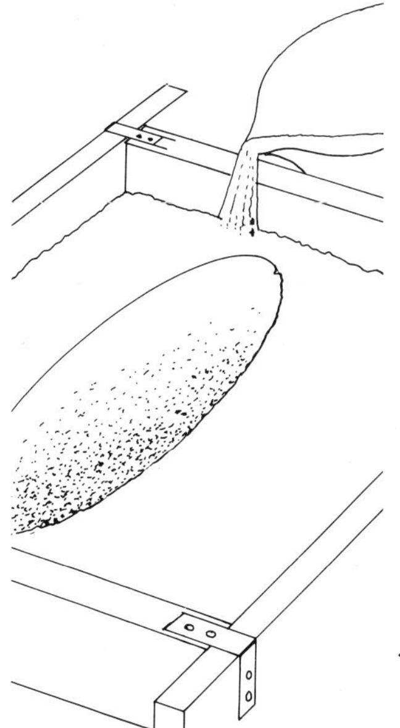

8 Leave the plaster to set. (It will grow hot as it sets, and then cool again). Remove the surrounding walls, clean up the edges of the underside with a steel scraper and an open-toothed plane called a surform. Round off any sharp edges that might otherwise chip off when the mold is used later.

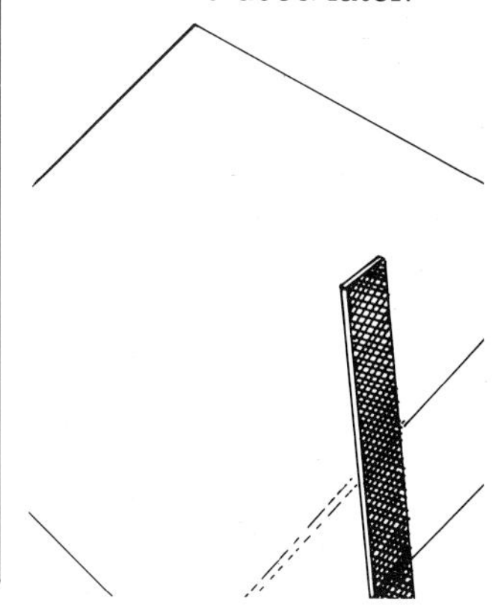

Casting surrounds

Different kinds of wall can be used to surround a model when a casting box is not suitable or available.
Roll out a flap strap of clay and place this around the model. Seal the join, leaving a suitable margin into which the plaster is to be poured. (This type is useful if the shape is not square.)

It is also possible to use a strip of vinyl floor covering held in place with coils of clay. Rubber bands or string tied around it will hold the two ends together. Seal all joints with clay before pouring in the plaster. Lino or corrugated card can also be used in this way.

Casting the two-piece mold (useful if shape required has convex *and* concave curves)

1 Stick the ball on to the wheel and mark a line around the center as a guide to the division of the two halves.

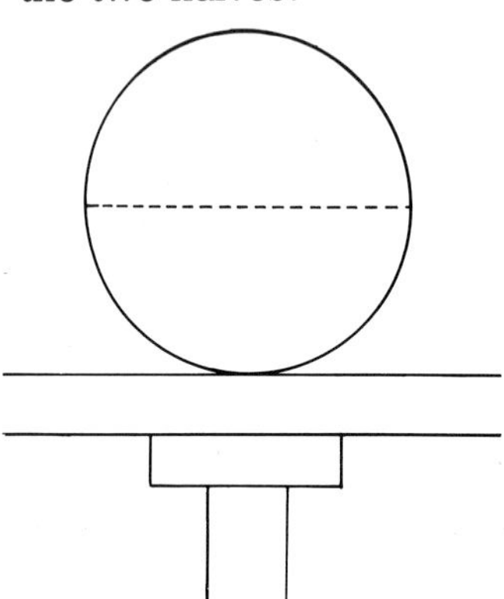

2 Make or find a box that is large enough to hold the ball and allow about a 2 inch (5 cm) space all around.

Make a cone of clay with the point cut off. Place this in the center of the base of the box.

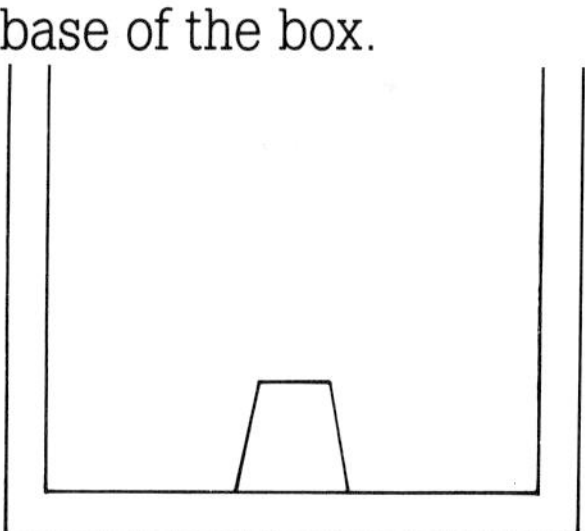

3 Grease the surface of the ball and place it on the cone of clay, with the marked halfway line horizontal.

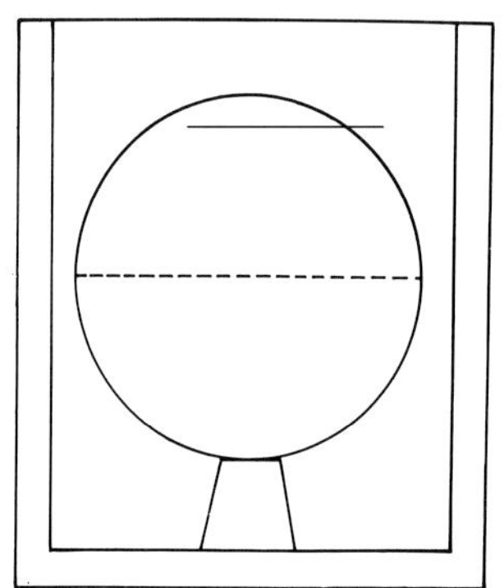

4 Mix plaster of paris and pour it carefully into the box around the ball up to the halfway mark.

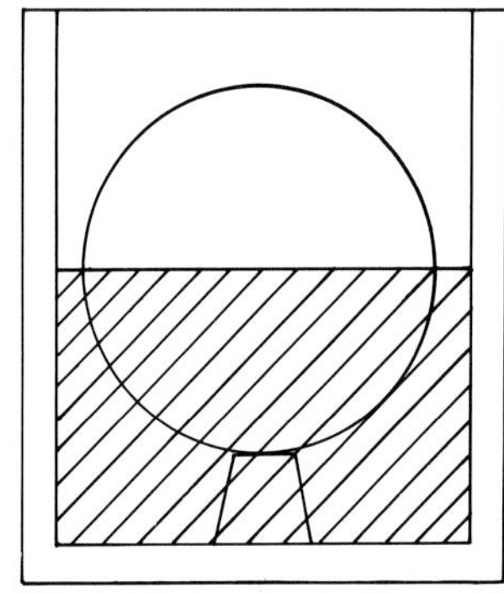

5 Wait until the plaster sets. Make two key marks on either side of the ball. (These key marks will create notches that will later enable the two halves of the mold to fit neatly together in exactly the right place.)

Cover the new plaster surface with clay slip to stop the second half of the mold sticking to the first.

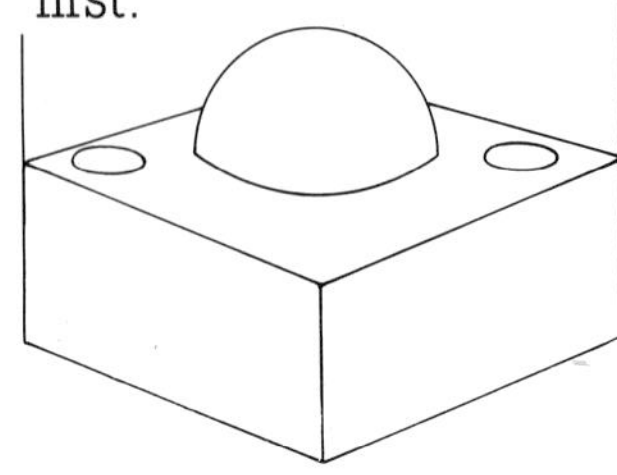

6 The ball is already greased so simply pour the plaster over the second half of the ball until the top is covered up by plaster about 2 inches (5 cm) deep.

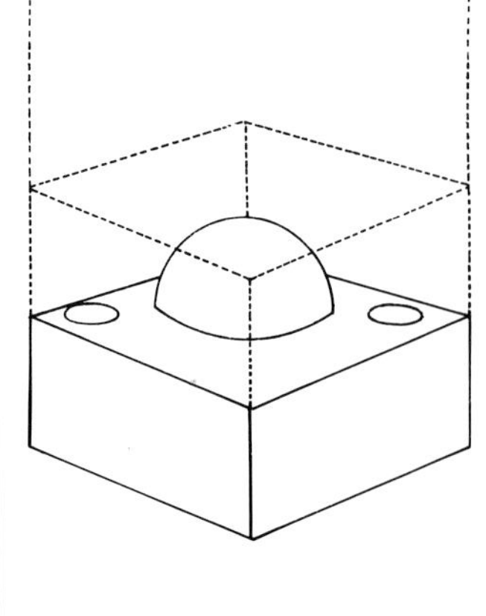

7 Leave until the second set of plaster is hard, then gently knock apart the casting box.

Clean up the sharp edges of the two-piece mold with a scraper and surform.

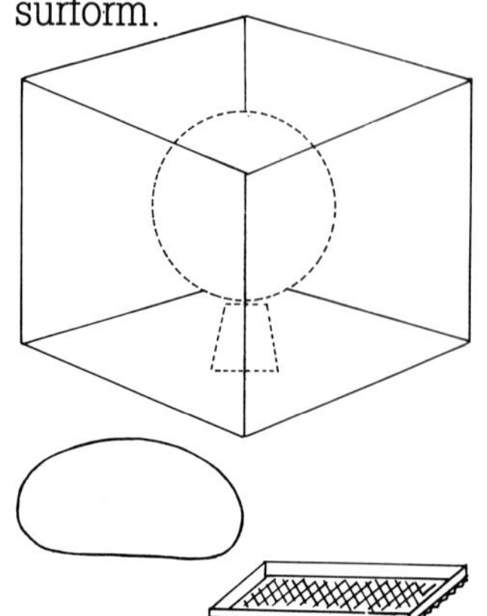

8 Divide the two halves of the mold and remove the ball and the cone of clay. Wash and clean the mold, making it as smooth as possible. Leave to dry.

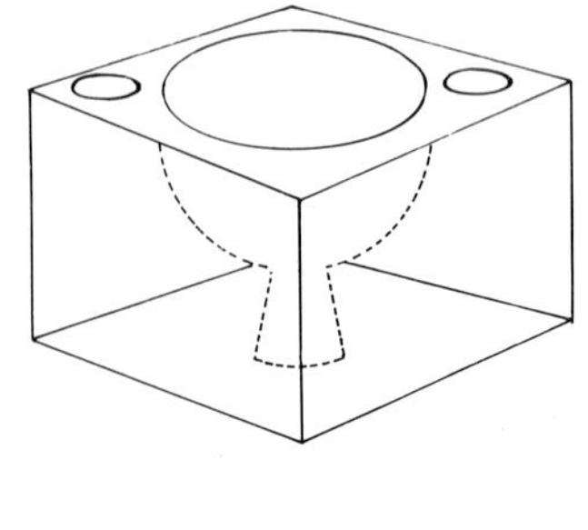

9 Cut strips of a car-tyre inner tube and use to bind together mold pieces (with hole at top).

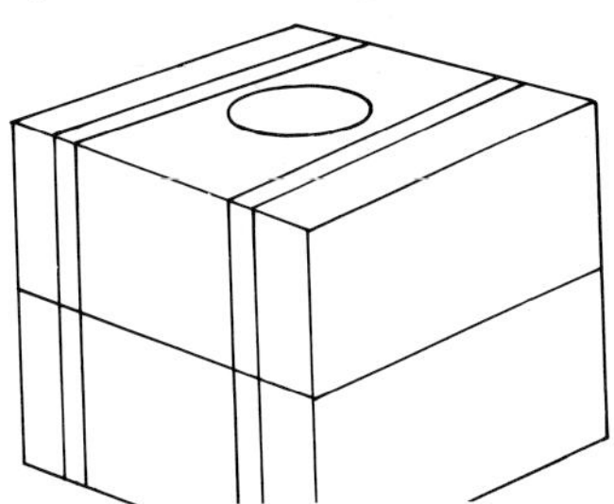

10 Pour casting slip through the hole and fill to rim. Leave for 15 minutes.

11 Upend and drain over a bucket. Leave till slip is leather hard.

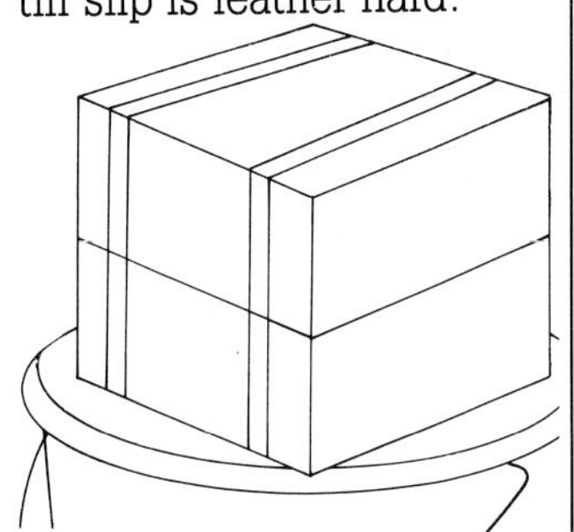

12 When slip is leather hard separate mold sections. Remove cast, trim neck and smooth rim.

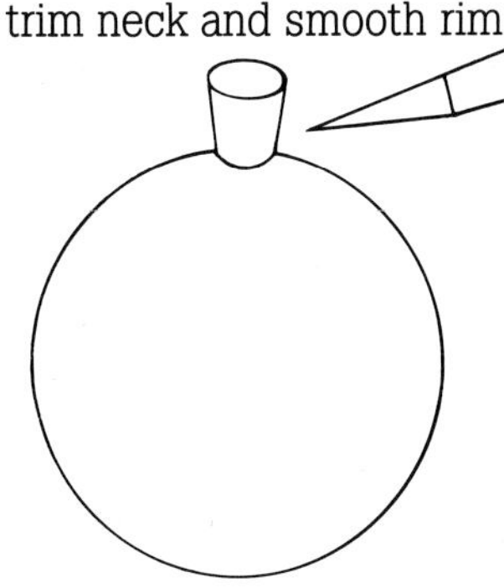

Making a mushroom or hump mold

1 Cover the inside surface of the mold with soft soap to seal the surface.

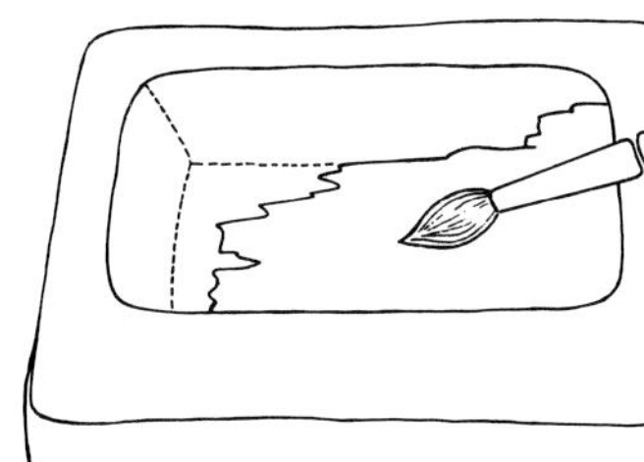

2 Mix more plaster and fill up the mold.

3 When it is nearly dry set in the center a tube of cardboard or vinyl and fill that with plaster to form the stand for the mushroom.

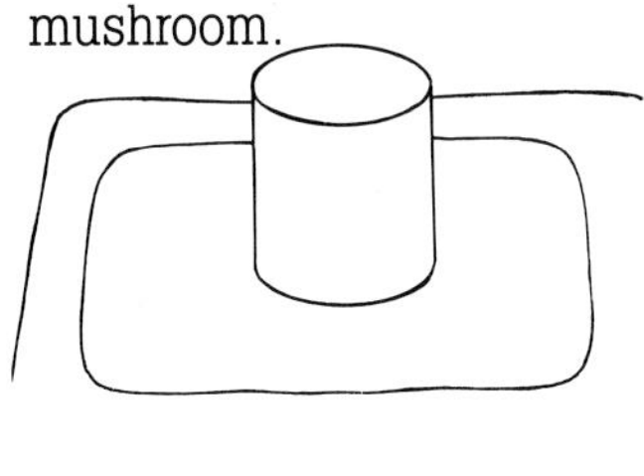

4 When fully dry and set, remove the hump mold from the hollow concave mold.

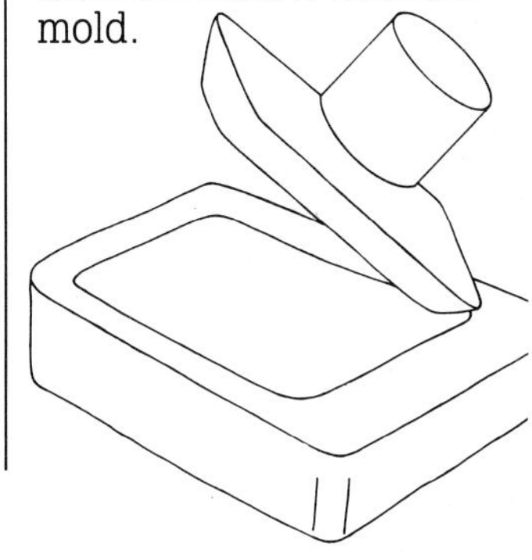

Making a dish on a hump or mushroom mold.

1 Roll out a thin even slab of clay as normal. Lift both the clay and the cloth and place them face down over a mushroom mold.

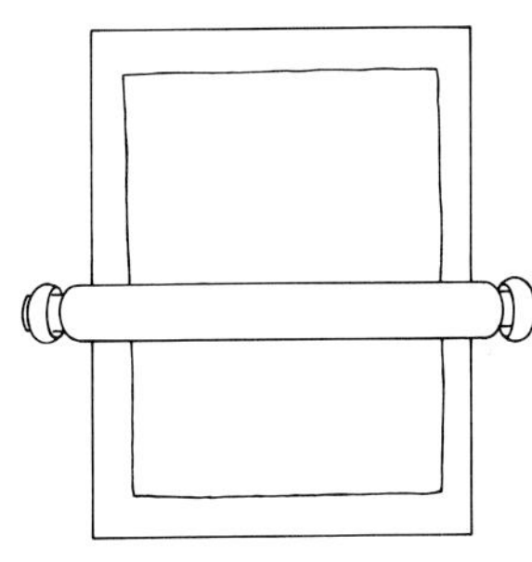

2 Peel off the cloth and flatten the clay on to the mold with sponge and water.

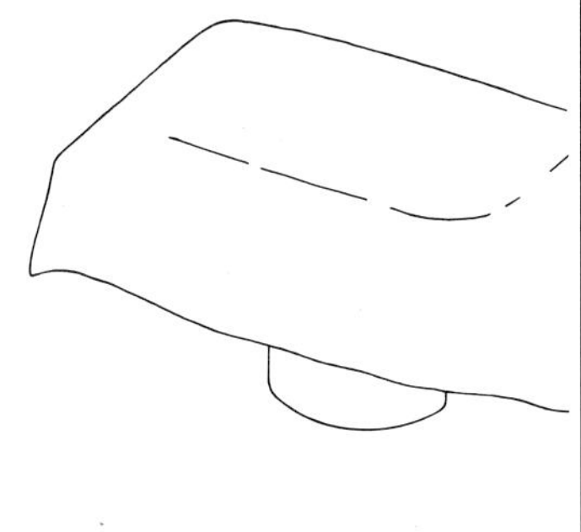

3 Trim off the spare clay round the edge with a wooden modeling tool.

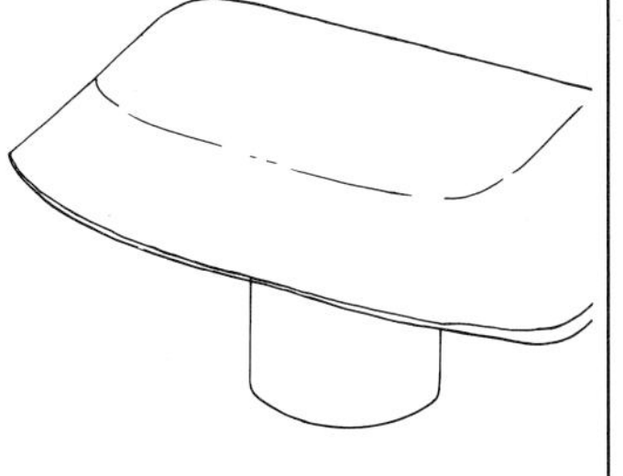

4 One advantage when using this type of mold is that a foot ring may be joined on to the base while it is upside down. This can be made from a strap of clay or coils.

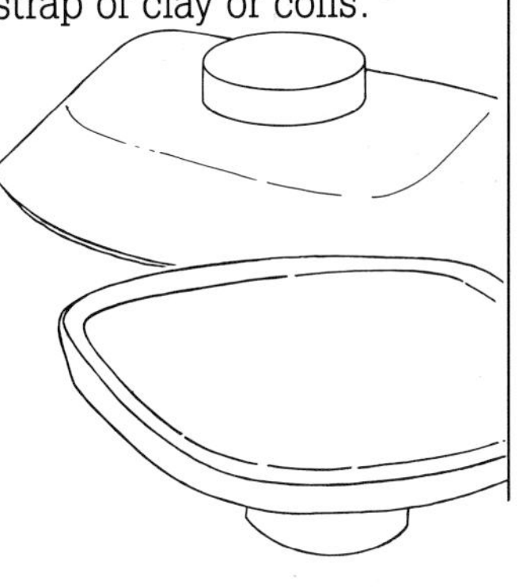

Slip casting

'Greenboots' by Tony Bennett (UK)
Made of slipcast earthenware, glazed, with underglaze colors, this piece has a lovely wrinkled surface texture.

Bowls and plates by Lisa Katzenstein (UK)
Each of these pieces has been 'double-cast'. Because they are hollow, a rounded massive appearance is achieved without the usual weight. The seam line is midway on the rim.

'Equestrian frog' by Tony Bennett (UK)
Using slip casting and molds offers the opportunity to explore different decorative and textural effects on the same figure. This unlikely but delightful pair are made of cast earthenware.

Slip cast porcelain by Sandra Black (Australia)

Making slipcast pots

Used primarily in industrial potteries to produce matching shapes quickly and easily with semi-skilled labor, slip casting also has exciting possibilities when utilized for one-off pieces and is being used increasingly by studio potters.

The process is based on the absorption of water into a plaster of paris mold. Clay and water are mixed into a smooth creamy-like state called slip. When a plaster mold is filled with slip, the water content is absorbed by the plaster. The clay in the slip is left as a deposit on the inside surface of the mold. When the excess slip is poured out, the deposit is left to harden and become the plastic clay form of the mold.

Ordinary clay made into slip requires a great deal of water to reduce it to a liquid clay. When dried out it shrinks excessively and cracks and warps.

In order to make a casting slip which will not shrink quite so much, a 'deflocculant' is used; this is a substance which disperses the particles of clay, making the slip fluid without the addition of more water. Generally sodium silicate or soda ash is used, to a maximum of 1 per cent — usually $\frac{1}{2}$ per cent will be enough to reduce soft clay to a slip.

Simple wheel-made mold

1 On the wheel, center roughly a fairly stiff smooth piece of clay.

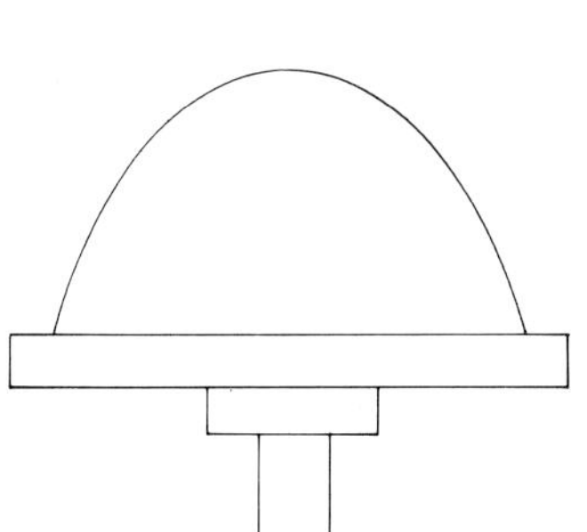

2 Shape it with tools and a scraper to the shape required — in this case a tapering open-mouthed bowl with a wider edge. It is made upside down.

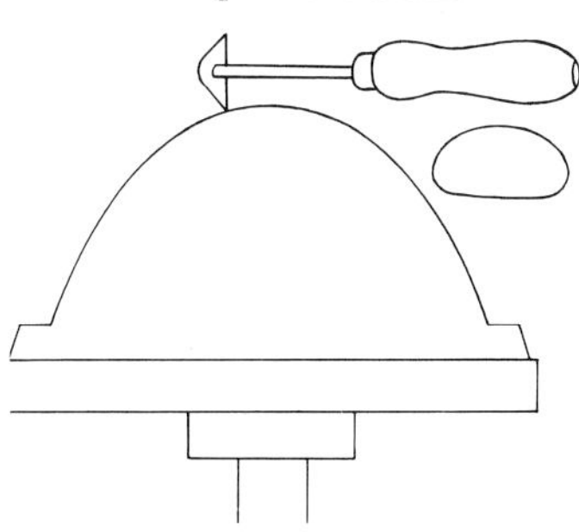

3 Surround the wheel head with a piece of cardboard, vinyl or lino. Seal all joints with clay and pour in the plaster.

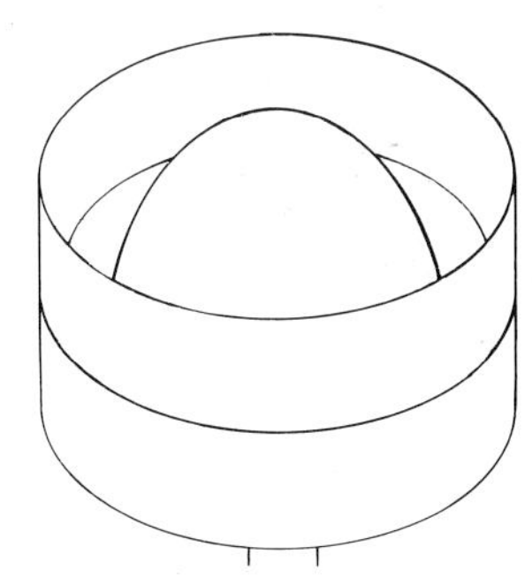

The wider edge of the mold will be removed as spare, leaving a smooth edge.

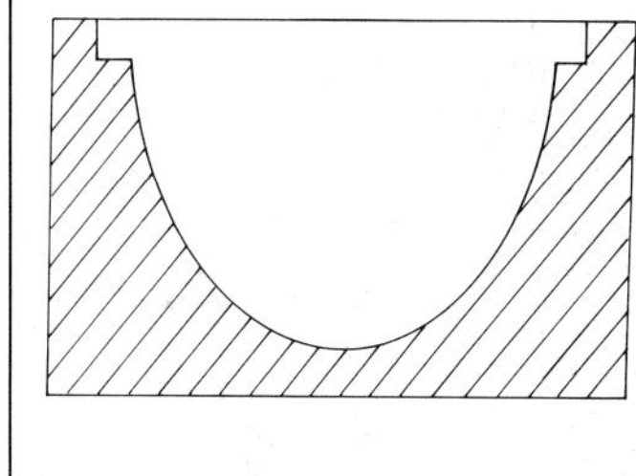

Casting the open mold in slip

1 Fill the mold with slip to the rim.

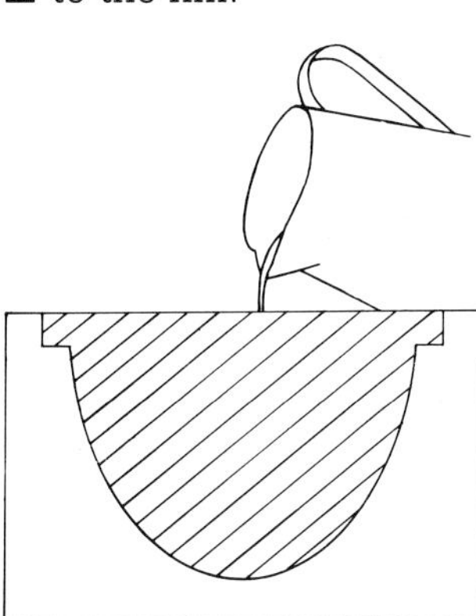

2 Allow it to stand in the mold until a thick coating of clay has been deposited on the inside surface. The level of the slip will drop as the water is absorbed by the plaster.

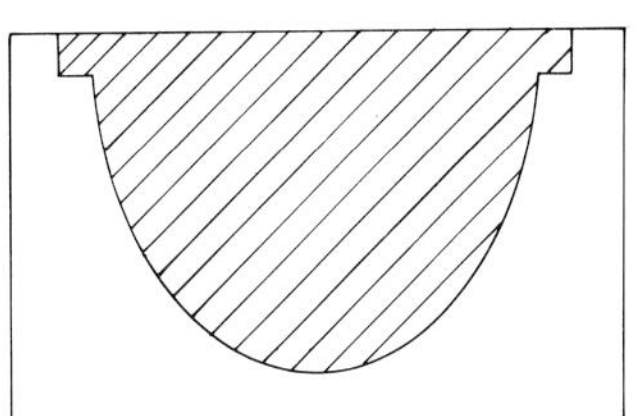

3 Pour out the excess slip and leave mold upside down to drain over the slip bucket.

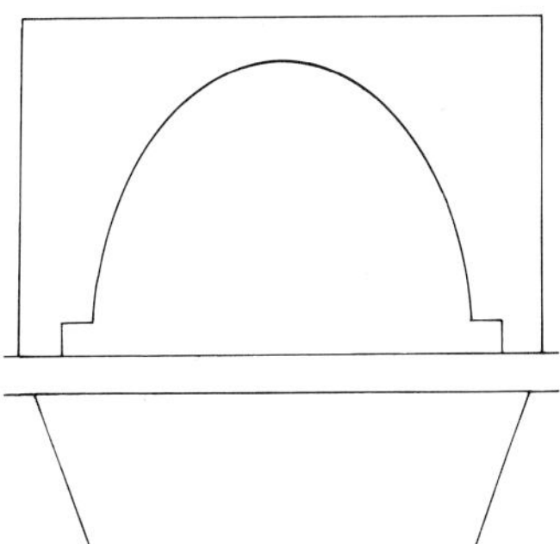

4 The bowl shrinks away from the mold and can be lifted out. Remove spare edge. Fettle ring with sponge and fingers.

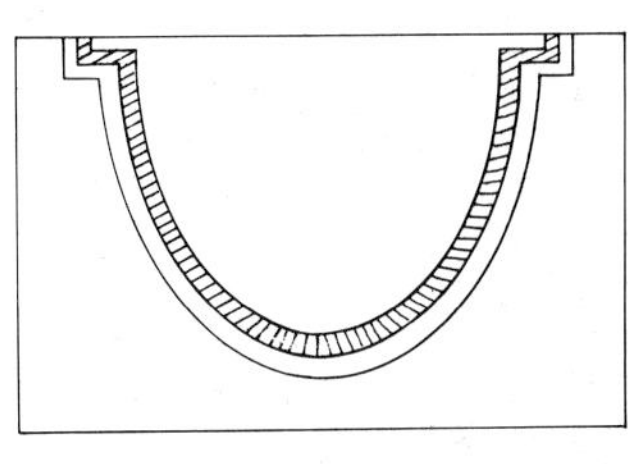

Molds

Molds for slip casting can be one-piece drop-out molds.

1 Make sure there is no undercut shape which would prevent the pot being tipped out.

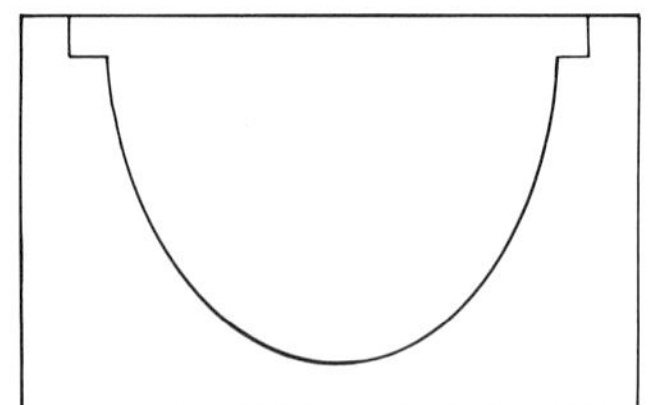

2 This simple egg shape can be made with two pieces.

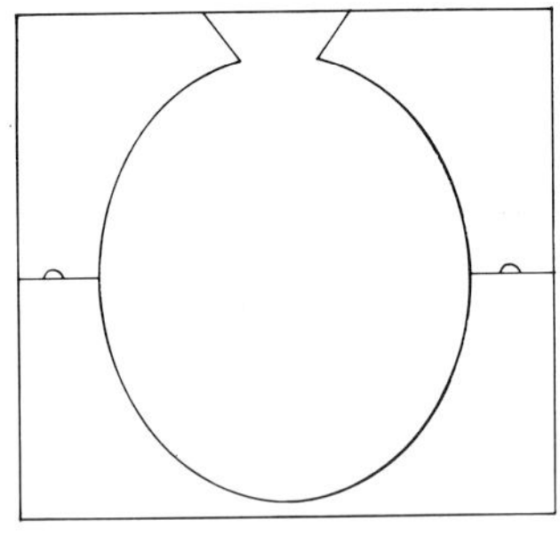

3 Three or more pieces can be used to create a more complicated shape.

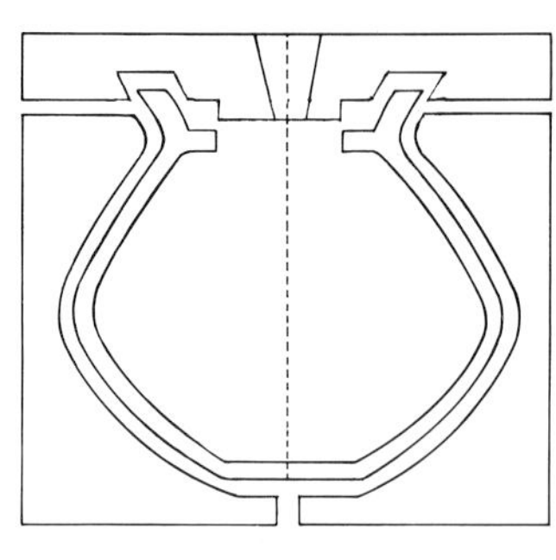

4 Remove the dish before it starts to shrink or it may crack.

Slip casting in a mold

1 Fill mold.

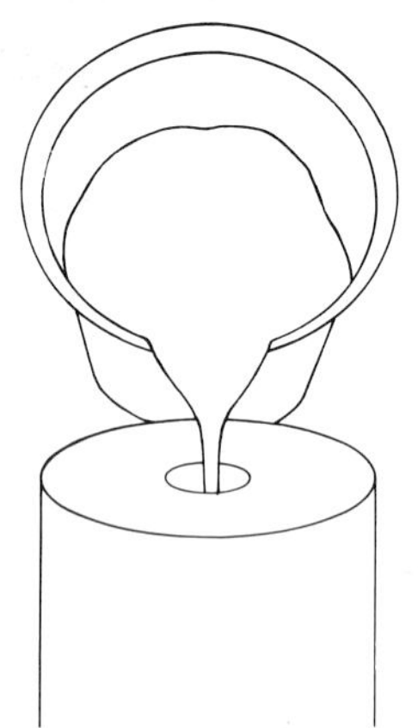

2 Leave 10-20 minutes; pour out excess.

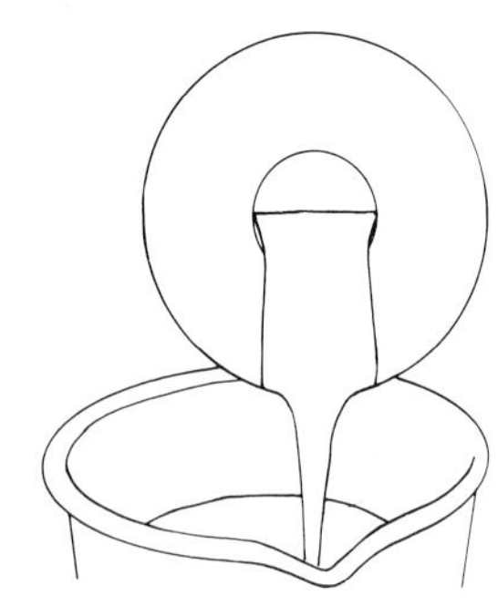

3 Leave to drain.

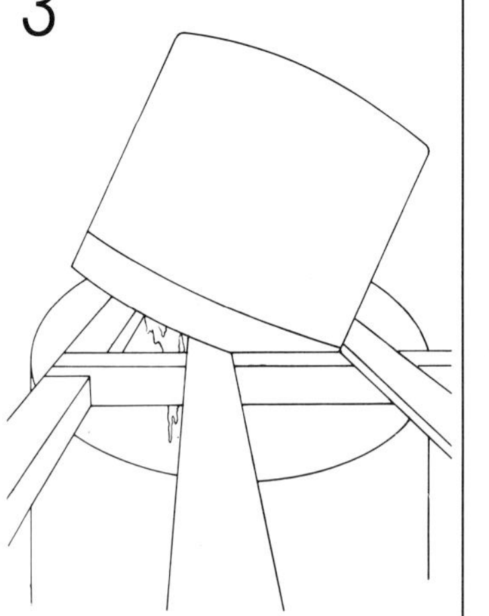

4 Cut off the spare around neck.

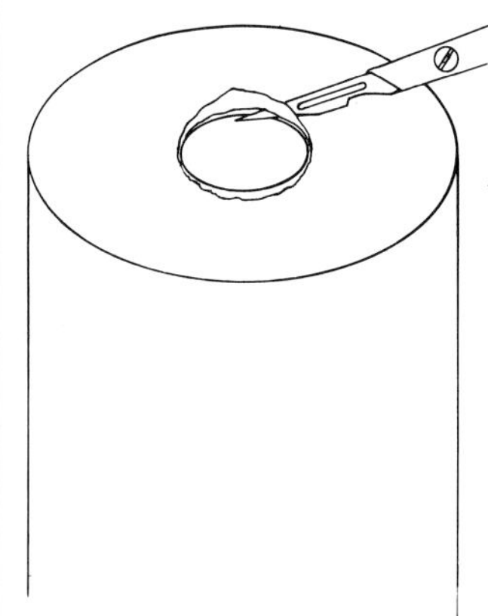

5 When dry, divide the mold.

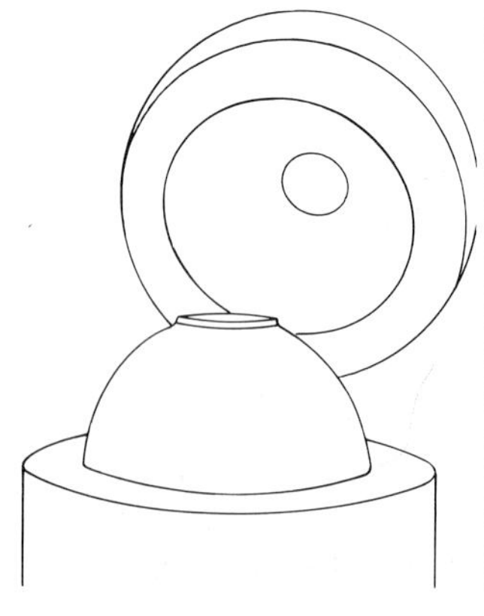

6 Fettle any casting marks with knife and sponge.

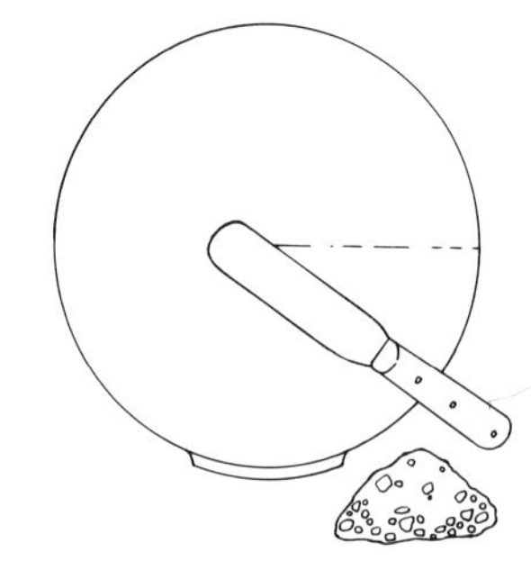

Cast ware by Lisa Katzenstein (UK)

Extrusion

The basic method

This is a method of producing clay shapes by passing the clay through a shaped die under pressure. This can take place in a wad box, a hand-operated extruding machine or through a powered pugmill.

Small amounts of work are usually produced from a wad box, such as coils and handles, pieces to be added on as decoration and sections that will be joined and combined later by handbuilding.

Pugmills are used industrially to make bricks, drainpipes, tiles and so on.

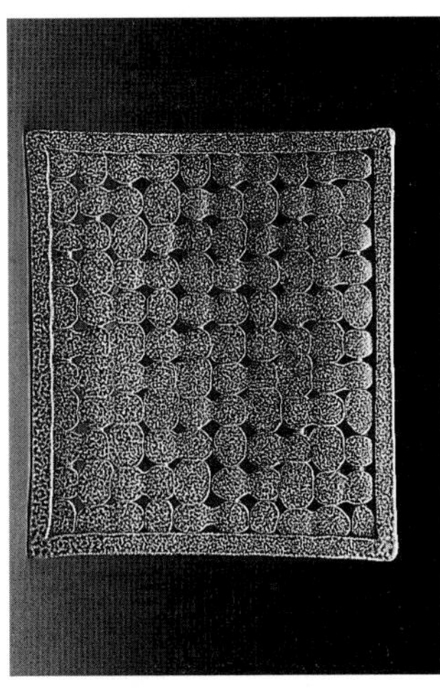

Woven plate by Jan Schachter (USA)
Extruded coils have been used to form a decorative plate.

Jigger and jolley

A jigger and jolley is a machine which shapes clay by pressing it against a plaster mold. Usually the term jiggering refers to the manufacture of hollow pieces and jolleying to flat ware.

Using a revolving plaster mold set in a wheelhead, the inside or the outside of the clay piece is shaped by pressing a template on to the flattened clay.

The clay is flattened roughly into the mold with fingers.

The inside shape of the dish is then smoothed and refined by the downward pressure of the template which scrapes and smooths the plate while the wheel is turning.

A variety of extruded shapes

Coils of any length or thickness can be produced quickly and cleanly — provided that the clay put into the machine is even in texture and plasticity and that any grog present in the clay is fine grained.

Hollow earthenware bricks are used for construction purposes, mainly in hot countries, as they keep buildings cool and insulated.

Pantiles, bright terracotta roofing tiles, can be handmade by using a mold or former or by throwing and cutting cylinders.

As a simple form of extrusion, clay can be pressed through a coarse sieve to produce a hair or furlike mass. This effect may be useful for a variety of purposes when modeling objects.

Pipes for drainage of land can be produced quickly and cheaply by moving the clay from the clay pit straight to a pugmill. This extrudes the pipes which are then moved and stacked in the kiln to dry and fire.

Pipes are cut off in fours or fives by a fixed length measure and a hinged cutter which can cut a number of pipes at once.

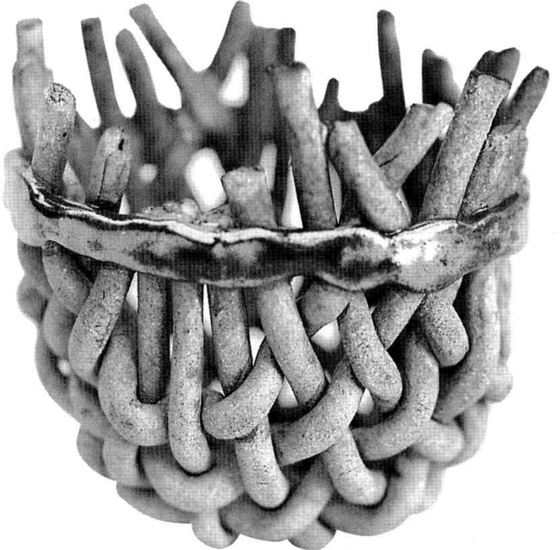

Woven vessel by Jan Schachter (USA)
Extruded coils of clay have been woven into a basket-like vessel.

Alternative creations

Developing alternative creations

Some artists discover that their interest leads them towards creating sculpture rather than vessels. They may still be using the same building, firing and decorating methods but are doing so in order to produce a particular form — not necessarily a pot.

For example, Paul Astbury's work during the early 1970s was mainly press molding, modeling and taking molds from objects which echoed technological development. However in the late 1970s he conceived a series of 'Reforms' — pieces built from already fired and glazed ceramics which were smashed and rebuilt, using other materials to bond the pieces together. This method of work has developed into the use of actual objects such as chairs, boxes, tables and tyres, all combined with clay.

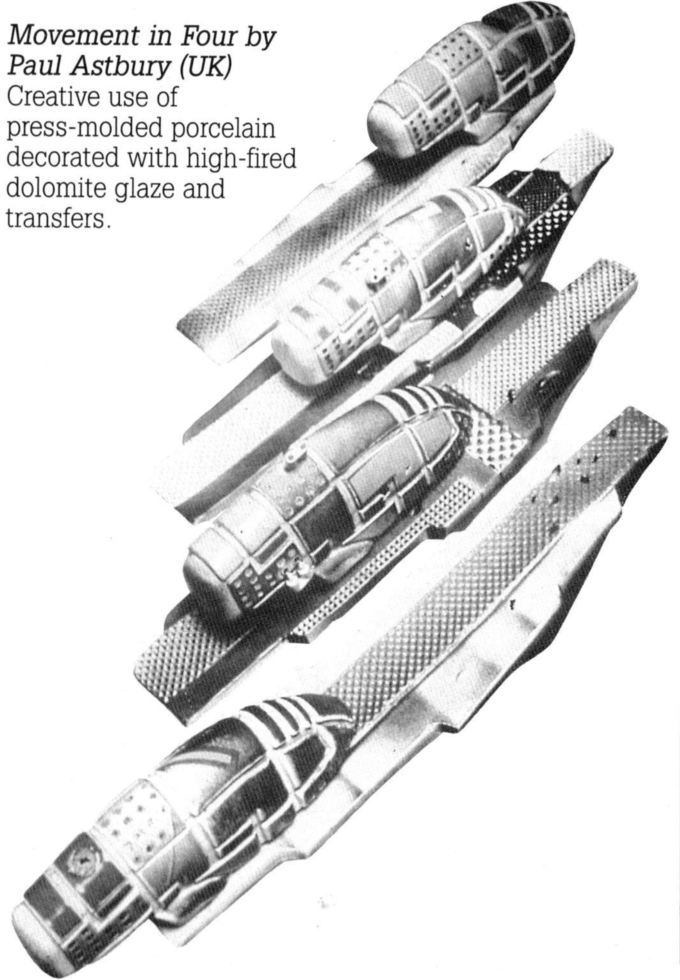

Movement in Four by Paul Astbury (UK)
Creative use of press-molded porcelain decorated with high-fired dolomite glaze and transfers.

Iconic form by Peter Simpson (UK)
Joined slabs with rolled-on relief decoration created into a sculptural form. Selected areas have been burnished with 'wet and dry' and wax.

'Jungle with Hipbath and Tiger' by Hilary Brock (UK)
An interesting combination of stoneware and porcelain clays, this ceramic piece was finished with stains, colored oxides, enamels, and in feldspathic glaze.

'Reform' by Paul Astbury (UK)
Here the mediums are an intermix of porcelain clay and cardboard from a cardboard box. Dolomite glaze and stains have been used. The ceramic work is slabbed and press molded with textures and bolted to a cardboard structure.

'Celtic Land III' by James Robinson (UK)
The stoneware columns, inspired by illuminated manuscripts and then interpreted in three dimensions were slabbuilt and handbuilt with stains and slips. They are 6 feet (nearly 2 meters) tall.

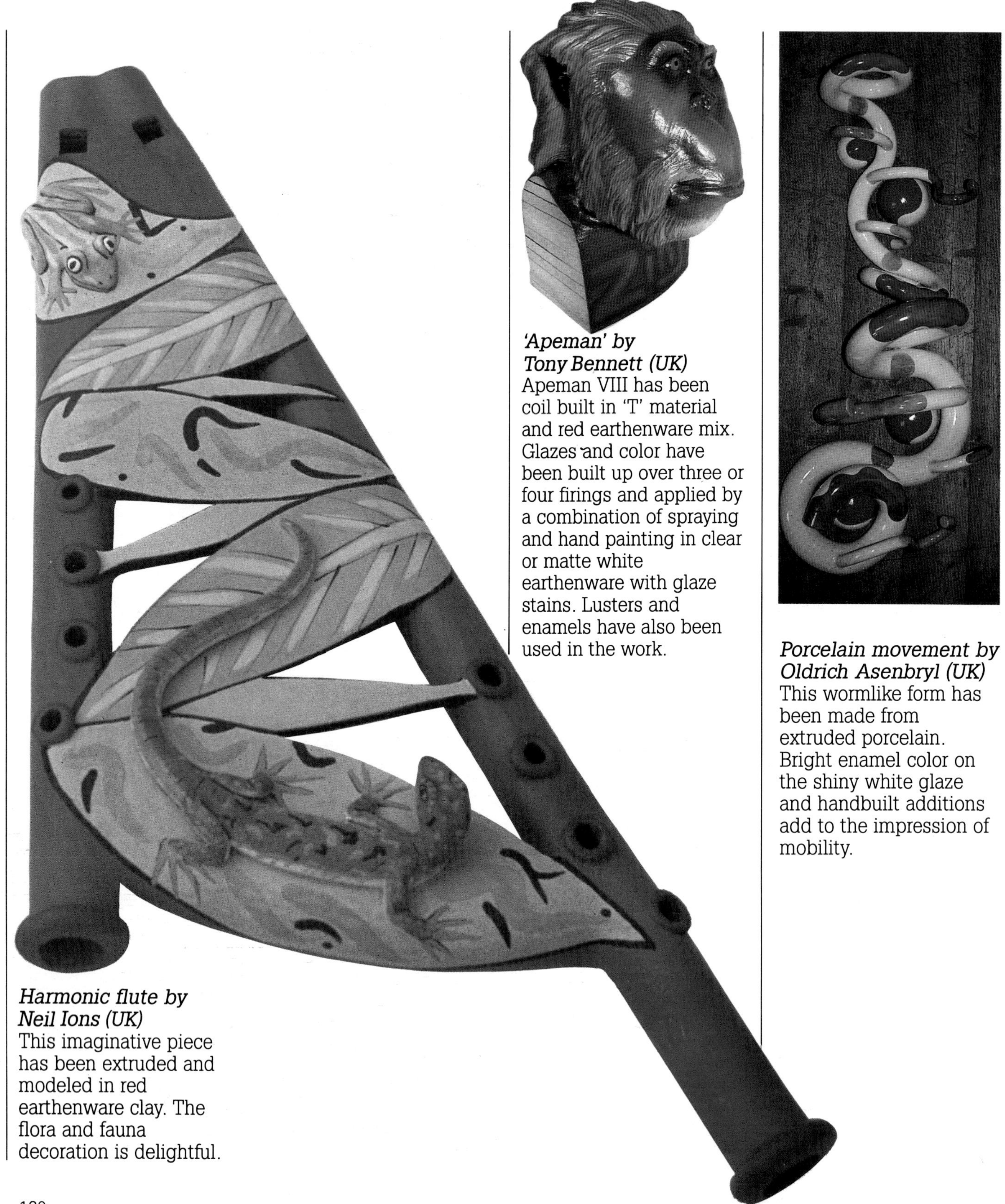

'Apeman' by Tony Bennett (UK)
Apeman VIII has been coil built in "T" material and red earthenware mix. Glazes and color have been built up over three or four firings and applied by a combination of spraying and hand painting in clear or matte white earthenware with glaze stains. Lusters and enamels have also been used in the work.

Porcelain movement by Oldrich Asenbryl (UK)
This wormlike form has been made from extruded porcelain. Bright enamel color on the shiny white glaze and handbuilt additions add to the impression of mobility.

Harmonic flute by Neil Ions (UK)
This imaginative piece has been extruded and modeled in red earthenware clay. The flora and fauna decoration is delightful.

Sandra Black (Australia)
Black bone china forms have been cast and then decoratedwith speckled gold and white.

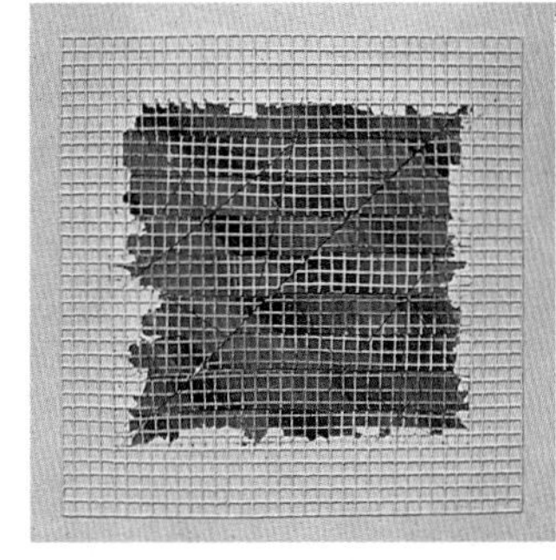

Porcelain wall panel by William Hall (UK)
This porcelain wall panel is approximately 10 x 10 inches (25 x 25cms). (Firing: 2066°F 1130°C oxidized)

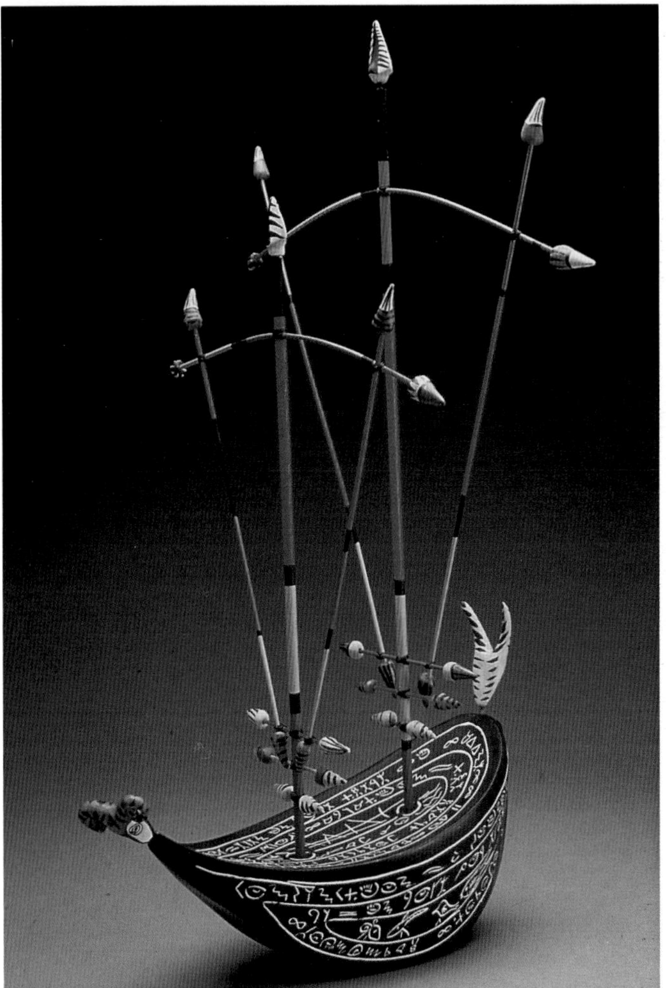

Convergence, 1988 by Roberta Kaserman (USA)
This is very much a mixed media piece which has made use of terra sigillata, wood, wax and gouache.

Textures by Paul Astbury (UK)
Plaster molds taken from found objects and illustrating the source of textures he used for his work in the early 1970 s. These three were derived from a carpet, a speaker grill and a toy puzzle, respectively.

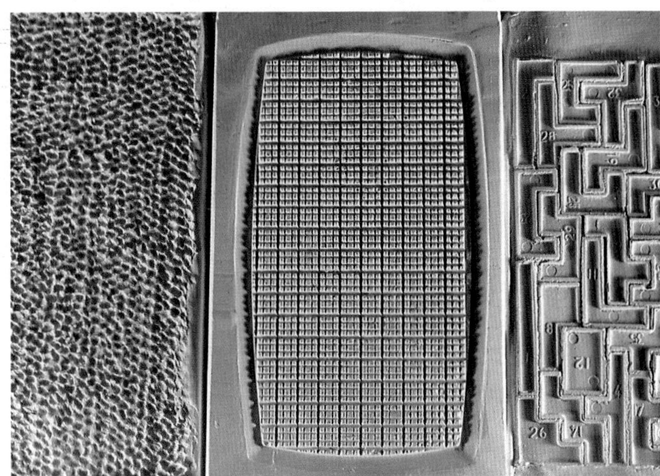

Surface effects

Chalice by Henry Pim (UK)
A striking and unusual chalice with its deeply contoured surface highlighted by skilful use of color. It is some 20 inches (52cms) tall.

Old Man of Port Grigor by Andrew Wood (UK)
White tin glaze, enamels and lusters complement the color and vigor of this lively pot.

Bird ocarinas by Neil Ions (UK)
Neil Ions' work results from a fusion of interests in the natural world, music and American Indian artefacts.

These were press molded and modeled in red earthenware clay and then painted with a range of earthenware slips. They were lightly burnished before firing and wax polished afterwards. (Firing: 1976°F 1080°C)

Treasure chalice by Sally Bowen Prange (USA)
Made in two sections this piece has been both wheel thrown and coiled, and then finished with silicon carbide slip, glaze and lusters to produce a gleaming surface.

Open bowl
by Peter Meanley
(Ireland)

5 Decoration

Decoration

An artistic approach: inspiration.

The production of a handmade object allows the artist considerable scope to create an individual piece which expresses his or her feelings — both for the subject matter and for the materials chosen for the work.

All potters develop their own approach and style; surface decoration is an integral part of this. It should be considered throughout the molding and construction process so that the eventual design enriches and enhances the ceramic form.

Decoration does not have to rely on colored clay bodies, slips and glazes; it can be achieved by manipulating and touching the clay itself. Plastic clay is an excellent material because it responds immediately to the pressure of the potter's hands. The marks so easily made by fingers, thumbs or nails may provide the basis for a wide variety of surface patterns or textures. The joins between pieces, strips or coils of clay can also contribute to the visual interest and, where the surface of a pot is patted or beaten into shape with hands or wooden paddles, those marks too may become an integral part of the design. Such apparently random marks can be used deliberately and, together with the character of the particular clay, might be all that is needed to complement the form. The kind of clay chosen for the work will also make an important contribution to the final result.

Visual awareness should be nurtured and developed so that you are able to make the best possible use in your work of those elements in the environment that interest you. The idea for an effective design can often come from drawings and photographs or from objects of all kinds. Textures, patterns and shapes collected and recorded with a camera and sketch book will provide endless sources of inspiration. The smallest detail of a leaf, a shell or a machine may become the foundation for a whole series of experiments in form and surface decoration.

Constant images surround us all. We may respond to pattern, texture, shape, form or color; or the size of an object, its symmetrical qualities or its originality. Trees, flowers, fruit, insects, rocks and bark are a source of inspiration to many artists: the roughness of bark or the smoothness of an eroded pebble found on a beach; a fungus growing on a tree stump, or a seedpod studied closely — perhaps cut in half.

Inspiration may also come from man-made objects: the clean precision of engine parts, the sleek finish of industrial pipes, the jagged structure of scaffolding, a house constructed in mellowed brick or stone. Any of these could be a starting point from which new pieces are formed and decorated. Collect natural and man-made objects, whatever can be kept in the studio to help in the approach, construction and development of a piece of work.

The choice of clay

Having absorbed the basic techniques of constructing a pot and with your ideas for decoration down on paper, the next matters to be considered are the color and texture of the clay to be used for any particular project.

As already explained, raw clay, when dug from the ground, can vary in color from a pure white to black, depending on how it was deposited millions of years ago and which minerals it collected on the journey to its final bed. Some can be used straight from the ground; others need to be combined with different clays and materials to make them into a workable body.

Traditionally, craftsmen did not travel great distances to find raw materials. Instead they made use of their own local clay. It is interesting to note the wide variety in color and texture among the many clay bodies used in the production of pottery around the world. The most common iron-bearing clays were often used in their raw state, making natural buff or terracotta bodies. Only in the last two hundred years, with the discovery of a suitable 'local' kaolin, were European potters able to produce porcelain pots similar to the ones that the Chinese had been making for hundreds of years.

Modern potters now have an enormous choice of ready-prepared materials and clays which can be mixed together or with one or more metallic oxides to stain the body. During the 1950s and 60s the mid-range of grays and browns were fashionable but from 1970 onwards there was increasing use of white high-firing porcelain bodies and brightly colored decoration. Traditional methods and techniques provide us with a rich variety of options in our own work but it is not sufficient to attempt merely to copy the style of other cultures or past masters. Potters should always approach their work with freshness and originality and, at the same time, try to reflect their own time and place.

In many cases an explosion of color is a reaction against the 'grays and browns' of the 60s. The availability of excellent and reliable ceramic stains in a wide range of bright colors — for painting direct, for slips and bodies, and for glazes — has enabled many potters to experiment in a contemporary idiom. The visual arts, especially painting and textile design, have provided source material to aid the swing towards color. The abstract expressionist movement in the USA (on the West Coast in particular) has also been an added stimulus.

Develop your visual awareness: collect reference material and ideas to implement in creative design.

Explorations in design

Inspiration for design and shape can develop from almost anything seen or handled, photographed, or drawn and painted. As you become accustomed to the clay, its feel, texture and colors, the endless variety of decoration techniques practised will spring to mind when almost anything is observed. Textures and colors of rusting metal may seem like an iron-rich, dry wood-ash glaze. Tree bark can be portrayed in a deeply scored and beaten coiled pot. Geometric patterns of seeds may suggest a slip-trailed design on a dish; the pattern of colors of sea and sky could reappear on a white porcelain ground. Painted slip, sgraffito and relief work can all be used to recreate the effects required. Wide landscape, or minute detail seen through a microscope — both may serve as material for ideas.

Try using other forms of artistic expression as a source of inspiration for shape, design and decoration. Historically the decoration of many ceramic pieces can be seen to bear a close relationship to the paintings of the period. It is important to always ensure that the decoration relates closely to and complements the form — which may also draw upon the potter's awareness of his or her surroundings and culture.

For example, initial thoughts can be triggered by the sight of colors in landscape, the tactile pleasure of texture, the shape and feel of a shell, a stone or a leaf, the form of a statue or style of a painting.

A small area of a water-color can serve as inspiration for color and style. In this case the color and form relate to the East Anglian landscape with its agricultural vistas and vast ever-changing skies. A section of landscape is masked around. The different areas become bands of color and texture on the surface of the pot, rising in a spiral design.

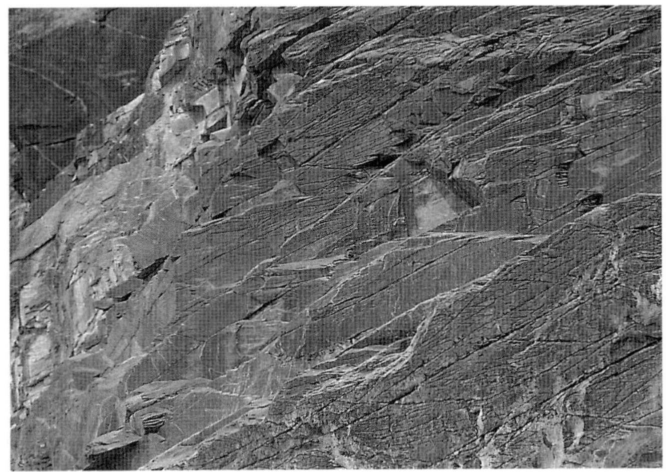

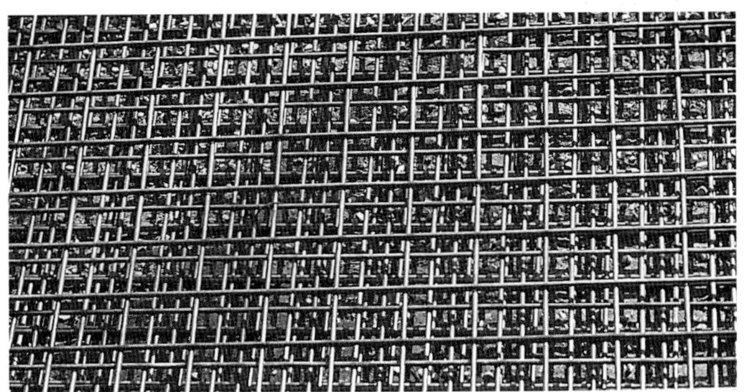

Natural and man-made objects provide an endless source of inspiration. Examine the shape and textures of all you see around you.

Experimental piece (and detail) by Maggie Barnes (UK)
Individual porcelain pellets have been joined by porcelain slip, sheet rolled and press-molded into a bowl form. The exterior has been given a black slip 'crust' to contrast with interior.

Open bowl
by William Hunt (USA)
Here the emphasis has been placed on the wonderful cratered texture of the surface glaze.

'Dieu des Mers'
by Monique Vézina (Canada)
Handbuilt stoneware sculpture, inspired by a fascination with the human form.

Oval bowl by
Peter Meanley (N.Ireland)
With vigorously profiled handles and rim and unusual decoration, this rich vessel is the result of long exploration into the aesthetics of the bowl.

Burnishing

Burnishing is nowadays used as a means of enhancing the surface of a pot and originally was also a way of making the pot more waterproof. The action of polishing the clay with a smooth pebble or the handle of a spoon compacts the clay — filling the pores of the clay texture. It provides a smooth surface for decoration, particularly for scratched designs which show up well on smoke-blackened pots. Animal fats were also used in the burnishing process, to help seal the surface. The Greeks perfected the use of ultra-fine particled slip (made by decanting a slip mixture and known as terra sigillata) which when painted on to a surface develops a shine on drying — before any burnishing is done — and which was thought until quite recently to be a glaze. **H** see page 260

If clay has been sponged or scraped it has the effect of revealing the grittiness in the clay. Burnishing pushes the coarser grains into the clay and smooths over the finer particles. Fireclays are most suitable for burnishing; coating of fine ground slip or oxide applied over the coarser clay will enable you to burnish to a fine gloss. The slip or oxide (iron and manganese work well) should be painted on to the leather-hard clay and allowed to dry slightly. Test a small area for the correct state of dryness; if it is too wet the surface will smear; if it is left too long to dry it will be chalky and will not polish evenly.

Any smooth object can be used for burnishing: a knife blade, a piece of hardwood, a pebble or bone or the back of a spoon.

To retain the shine, the temperature when firing should not be above 1832°F (1000°C) but the best results are often fired as low as 1470°F (800°C). Interesting mottled effects appear on the smooth surface if the pots are fired in a simple saw-dust kiln, as a result of smoke and uneven heat. It is important to remember that high firing usually destroys the gloss — due to shrinkage and the subsequent appearance of harder particles on the surface.

Burnished bottle by Robyn Stewart (N.Z.)
Swirls carved into the highly burnished bottle serve to emphasize the gleam of the polished surface surrounding them.

Open vessel by Petrus Spronk (Australia)
This lovely spherical form has a highly burnished and carved surface.

A selection of burnishing tools

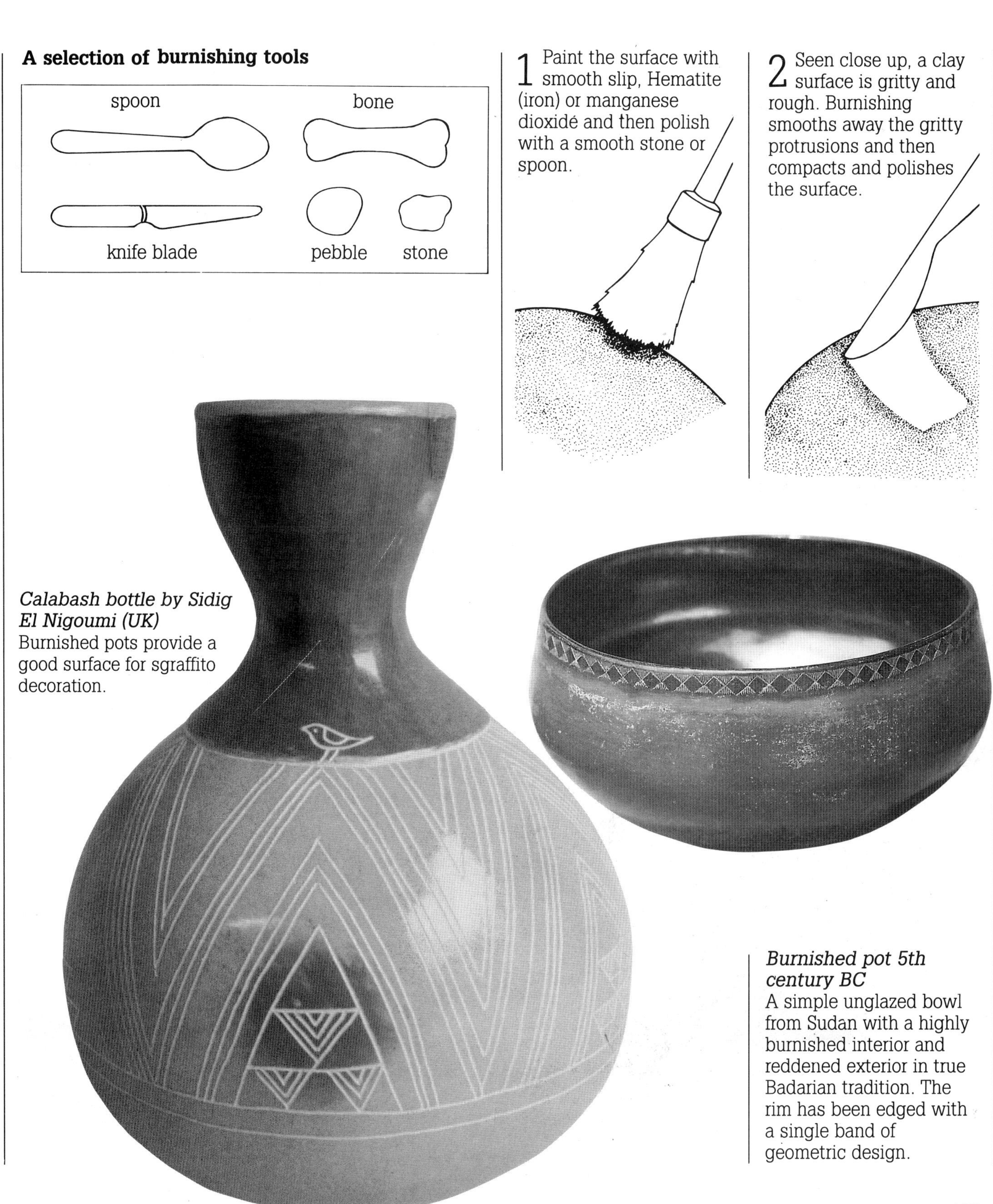

1 Paint the surface with smooth slip, Hematite (iron) or manganese dioxide and then polish with a smooth stone or spoon.

2 Seen close up, a clay surface is gritty and rough. Burnishing smooths away the gritty protrusions and then compacts and polishes the surface.

Calabash bottle by Sidig El Nigoumi (UK)
Burnished pots provide a good surface for sgraffito decoration.

Burnished pot 5th century BC
A simple unglazed bowl from Sudan with a highly burnished interior and reddened exterior in true Badarian tradition. The rim has been edged with a single band of geometric design.

Piercing

Pierced decoration

By cutting right through the surface of the pot you can add a new dimension and fragility and allow light to contribute to the design. When the clay is leather-hard cut through it carefully with a narrow tapering blade, using a rotating action rather than forcing the blade through the clay. If too much pressure is applied cracks may appear later during firing. Clean and smooth the edges of the holes with a damp sponge, a nylon pad, steel wool or a piece of coarse linen stuck to the end of a small stick. Remember to wear a protective mask when rubbing down dry clay and sponge the dust from the bench. Steel wool (of a fine grade) is often the most effective means of cleaning burred and chipped edges when the clay is dry. However, apart from the health hazards of airborn dust thus created, all fragments of the steel wool together with the dust should be discarded to avoid contaminating other clays. For porcelain cleaning use a nylon pan-scrubber as any fragments of steel wool inadvertently left behind may go rusty and will show on the brilliant white surface. ◇

Scalpel blades and hole cutters can be bought easily but a little experimenting will expand your own source of tools and needles. Try sharpening metal tubes or use twist drills in various sizes for symmetrical designs.

Tools

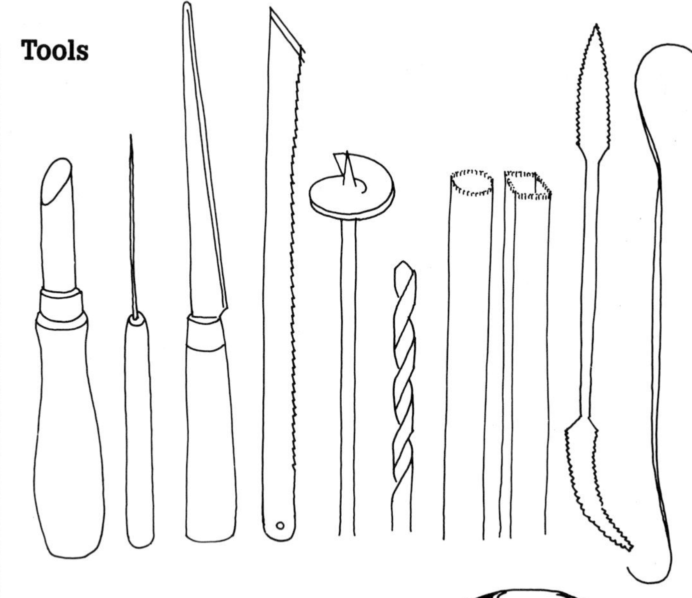

curved blade hole cutter
thin-bladed knife
drill bits
ground thin hacksaw blades
sharpened metal tubes
twist drills

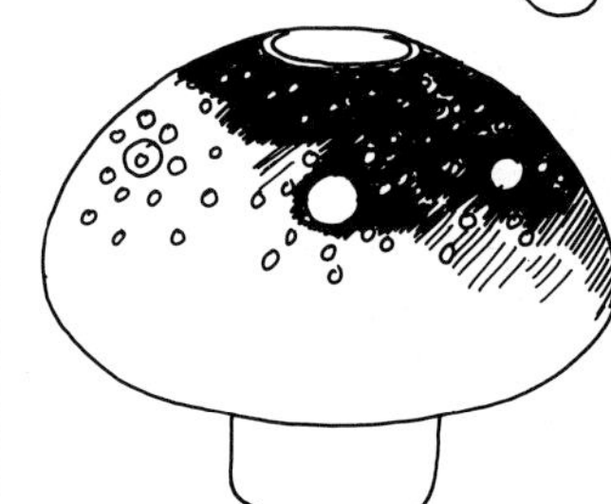

Lamp-shade type pot with pierced decoration (made by using twist drills).

For a fine bowl, support the wall on the inside while applying pressure.

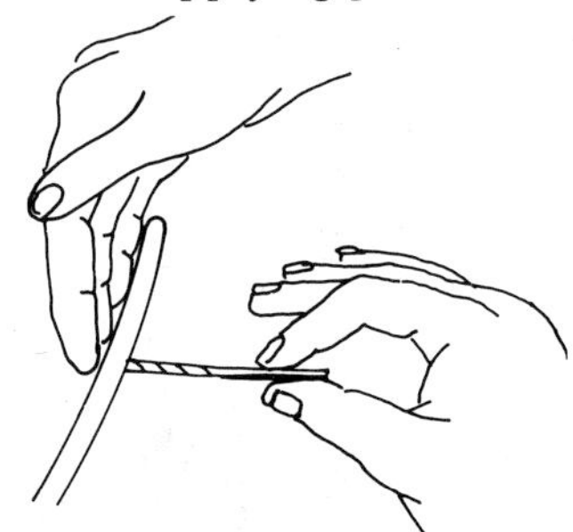

Domed forms by Maggie Barnes (UK)
Porcelain 'toadstool' forms have been given a lacy crocheted appearance by the intricate pattern of piercing.

Porcelain sphere by Maggie Barnes (UK) Carved, pierced and drilled, this delicate piece has been airbrushed with copper oxide decoration.

Bone china form by Angela Verdon (UK)
The finely cast bone china has incized and pierced surfaces which enhance the translucency of the medium. The piercings exploit to the full the fine delicacy of the bone china as well as its surprising strength.

Cast and pierced bone china by Sandra Black (Australia)

Hills and trees by Peter Lane (UK)
Translucent porcelain has been thrown and carved with the white glaze shading to delicate shades of gray. This is a good example of how piercing and carving can exploit the effects of light. The colors have been obtained by using copper carbonate under a dolomite glaze.

Detail of pierced work by Maggie Barnes (UK)
Carved into swirls and folds, the textured slip decoration and cut work create fascinating surface effects.

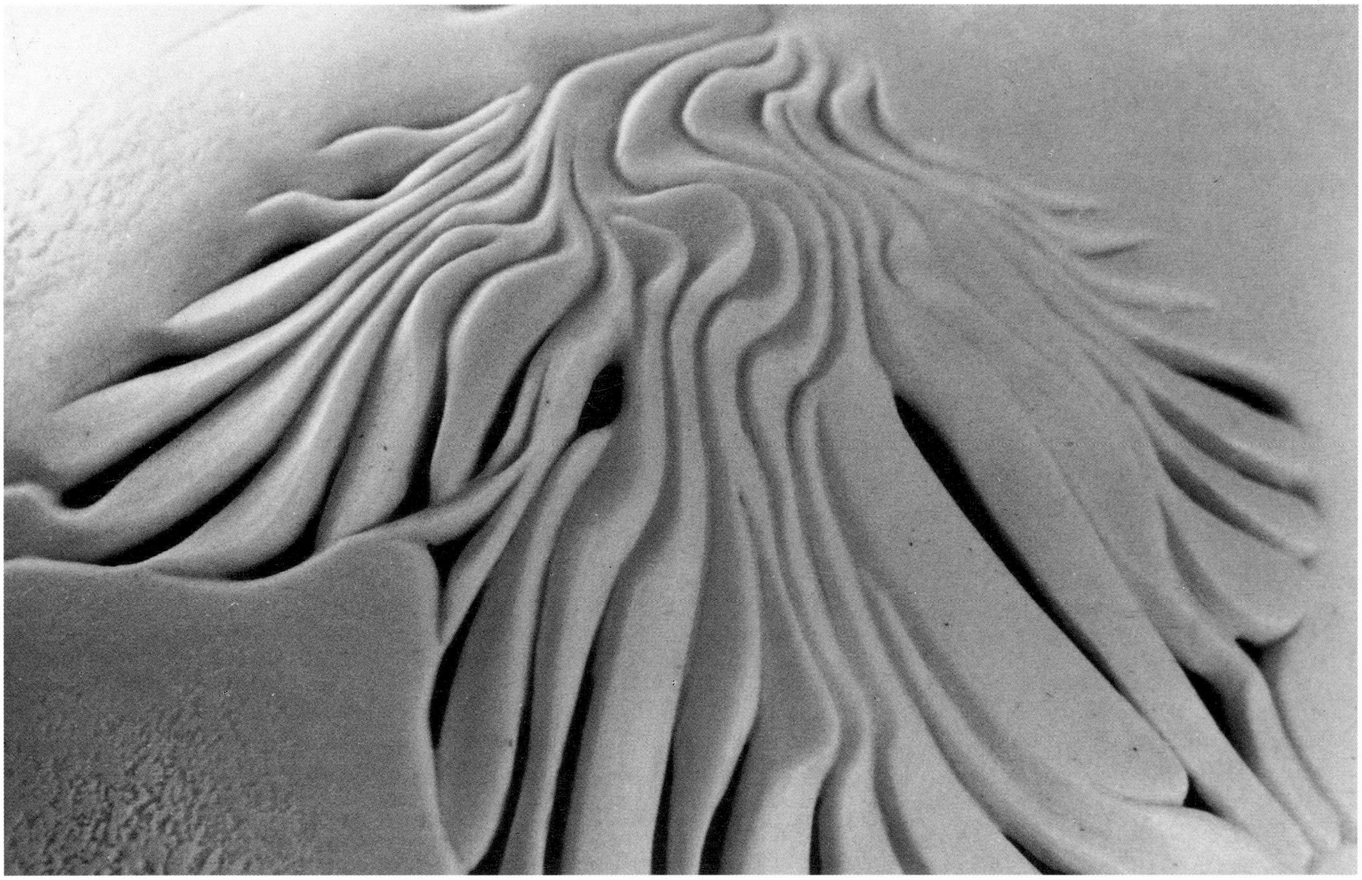

Incising

This technique involves cutting or scratching into the clay surface with a single blade or point. The lines can be deep or shallow, depending on the tool used, and can be made when the clay is at any stage from wet to dry. For variety you can build up the surface in some areas and engrave fine detail in others.

Leather-hard clay can be crisply carved. Clay in this state can be handled without damage and will not chip or distort. A sharp point will lift the surface of the clay as it cuts and a wire tool may be more effective. Sharpen various pieces of wood and practise making lines on a spare piece of clay of the same consistency as the pot. If the clay is soft enough a loop of wire will produce a smooth incision. A sharpened wire or narrow strip tool cuts clay cleanly even when almost dry. As the clay dries firmer lines can be cut but if the clay is too hard this can result in chipping at the edges of lines. Fine patterns can be scratched on to a biscuit-fired pot, and the glaze will gather here to create a deeper color, or the lines can be pricked out with an oxide by sponging it on and then wiping the surface clear. However, this practice is not easily controlled and can be hard work. It is better to work on the surface of a firm leather-hard pot. Color can be applied immediately to fill the fine lines (the top surface of the pot being wiped clean with fine steel wool) or it can be added after bisque firing.

A variety of different tools can be used to make interesting and decorative marks in the clay.

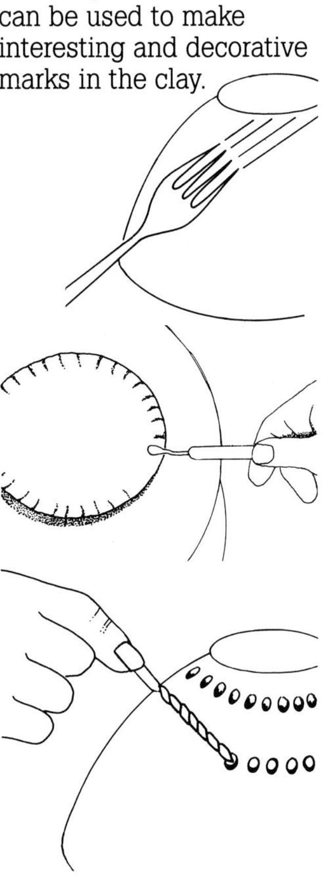

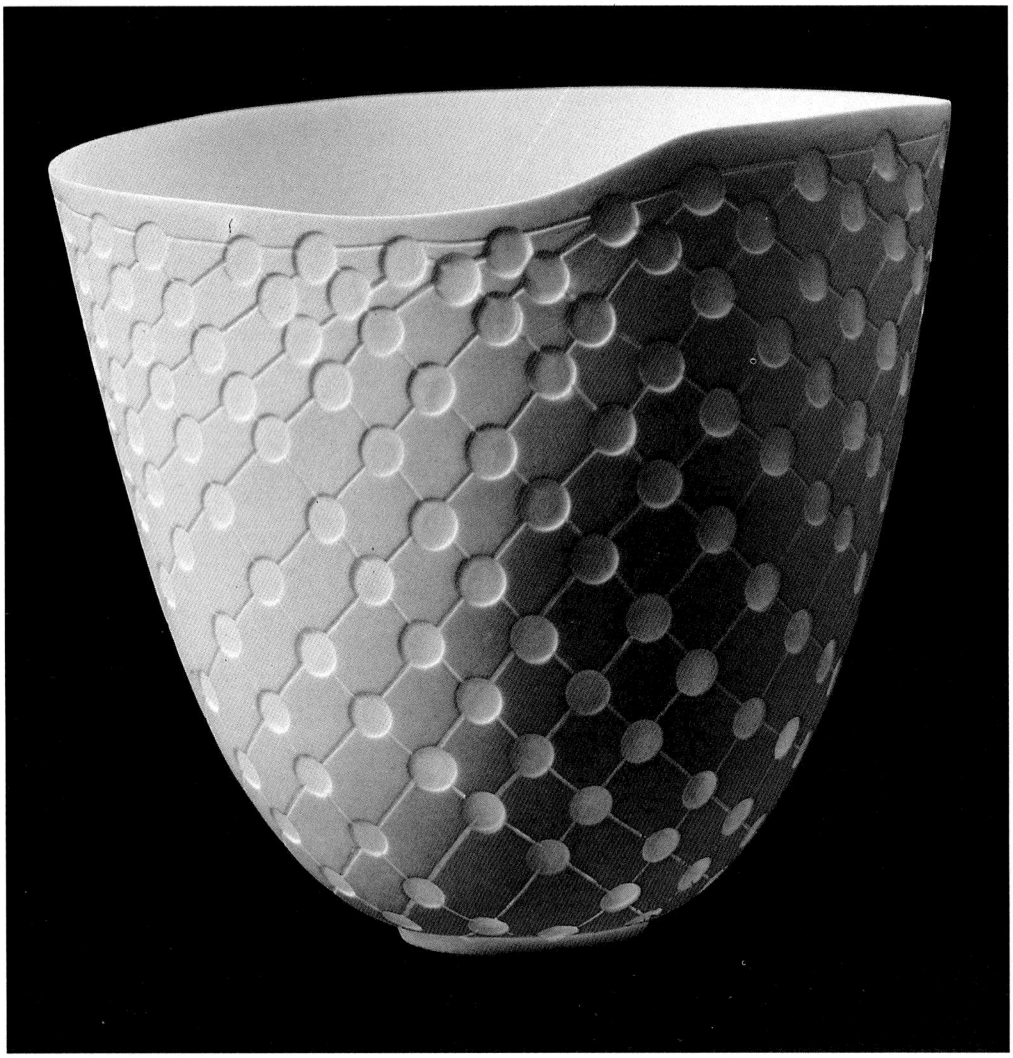

Bone china form by Angela Verdon (UK)
The semi-pierced and incized surface enhances the translucency of this simple pleasing form.

A fettling knife can be used to cut areas of clay that have been outlined within the original incized design.

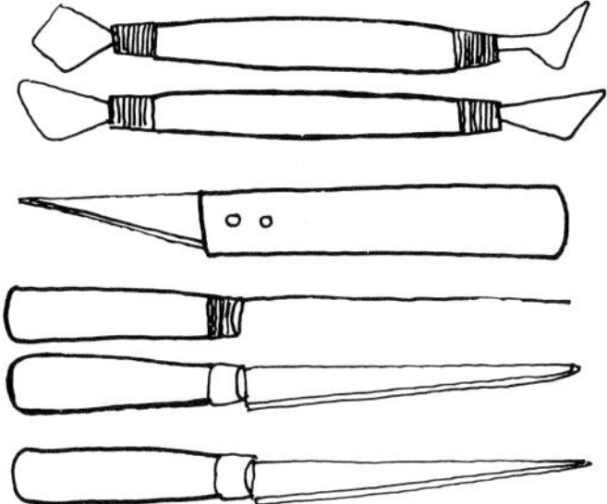

Loop tools can be used for delicate incizing.

Sharp clear lines can be incized into the surface of a pot with a sharp tool. Effective surface decoration can be achieved this way.

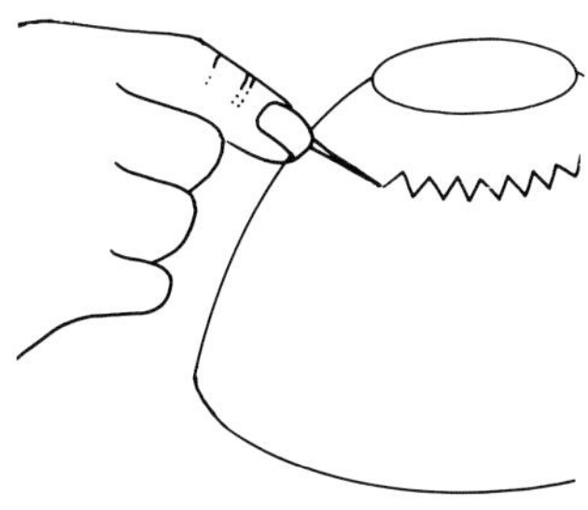

Swirls have been incised to create an interesting surface effect on a pot by Peter Lane

Fluting and faceting

Fluting

Fluting is the use of grooves, usually regularly spaced, that are cut into the surface of a pot. Normally, the ware has a plain monochrome glaze and its effectiveness relies on the surface texture and the play of light and shade on the different levels. Certain glazes flow away from the ridges and sharp edges to reveal some of the underlying clay color and these are often preferred because they help to accentuate the design.

Grooves can be made into fairly soft clay with loops of wire.

Cut into the surface with horizontal, vertical or spiraling lines to create a pattern of shadows. The grooves should be carved when the clay is in a fairly firm leather state. A loop of fine wire is usually sufficient to cut through the clay rather than scraping the clay away with a solid round-ended tool but it depends on the effect you want; the end of a wooden or metal ruler may be suitable. Whenever possible, support the wall of the pot with the other hand while putting pressure on the surface. Experiment with different shapes and sizes of tool to discover your own range of marks and textures for surface treatment. Each stage from wet to dry will give a different effect but a cleaner cut can be produced at the leather-hard stage.

Tools

hooped tools
hoops of wire
sharpened tube

1 Throw a fairly thick cylinder and mark with a sharp point the intended position of grooves.

2 Cut grooves from the bottom upwards with a ruler as a guide if required.

3 Rotate wheel to create final bowl shape, being careful not to touch exterior surface.

4 Fluting lends itself to a variety of forms.

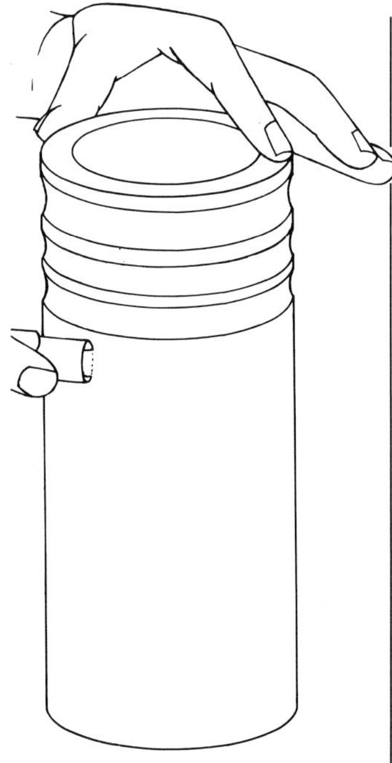

Fluted porcelain vases by Peter Lane (UK)

Faceting

Faceting involves cutting a number of flat sides into a round pot. The resulting contrast between the round form of the pot's interior and the sharper lines of its exterior cut surfaces can be most effective.

If you intend a pot to be faceted make sure it is thrown with fairly thick walls to allow for the later cutting. This cutting can be effected quickly while the pot is still on the wheel. However, more even cutting can be achieved if the pot is removed and allowed to dry to the leather-hard stage.

(Of course a pot can also be molded or beaten to create a number of distinct faces on the exterior surface.)

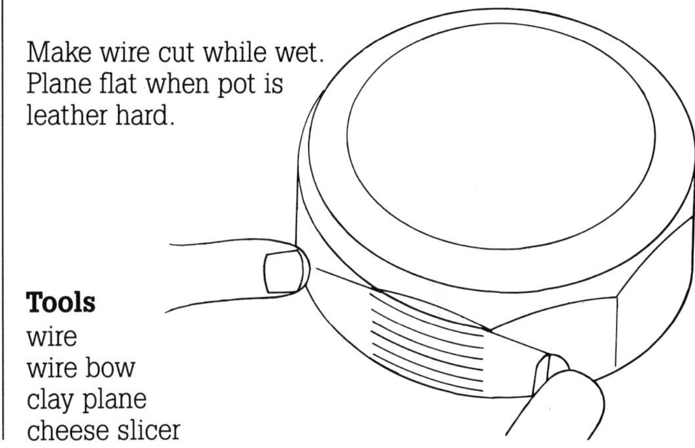

Make wire cut while wet. Plane flat when pot is leather hard.

Tools
wire
wire bow
clay plane
cheese slicer

Bowl by Sasha Wardell (UK)
Fluted surfaces have been carved by hand into slip-cast bone china.

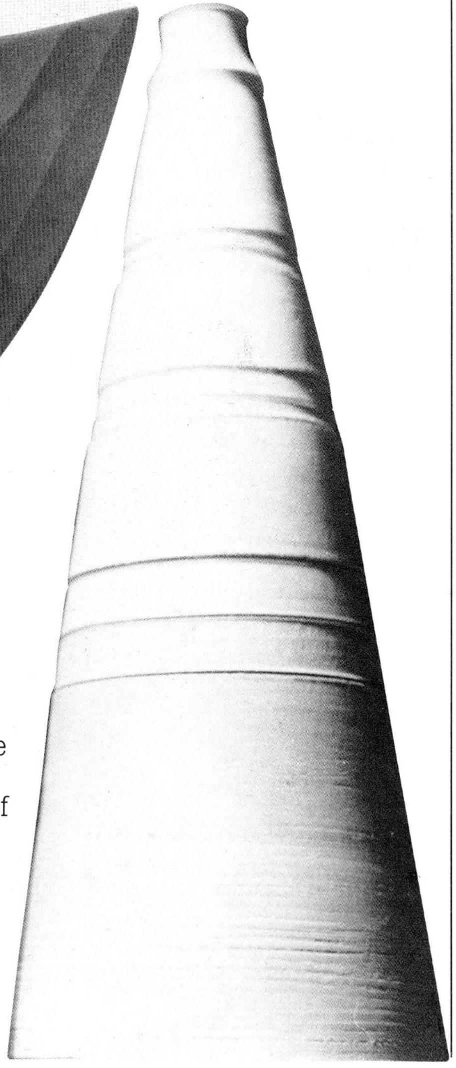

Lampshade shapes by Ian Sprague (Australia)
Stoneware bases have been given surface interest by a variety of grooved effects.

Jug by Jim Malone (UK)
The jug has been cut with wire while still on the wheel. A local granite ash and clay glaze has been used effectively.

Impressed decoration

Any object will make an impression in soft clay so the possibilities are endless—from stamping and embossing to rolling and beating. However, applying pressure to the surface of a freshly thrown pot to produce an instant design can create distortion although this too can sometimes be used as a decorative device. Even if the pot is left to harden the pressure required may still be enough to distort the shape of the pot. So the technique is best used on simple forms and shapes. Flat objects, with support behind, are especially suitable: tiles, molded dishes and slabs (before they are assembled into one piece) all make ideal surfaces for impressed decoration. Over-enthusiasm can lead to disaster so be careful. If too many patterns compete with each other, their impact will be reduced and the overall shape of the pot will be masked, rather than enhanced.

Modeling tools, saw blades, wood, metal, coarse corrugated card; all can create interesting patterns. Repeating patterns can be produced using decorative rollers.

Roulettes can be made from plaster or clay. These are circular discs with patterned edges which are held between thumb and finger and rolled across the clay. Wooden battens can be used to beat the surface of a slab into areas of rich texture. Fruit can be used to create fascinating repeat patterns — and fingers may often prove the best tools of all. Different textures can be built up on the surface by hitting leather-hard clay repeatedly with a wooden or metal implement such as a spoon, wooden batten, fork, kitchen grater or brush. A wire handle slotted through the center can also be used. Metal or rubber type face makes good impressed lettering, often used on wine vessels and beer jars to indicate the contents or maker.

Many potters use a stamp when marking their work. Roll out clay coils of various widths and cut them into sections 1-1½inches (3-4 cm) long. Cut your design on to the end of each coil — a raised pattern will result in an indented decoration and vice versa. Your personal monogram or symbol should be incised in reverse, and can be stamped directly or embossed on to a soft pad of clay (or sprigged) and then attached with slurry to the surface of the pot.

Impressed design

Roll out several slabs, some soft and some stiff. Lay a variety of objects over them, covering these with a clean

Neolithic beakers
The Beaker Folk used simple incized designs and twisted cord or string to create impressed patterns in their pottery drinking vessels.

cloth and then rolling them into the slab with a rolling pin. Try cloth, lace, leaves, twigs, glass, net, string and rope. Try cogs and gears from clocks and watches, or machine parts. You will find that they will make a clean impression on harder clay but will cling to soft clay.

Paddles of wood with saw cuts can be used to beat a slab and so give it a general texture. A piece of wood bound with coarse string will also do this effectively.

Pick up patterns on the clay by rolling, beating or pressing it into a fixed object. This will produce exciting unusual textures and patterns. The slab may even be taken outside and then returned to the studio for construction. For example it could be taken out and pressed on to a coarse-grained plank, rough stone, brick wall or manhole cover.

Making clay stamps and casts

If these fixed patterns need to be repeated then the pattern in the clay can be cast in plaster. Surround the object to be cast with a clay or card wall, sealing any cracks before pouring plaster of paris over it. Some useful clay stamps can be made by taking a box of small balls of clay outside and pressing these into any surface or object which appeals to you, such as a house wall, paving stone or tree trunk.

The clay stamps can then be bisque fired. You can also make your own pattern stamps in clay by cutting away part of the surface into a design and firing the finished piece to 1830°F (1000°C). The simpler the design of the stamp, the greater the variety of different combinations of its impression that can be achieved.

Stiff clay blocks are beautiful to carve into. Use gouging tools, knives and needles to score fine lines. (Line cutters are excellent but will lose their sharp edge quickly.) These must be bisque fired before use or surrounded with a wall of clay and cast in plaster.

The Ancient Egyptians used fired clay molds for making impressed patterns, as well as cylindrical rolling stamps or roulettes. These are made from coils of clay cut into short lengths, and may have a hole through the center where a wire handle can be fitted for easier handling. (Failing this, the roulette is held between forefinger and thumb and rolled across the clay surface). They are often used on a newly thrown pot, being held against the rim as the pot revolves, and were common in early salt-glazed ware when the glaze picked out the pattern without the use of color.

Plaster rollers can be made by casting a shape in a tube of card embedded in clay. (Often packaging in the shape of a tube can be found.) These rollers should be carved after setting and before they become too hard and easily chipped. After carving, roll them on a piece of old clay which can be discarded later. This will remove any particles of plaster which otherwise may embed in the surface of the pot, causing blow-outs and cracking.

Plaster blocks can be carved into for repeat patterns, or a prepared clay design can be cast in plaster. Small plaster stamps of any shape can be carved and scratched.

Cast some in small cardboard or clay boxes. Discarded packaging often has small clean compartments which can be filled with plaster and used as a stamp.

If you wish to make a cast from some object or surface texture which is metal or stone, the surface must be painted with oil, grease or soft soap to prevent the plaster sticking.

Using mushroom molds

The Romans developed very fine quality small bowls. These were made in carved and incised clay molds and then biscuit fired. The clay was pressed into the mold, making an impressed pattern on the outside while the inside was smoothed on a wheel. Foot rings, handles and spouts were added after removal from the mold. These pieces were unglazed. **H** see page 262

This style can be emulated (although in reverse) by using a hump or mushroom mold made in plaster. First carve and scratch a design on the plaster. Lay a soft sheet of clay over the mold, pressing firmly to produce a strong pattern to the inside of the dish — which will need only a colored transparent glaze to pick up the hollows and high points of the design. This method was used to full effect by Chinese potters in about the 12th century AD. **H** see page 268

Larger cut impression makers

Soft insulating bricks (as used in kilns) are easily cut but the nature of these blocks makes finer details impossible. Sawn or filed line patterns work well. The brick can be pressed into the clay slab or the slab can be laid on top and rolled into the pattern.

Cast a block of plaster in a box. Carve into the surface, and roll clay into those patterns you have created.

Try out as many tools and objects as you can find; use them first on a sheet of soft clay, and then on a sheet of hard clay. Compare the differences.

Some objects will make a sharp mark or cut when pressed into hard clay, and a smudged shape when pressed into soft clay.

Slab construction is the most successful form on which to try impressed and applied decoration. The flat clay can be decorated before assembly and then allowed to harden before cutting.

Found objects, carefully chosen, can be an endless source of supply. Examine both natural and manmade shapes and textures and the results they produce. Be prepared to experiment and always be alert to potential sources of inspiration.

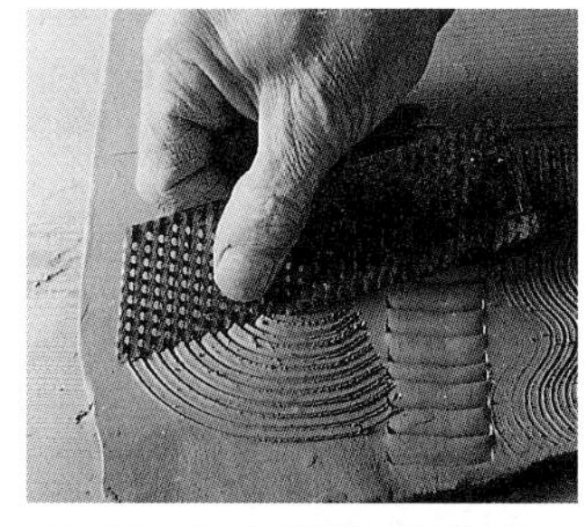

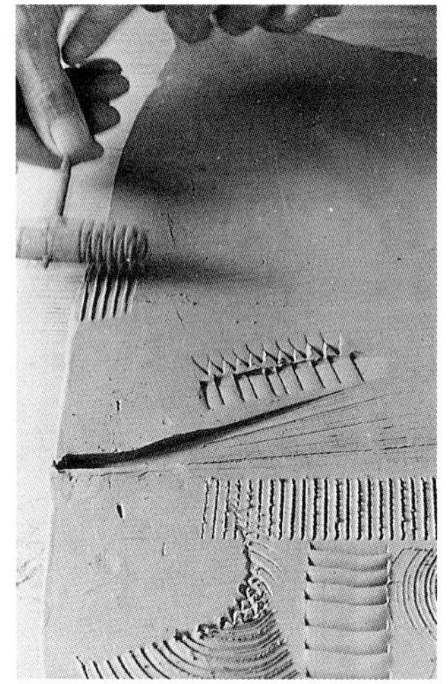

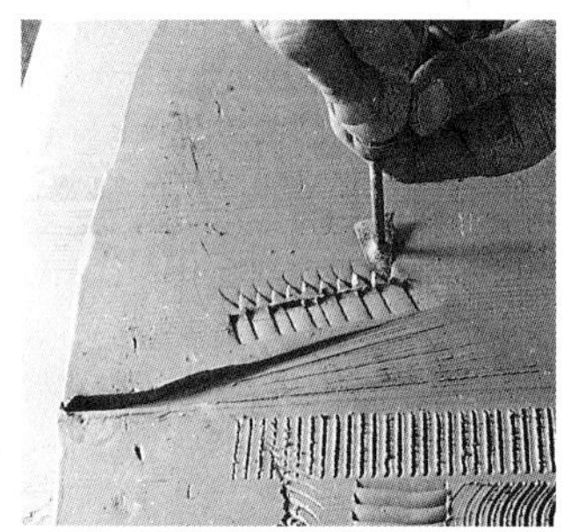

Materials for impressed decoration

Use fine materials such as cloth, lace and linen as well as leaves, twigs, net, string, rope, sacking and metal gridding.

Impressed decoration can be carried out on the flat slabs of clay before assembling them together into their final form.

Cover the chosen materials with a clean cloth. Then roll them into the surface of the clay.

Slab with repeated leaf impressions ready for assembly.

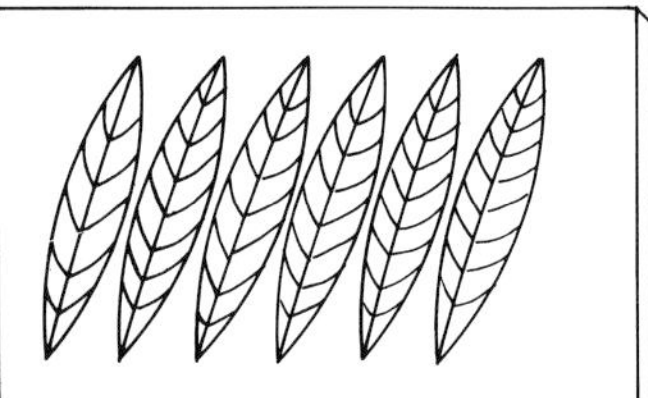

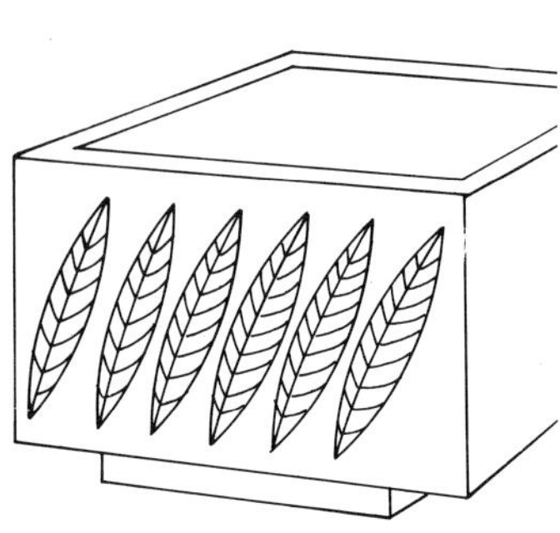

This slab with grass-blades impression has been constructed around a cardboard tube.

Stamped impression

After experimenting with found objects, try making impressions with designed patterns.

Clay stamps

These are made by modeling and carving a stiffened piece of clay. Make sure there is enough clay to provide a hold for the fingers. Sharp knives, gouging chisels or lino-cutting tools are good for carving into stiff clay.

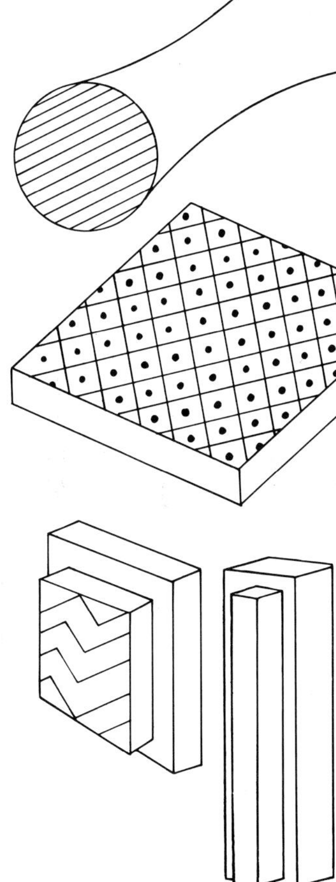

The clay stamps must be biscuit fired before they are ready to be used continually.

Plaster stamps

Make plaster casts of found objects to produce permanent stamps. Alternatively, plain plaster stamps can be cast and then carved into a chosen design.

Casting a disc or coin

1 Cover the metal with oil to prevent its sticking to the plaster

2 Surround coin or disc with a tube of card or vinyl which is held in place by a coil of clay. Pour in the liquid plaster.

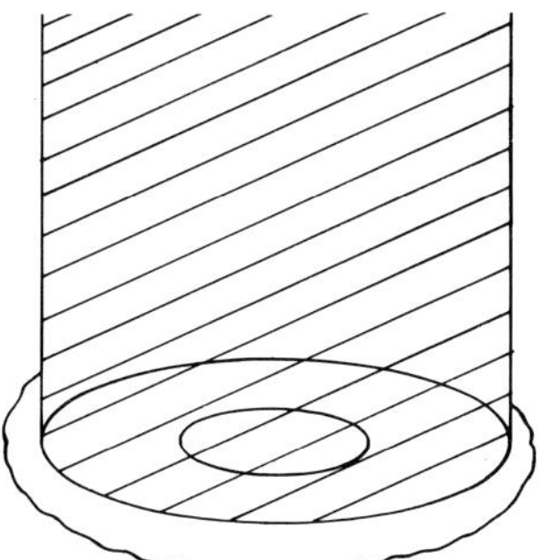

3 Allow plaster to harden, remove the tube and disc and then wait until the cast is completely dry before using. A raised pattern of discs can be created with this stamp.

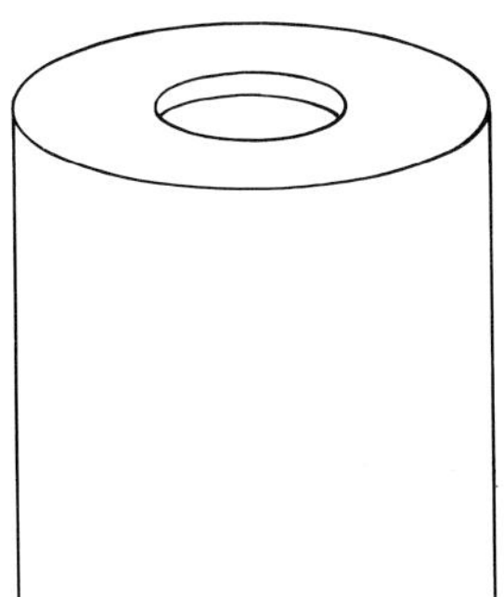

Carved stamps

Blocks or tubes can be cast and then carved or incised to make stamps.

Use boxes and tubes of card, clay boxes, or compartments in polystyrene packaging to cast handy-sized pieces of plaster.

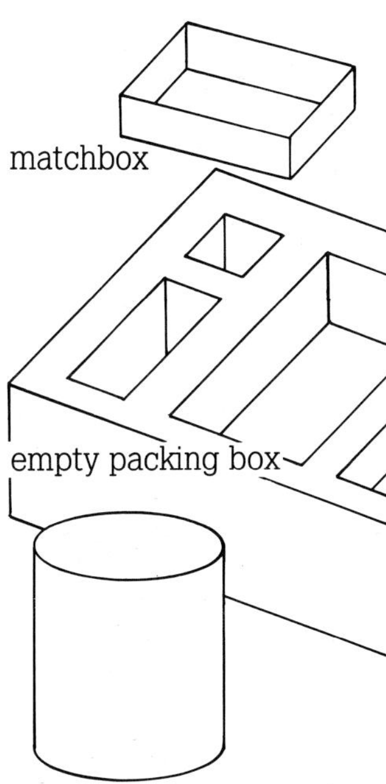

cardboard tubes

A strip of clay joined in a circle and then placed on a sheet of glass will make an ideal casting surround.

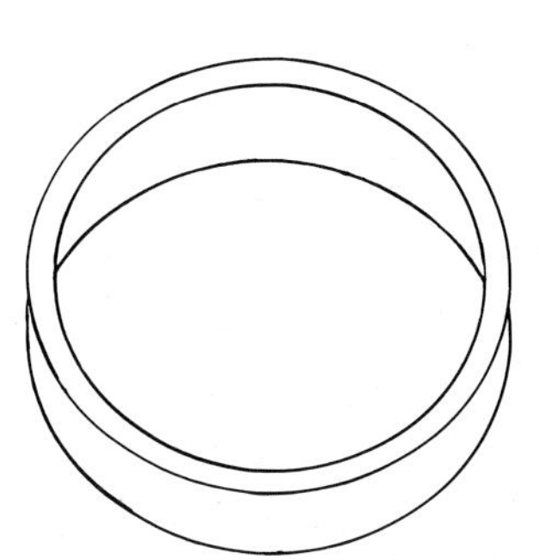

Larger impressions

Soft insulating bricks (like those in kilns) are easily cut—provided no fine detail is required. Sawn or filed line patterns work well.

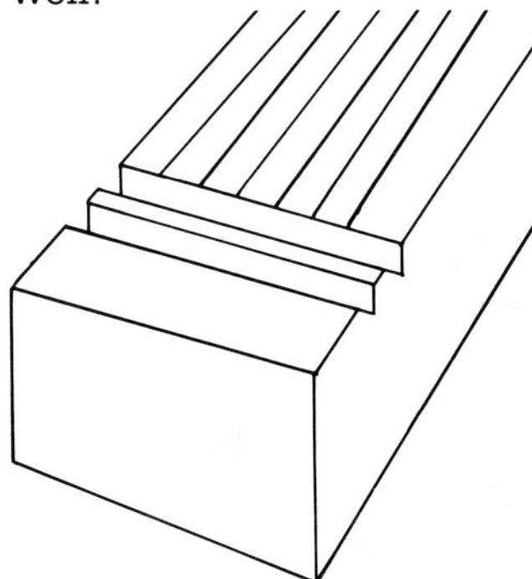

Cast a block of plaster in a box. Carve into the surface and roll clay into the patterns created.

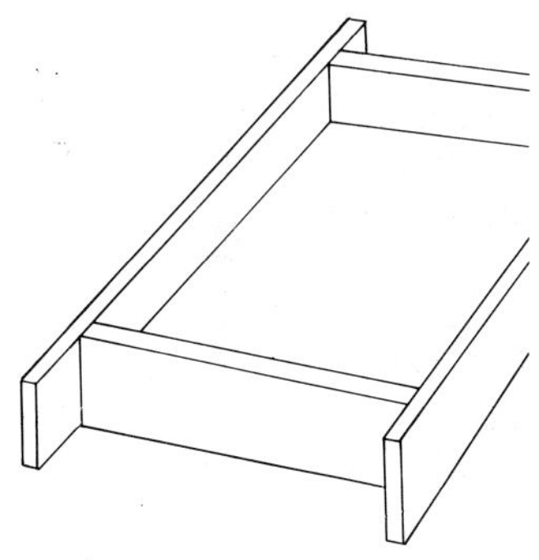

To make a dish with decoration on the inside surface, use a hump mold with a carved design.

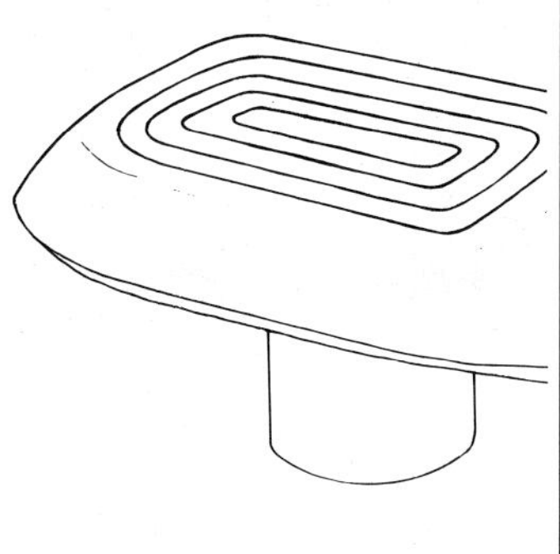

Samples of impressed and applied work

Slab-made vase by James Robison (UK)
A multiple slab construction, this vase has been given textural interest with porcelain slip used over a cloth stencil.

Lidded pot by Jan Schachter (USA)
stoneware pot with interesting texture and an applied handle made from extruded coils.

Teapot by Mary White (Germany)
This stoneware pyramid vessel makes good use of larger, applied decorative pieces.

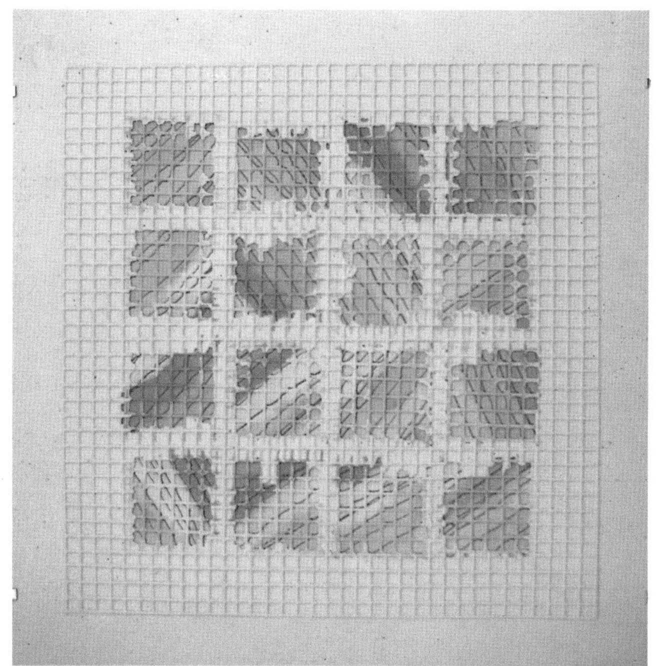

Porcelain panel by William Hall (UK)
William Hall uses plaster molds or low-relief patterned plaster bats to cast textured surfaces.

Drought panel by Ian Sprague (Australia)
A fascinating mixture of textured surfaces and applied relief pieces has been accentuated by the use of an iron pigment.

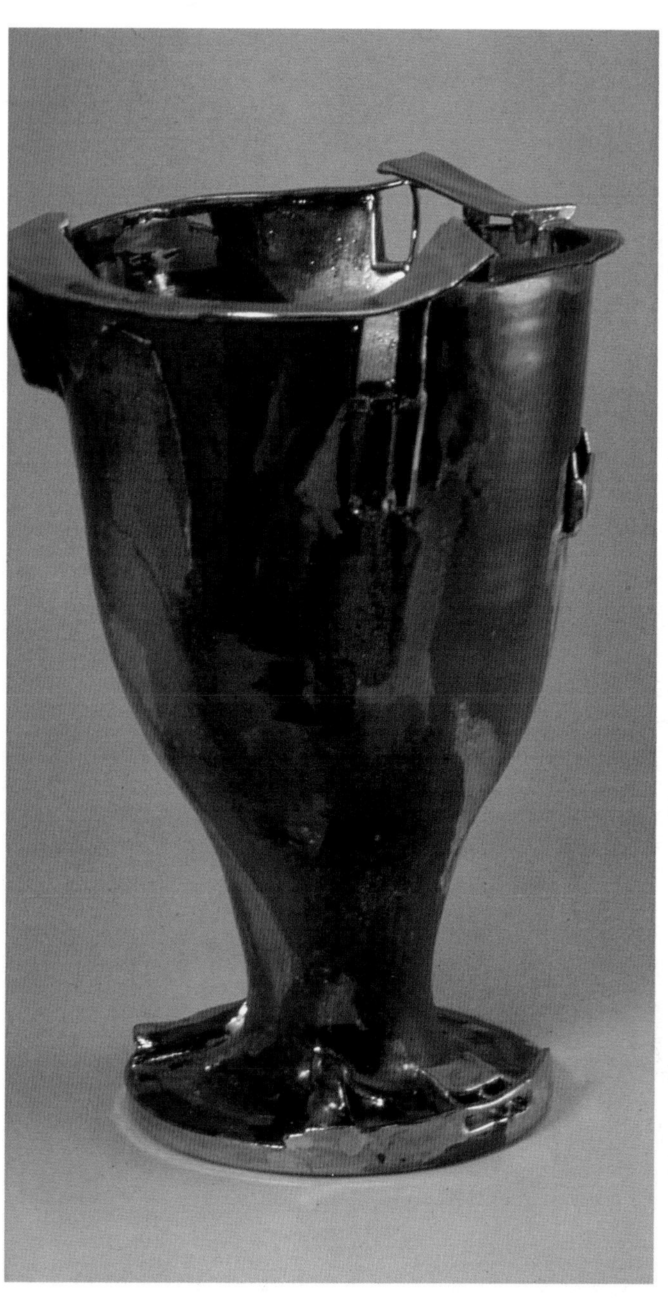

Treasure chalice by Sally Bowen Prange (USA)
This is an example of the 'Wreck Vessel' series with celebration glaze, lusters and added decoration

Oval dish
by John Maltby (UK)
This pressed molded dish exploits the different colors of clay — applied through a stencil and complemented by enamels.

Color

Color in the body is a combination of the natural colorant in the base clay and any added metal oxides of commercially prepared body stains. Trials with progressive amounts of colorant in the clay from 10 per cent upwards are the best way to assess the amount of color you will need. Bear in mind that large amounts of oxide added to a body will lower its maturing temperature — the oxide acting as a flux in the mixture. The nature of the base clay, the firing temperature and the atmosphere in the kiln will all serve to make changes in the final result.

White or pale clay bodies are best used for the brighter colors but they may need large amounts of oxide to produce the dark colors, whereas a red clay, already naturally rich in iron, will need only a small amount to produce a dark tone, albeit of limited color range. The same amounts of oxide added to white, gray and red clay bodies will give startlingly different results, rather like painting the same color on a white, gray or red paper. The white body will make the colors bright and strong, the gray body will make muted shades and the red will produce dark colors.

Changing the body color

Staining the body a different color may be done by weighing out clay and color in dry powder form. This method allows for accurate measurement and gives greater control if you wish to make a repeat batch of a particular hue or to adjust the strength of color. If your clay is plastic, roll it out into thin sheets, allow it to dry and then crush it into a powder. Mix the dry clay and oxide thoroughly after weighing before covering it with water and allowing it to stand and soak. Sieve the mixture through a 60-mesh and then through a 100-mesh sieve to produce a fine finished slip. Allow this to settle, pour off the excess water and leave the mixture to dry to a plastic condition. It can then be stored in plastic bags until required.

Alternatively, add water to the powdered, weighed clay to make a slip and then add the oxide (mixed with water) to the base. Some oxides may need to be ground in a pestle and mortar with water to reduce the particle size before they are added to the base.

If only small amounts are required, the oxide can be added in powder form to the plastic body by sprinkling the oxide on to a table and wedging it into the clay. Another method is to slice the body into thin slabs with a wire and then place the oxide between the slabs like a sandwich. If the clay is unevenly wedged this can produce some unusual and attractive effects. Some potters throw this mixture, creating a spiraled pattern of color on a wheel-thrown pot.

Drying the mixture

A quick method of drying the mixture is to pour it over a finely woven nylon cloth placed in plaster of paris mold or slab or, for large quantities, a box made of loose-laid bricks. The plaster or the bricks will absorb the moisture while the nylon retains the clay and prevents it sticking to the container. When the soft plastic clay is ready it must be stored in airtight polythene bags or bins and left a while to mature and to improve plasticity.

Points to remember

Always wear a mask if you are mixing dry ingredients, and have plenty of ventilation in the room. Wipe up any spilled powder with a damp sponge; do not brush it away as this will create dust which could be inhaled. ◊

For further details on the use of oxides and enamels see page 188-9.

Vase
by Greg Daly (Australia)
Gold leaf and enamel decoration have been boldly applied to enhance a simple thrown form.

Roulettes

These are made from a disc of clay, plaster or wood with a design carved into the surface. Using a piece of wire through the middle for a handle, the roulette can be rolled across a slab or around a newly thrown pot.

In this way a repetitive pattern can be created over the surface of the clay while it is still reasonably soft. If too dry, the clay may crack; if too wet the stamp will stick, so it is important to check that the clay is just right. The pattern can take the form of a straight or a wavy line and is effected very quickly indeed compared to many other decorative techniques.

It is not always essential to use a handle, especially if raised areas or indentations in the pot provide support and guidance for the stamp roller held firmly between thumb and forefinger. In fact, if the roulette is very narrow it is sometimes less easy to control with a handle and may wobble on its axle from side to side. Generally though, a handle will help you to keep the roller at an even pressure — firm but gentle.

Decorative roulettes can be made by setting plaster inside a cardboard tube to form small 'pillars' which can be carved when dry or from thick coil sections carved when the clay is leatherhard and then biscuit fired.

Even the most simple of line patterns can be very effective. Alternatively these patterns can be built up into quite complex surface textures. This is especially useful for creating decorative detail on slabs before they are constructed into their final form.

Roulettes can be made from clay, plaster or wood with a design carved into the surface which is then rolled over the clay.

1 Use a strip of clay for the wall. Hold a thin stick in the center when casting to make a hole through the middle which will eventually accommodate the handle.

2 Carve the pattern into the dry plaster before it becomes too hard (2-3 hours later). Hacksaw blades, sharp knives, scalpels, tiles; all these make good tools for carving plaster.

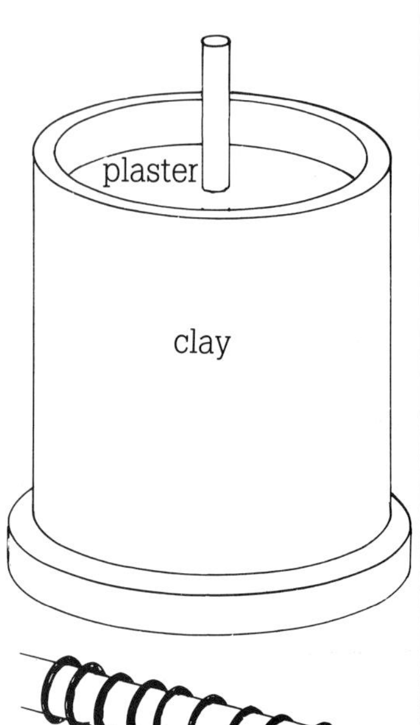

String-coated pencil for rolling a shallow pattern or texture

H see page 248

Vessel
by Jane Hamlyn (UK)
The ware has been salt glazed with polychrome splashes and the detailed decoration created with roulette impressions.

Various decorative processes

Sprigging

A well-known form of applied decoration is the one used on Wedgwood pottery and china, in which a molded piece is fixed to the side of a pot to create a raised pattern. In the Wedgwood pots it is usually a series of beautifully modelled figures or leaves which are taken from a mold, made in a contrasting color to the form of the pot, and then fixed to the pot with slip. This is called 'sprigging'. Simpler forms of applied decoration have just as much effect. By adding a coil of clay and thumbing it into the pot a strong surface texture catching light and shade will appear. Coils, thin rolled slabs cut geometrically or applied randomly, or a collection of dots and balls of clay will all add texture to the surface.

The practice of pressing a clay frit mixture into a hardened clay mold was used by the early Egyptians to make seals, beads and jewelry, the soda and oxide color in the mixture forming a glaze on the surface when fired — usually a rich turquoise blue.

The Romans used carved molds to make relief patterned bowls. The practice of molding carved sections alone, before making the bowl, was possibly first used accidently. Clay of a slightly different color was pressed into the carving and dampened, then the main body of the bowl was pressed in on top. When the bowl was removed from the mold with the attached decoration, a slight difference in color between bowl and decoration would show.

This can be achieved in a rough way by coating the surface of the carved sections with a colored slip before pressing in the clay. Molds were made from fired clay which had been carved while leather hard, dry and porous after firing.

Sprigging was developed as an industrial process in the 18th century as a style of decoration. The model and making of the sprig mold could be done by a skilled craftsman. However, the production of the press molded relief and its application to the pot can easily be achieved, with practice, by an unskilled hand.

A particular motif can be applied to the surface of a pot in the Wedgwood manner by attaching small, separately cast sprigs with slip. To make a sprig mold, the motif is modeled as a negative image and then cast in plaster of paris and allowed to dry. When the new mold is dry, press soft plastic clay into it and scrape off the excess clay until it is level with the top of the mold. The molded piece can then be removed from the mold with the flat blade of a damp knife pressed on to it. Draw the sprig cleanly out. Put it to one side and repeat the process until you have enough sprigs. The clay used for the sprigs must be the same clay as the body of the work or a stained version of the same clay.

The sprigged design should be attached to the pot with slip when both are leather hard. You can apply all kinds of shapes, strips, pellets, coils and so on to the surface. They need not necessarily be made in a mold.

A decorated slab can be used to create interesting bowls. Use a slab with rolled-in inlays of different colors and apply this over a mold as shown on page 113 . Alternatively, the decoration can be first designed into or on the mold itself. Lay pieces of design in contrasting textures or colors on the surface of the mold, then cover this with a slab of clay and apply pressure smoothly to ensure complete adhesion.

H see pages 255-7 297 **C** see page 201

Decorative variations of dish molding

Clay coils and pellets can be pressed into the mold to make a dish.

Laying strips of various clays into a mold will create an interesting surface texture. In both these examples the clay must be smoothed over on the exposed surface as soon as the design has been completed. This will ensure that the separate pieces become fully integrated and thus preserve the pattern created on the outer molded wall.

Different colored clays that have been mixed with body stains may be laid in strips or in overlapping patterns. They should then be stuck together with slip as they are arranged in place.

Inlay various colors and patterns on a slab of clay before placing it in a dish mold.

Agateware

This is the mixing of two or more colors or shades of clay to form a controlled or random pattern of contrasts. Unlike marbling, which is a surface decoration, agate patterns run right through the clay and show up distinctly when the mixture is used for throwing as the different clay spirals up through the pot.

Using similar clays or the same clay with some parts stained to give a contrast, the clays are layered and stuck together rather like a sandwich, and are then cut through and layered again. This process is repeated until a pattern of fine lines appears when the clay is cut through. These strips are rolled out as slabs and are used in the construction of molded dishes, multi-piece molds and slab pots.

Laminating

A quite difficult process to control, lamination involves applying shapes of clay of different colors to the surface of the pot, pressing them and scraping the whole to give a smooth surface finish. The design for a laminated pot has to be planned carefully, for the thin sheets of clay will dry quickly when handled. The shrinkage of the different clays can cause drying problems. The whole pot is scraped and cleaned at the leather-hard stage and at the first and second firing stages. The contrasting clays may be beaten on to the surface of the pot to make sure they are attached well and that no air is trapped between the two surfaces. Another method is to place the laminates on to the surface of a mold in the required design and then cover them with a sheet of clay, pressing the pattern into the molded sheet.

Care must be taken when sticking the pieces of clay to each other. Coat both surfaces to be joined with a slurry of clay and water and press them together to eliminate any air trapped in between. Make sure the pot dries slowly, covered in a plastic sheet. This allows the different colored clays to dry without pulling away from the body or cracking.

Inlay

Inlay has been practised by potters for thousands of years. It is an effective method of inserting — within the main body color — decoration of a contrasting color.

A design is carved into the surface of a pot and infilled with clay of a contrasting color to produce a clean, sharp-edged effect, or flowing contour lines. Cut out the design when the pot is soft but firm enough to handle, using a square-ended tool or loop of fine wire. Dampen the carved areas and fill the channels with a clay of a similar type, stained with oxides or body stains. (If clay types are not compatible there may be a risk of the infill shrinking too much and dropping out of the inlay later when it has dried.) Over-fill the cavities, letting the filling clay overlap the pattern. It will look a little messy at this stage but the clean lines of the design will be revealed once the whole surface has been scraped flat. Do not clean the surface until it is almost dry or it will remain messy. This is a good way of producing well-defined lettering flush with a smooth surface.

Slab inlay

Inlay of another sort can be achieved by rolling contrasting pieces of clay into a flat surface using pressure. This is a form of applied decoration but the pattern is rolled completely level with the surface rather than initially standing out from it. A slab of clay rolled ready for making a molded dish or a slab pot is an ideal surface on which to lay a pattern or picture.

Cut out sections of clay of different colors and arrange them on the slab. When the design is to your satisfaction, dampen the pieces with a brush and water , place a clean cloth over the inlays, and roll them into the slab with a clean, dry roller. Some pressure will be required to incorporate the new pieces and this will cause an amount of spreading which must be allowed for in the design and will give a soft fluid effect. More pieces can be added if necessary by rolling them in carefully or pressing in individually.

For a different effect, drier pieces of clay can be used. These will not spread but will leave a slight indentation around their edges after being pressed in. This often works better if the dried fragments have the plastic clay slabs placed on top of them before rolling in.

Clay that has been forced through a dieplate, using a pugmill or hand syringe, to produce anything from wide straps to fine threads of clay is called extruded clay. This can be rolled into the surface but you will find that smaller pieces will take more easily to the slab. Thin sheets of clay, not necessarily colored, can be overlapped and rolled in. The slightly raised edge created by the overlap will catch a colored glaze during firing and make a pattern of tones rather than colors.

Dorothy Feibleman (UK) preparing laminated sheets for handbuilding

Sprigging

1 Model a simple design in clay. Make sure there are no undercuts so that it will remove from the mold without catching anywhere.

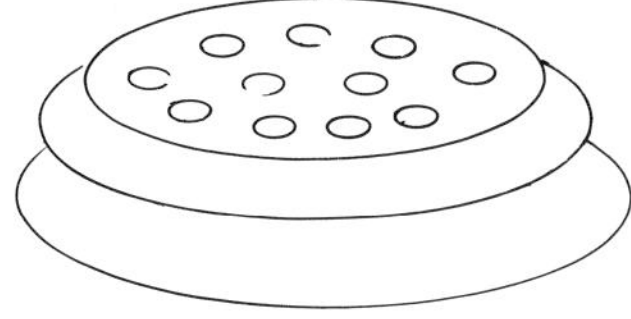

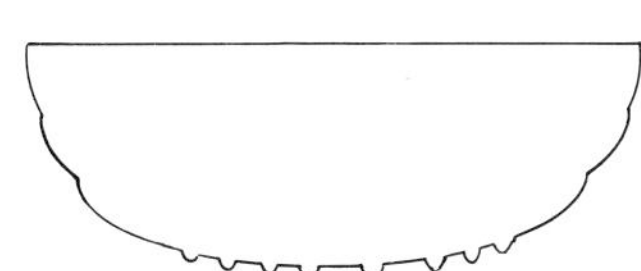

2 This shape would not be suitable for molding. Its sharp angles would prevent the clay being removed.

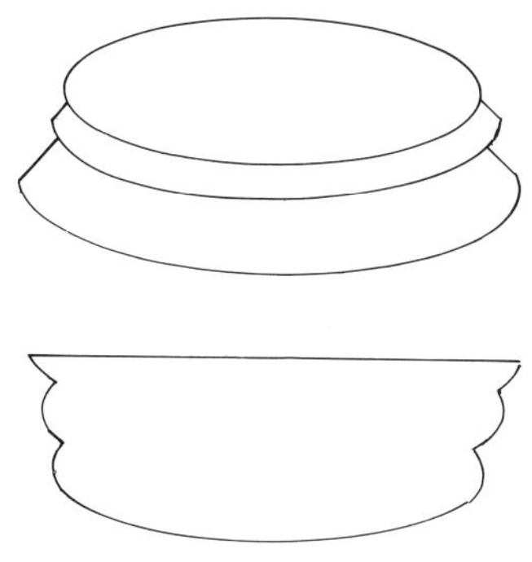

3 Surround the molded shape with a wall of clay and pour in plaster of paris.

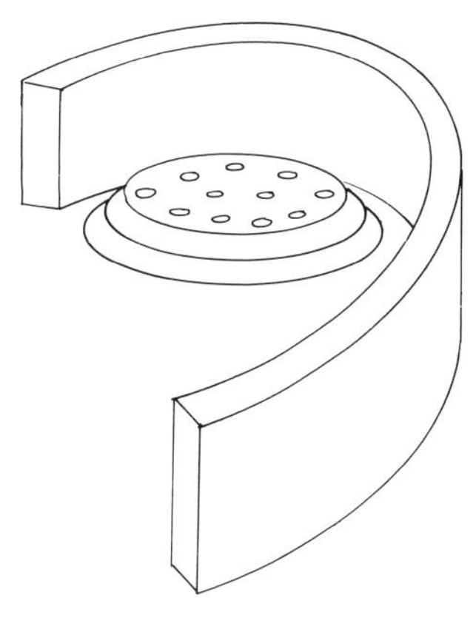

4 When the mold is dry, use your thumb to press on to the surface of the sprig a piece of plastic clay. Scrape any excess off with a wooden tool. This piece of clay you have attached will now act as a handle. The mold can then be left to dry.

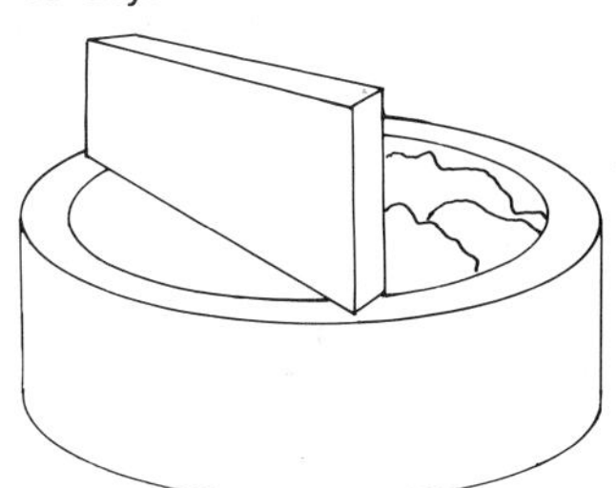

5 To remove the sprig from the mold, use a damp palette knife pressed firmly on the flattened piece of plastic clay, and simply lift out the sprig. The sprigs can then be applied to the pot with slip.

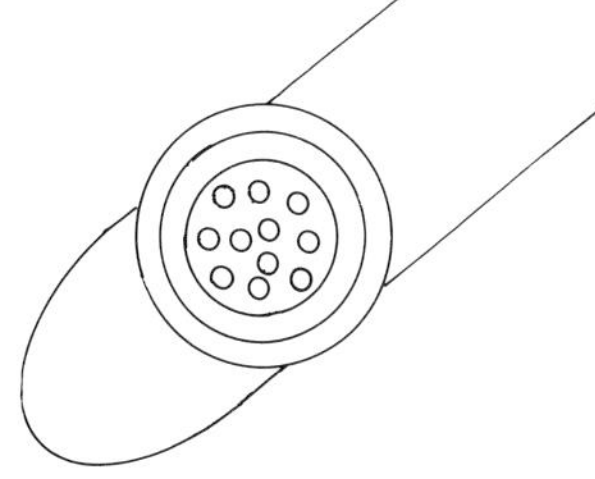

Wedgwood Jasper teapot early 19th century
Solid blue Jasper cabinet teapot with white bas relief figures from the 'Domestic Employment' series. This fine example exhibits a 'stippled' effect on the main body of the teapot.

Agate

Agate is a method of sandwiching together clays of contrasting colors to form a marbled effect. The marbling goes right through the clay.

1 Sandwich together several layers of contrasting colored clay. Roll and press firmly.

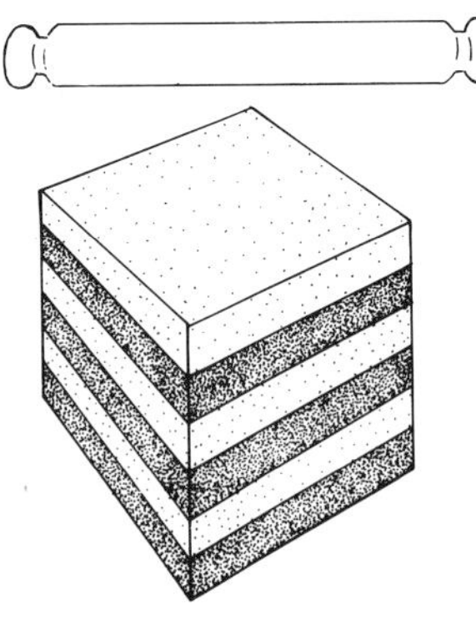

2 Cut in half and place one section on top of the other. Now roll and beat out the clay again.

3 Keep repeating this until the contrasting lines are very fine.

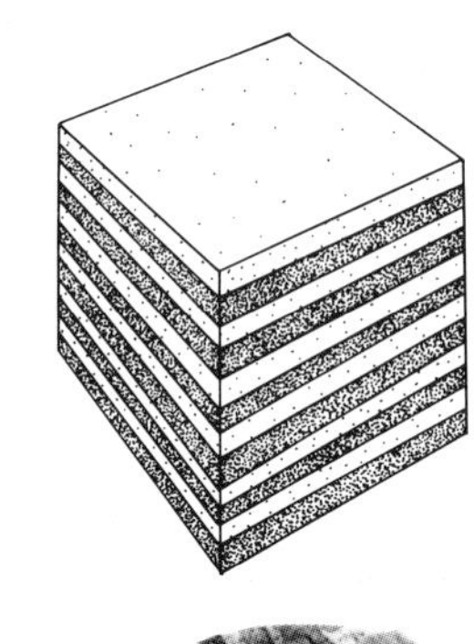

4 Roll out the clay into a slab ready for forming or decoration if required.

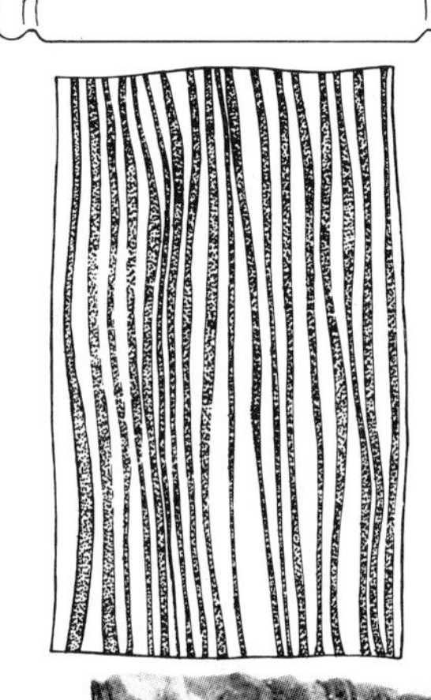

Millefiore

Millefiore employs a similar method but the form is built up from small sections cut from prepared laminations.

1 Different colored clays are rolled around each other and/or joined together so that when they are cut through and rolled they make small multicolored cross-sections.

2 Construction is slow and requires great care. Each piece must be joined to the next with slip, keeping the whole damp to allow maximum penetration of moisture to each piece.

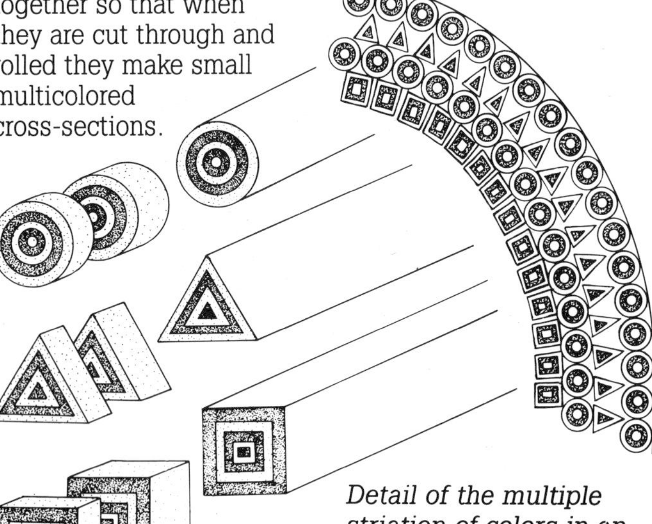

Detail of the multiple striation of colors in an agate piece by Virginia Mitchem (USA)

Agate and laminated techniques

Light and dark clays or differently colored bodies can be part wedged ready for throwing.

The pattern appears as the leather-hard surface is scraped and turned.

More controlled patterns are achieved by the careful assembly of different colored clays.

1 The basic bodies need to be the same if you are to avoid differences in shrinkage between the slices.

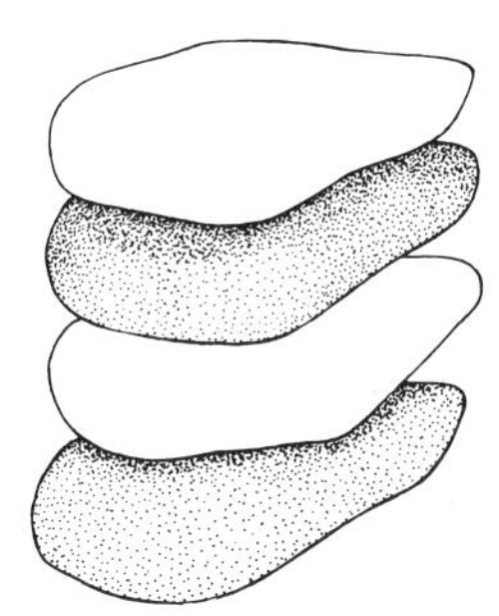

2 Roll out thin sheets of clay and cut them into narrow strips.

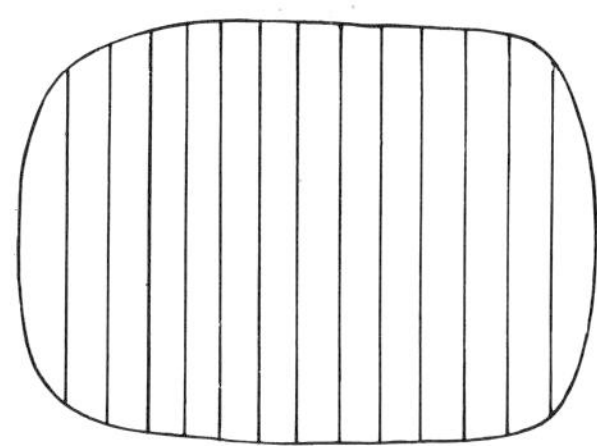

3 Layer these different colors, sticking them together with slip.

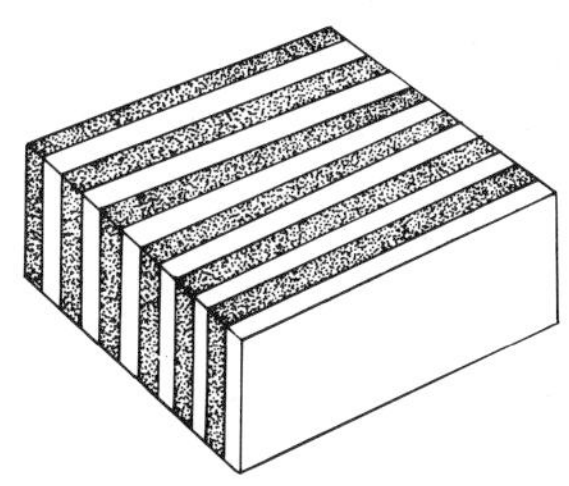

4 Roll the laminated slices into a slab.

5 The laminated slab can be placed over or into a mold to make a pot. Dampened biscuit molds are best for this kind of building; plaster is very absorbent and dries the clay too quickly. Drying must be accomplished in an airtight plastic box to slow down drying and shrinking. In this way the sections will be able to soak evenly, mature, and become fully integrated.

Bowl
by Robin Hopper (Canada)
Multiple colored clay creates fascinating swirls on this neriage bowl

Inlay on a flat slab

This is often used prior to slab building or dish molding.

1 Roll out a slab of clay of one color and thin slabs of another color (or colors). Cut shapes from the colored slabs and lay them in the pattern required on the main slab.

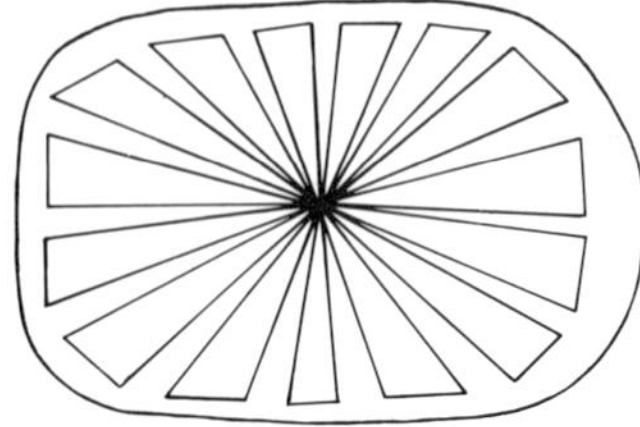

2 Cover with a clean cloth and roll the pattern into the slab until the surface is level.

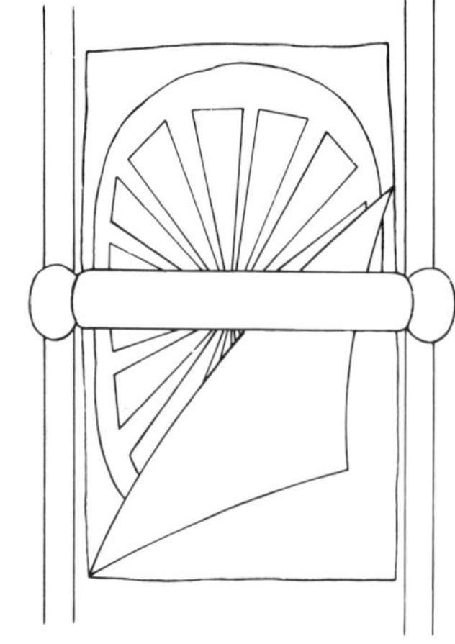

3 Press in small additions afterwards if needed. The decorated slab can then be formed into a pot or dish.

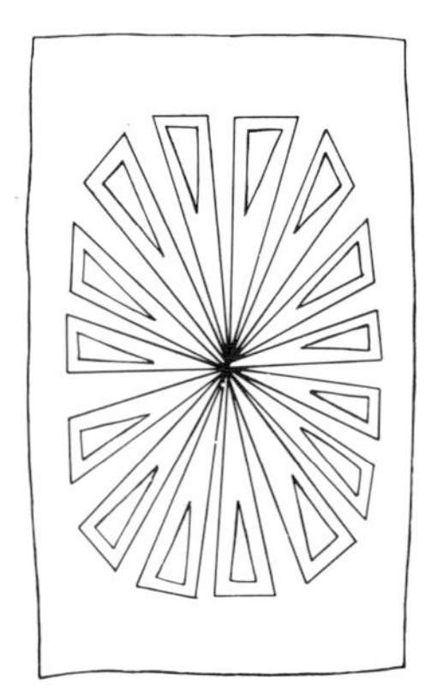

4 Yet finer designs may be made using the small syringe to extrude fine lines. These also can be rolled into the surface.

Molds for decorative coiling

There are many possibilities. Here are just a few suggestions:

Plaster mushroom or hump mold.

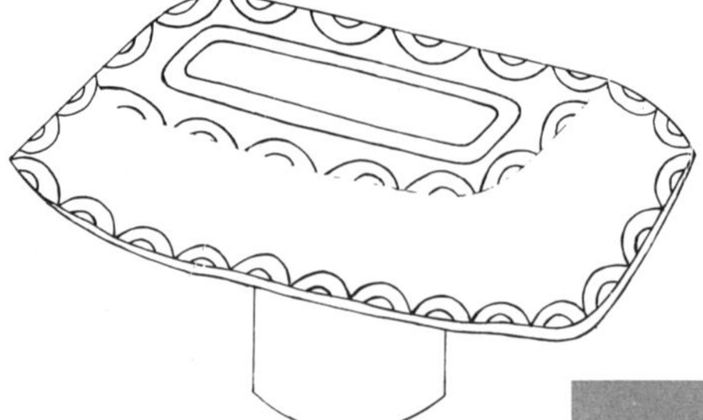

Plastic bowl lined with damp paper.

Pattern of coils built up on the mold.

Metal scraper to smooth the inside.

The dish must be removed from the mold while it is still damp and the coils smoothed together on the inside surface to form a tight seal.

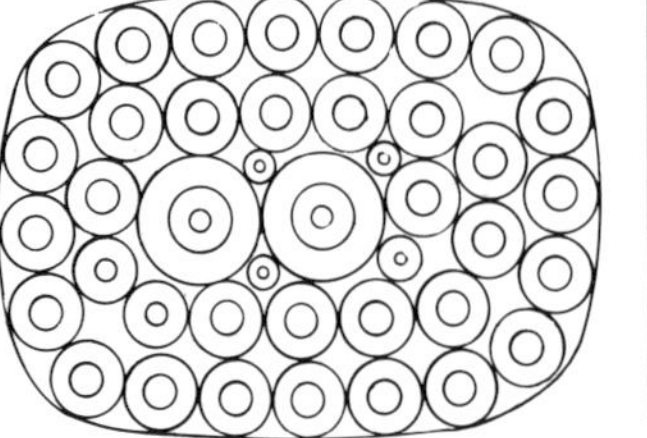

Teapot and cup by Yvonne Boutell (USA)
Color stains and oxides have been blended into porcelain clay. This is then inlaid into the surface of slabs prior to construction.

Applied decoration

Shapes cut out of thin sheets of clay can be beaten or textured before being applied to a pot with slip.

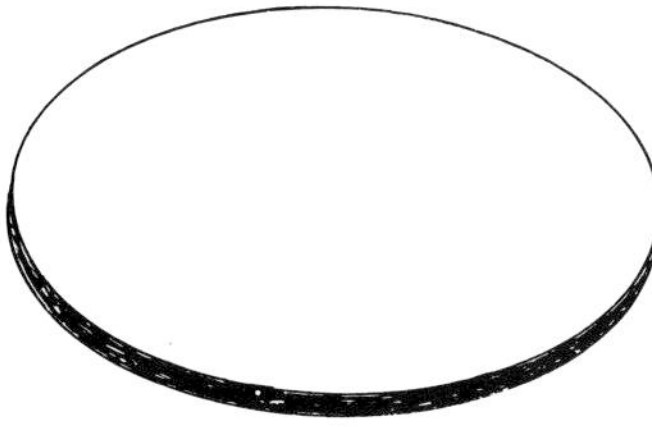

Thin sheet of clay is beaten with wood before being applied to the pot.

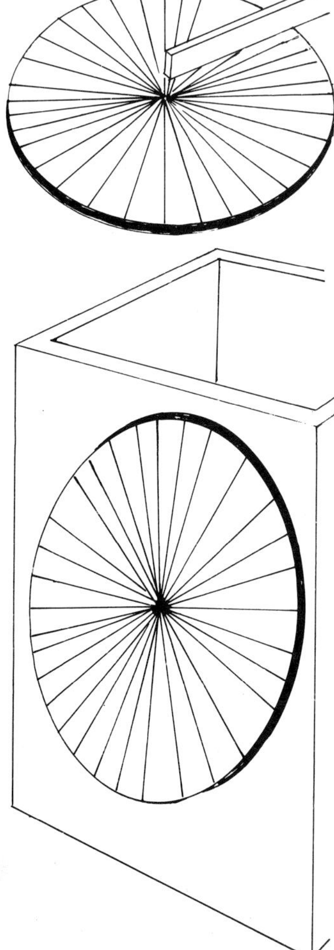

Thrown disc of clay is stuck on to pot.

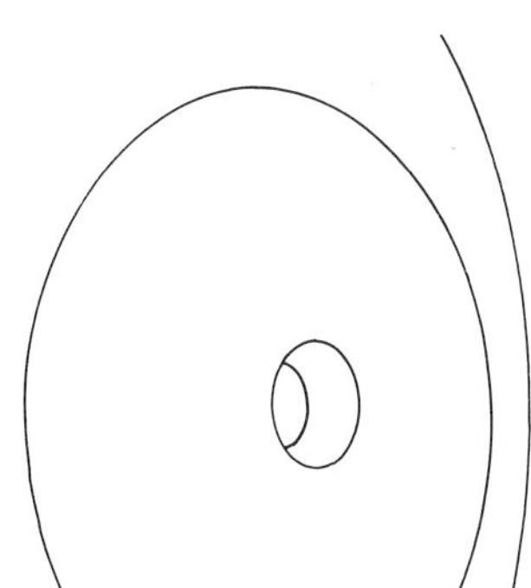

Texture scratched into surface with a fork.

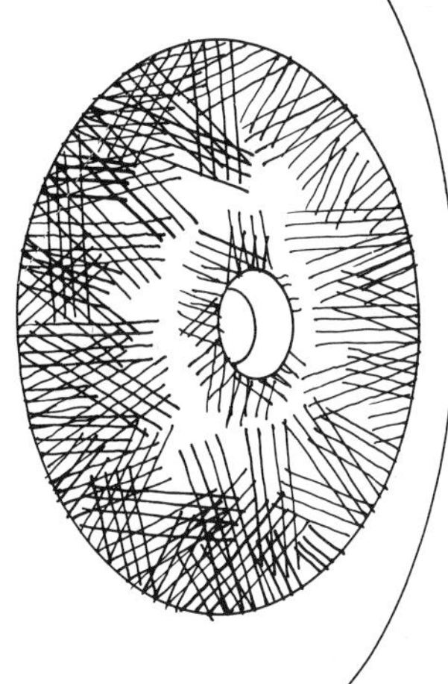

Color is added by rubbing in dry oxides at the bone-hard stage and then cleaning the highlights (or raised areas), leaving the color in the texture.

Teapot by Jenny Beavan (UK)
Hand-modeled decorative pieces have been applied to the surface of both teapot and base.

Coiled pot by Ian Sprague (Australia)
This thrown and coiled pot has been decorated with applied swirling coils of clay, their simplicity reflecting the 'primitive' feel of the pot's style.

Inlay designs

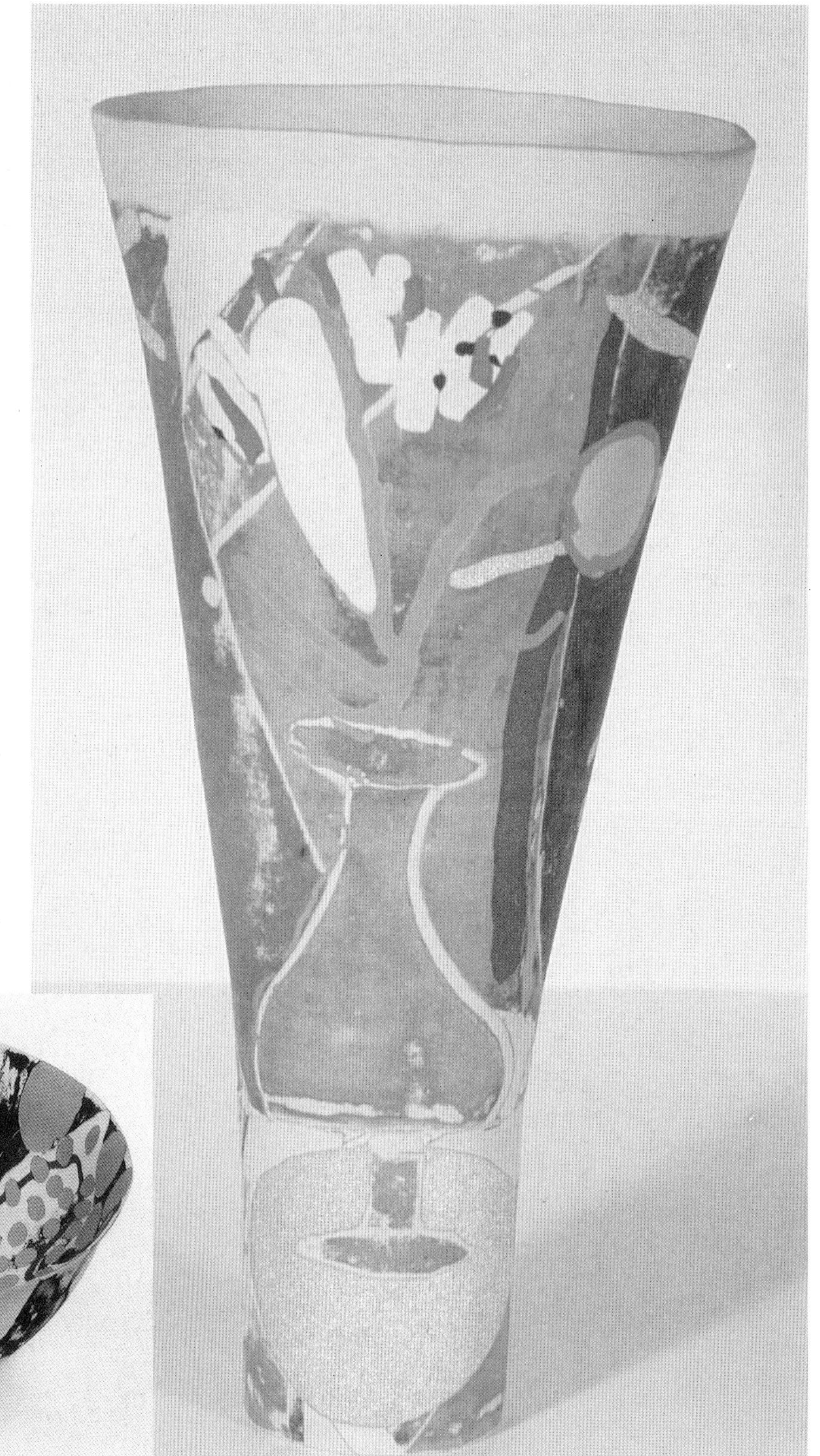

Vase and bowl by Susan Nemeth (UK)
The simple lines and delicacy of these pieces contrast with the bold decoration; this has been created with different slips and thin layers of colored clay used to form striking designs.

Folded bowl by Virginia Cartwright (USA)
Inlaid colored clay has been used to great effect on this brightly colored bowl. The striated pieces overlap in an irregular patchwork harmony.

Inlaid piece by Hiroe Swen (Australia)
Fine strips of contrasting clay sweep into fascinating 'liquorice' swirls inlaid into the surface of the pot.

Design and decoration options

Open porcelain bowl by Yvonne Boutell (USA)
A slab has been inlaid with oxide-dyed clay and then assembled into the bowl form.

Bowls by Peter Meanley (Ireland)
Semi-vitreous slips have been used to create a three-dimensional decorative effect.

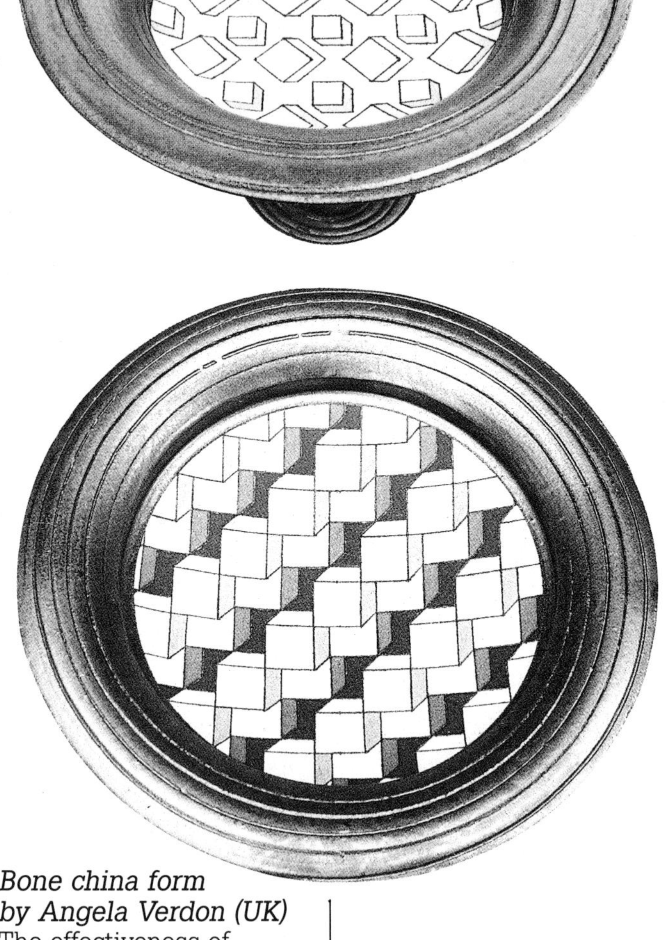

Bone china form by Angela Verdon (UK)
The effectiveness of applied black body stain and piercings is emphasized by their being restricted to the upper half of the pot.

Parrot by Tessa Fuchs (UK)
Choice of color is one of the most important decorative options and can greatly affect the impact of the finished piece.

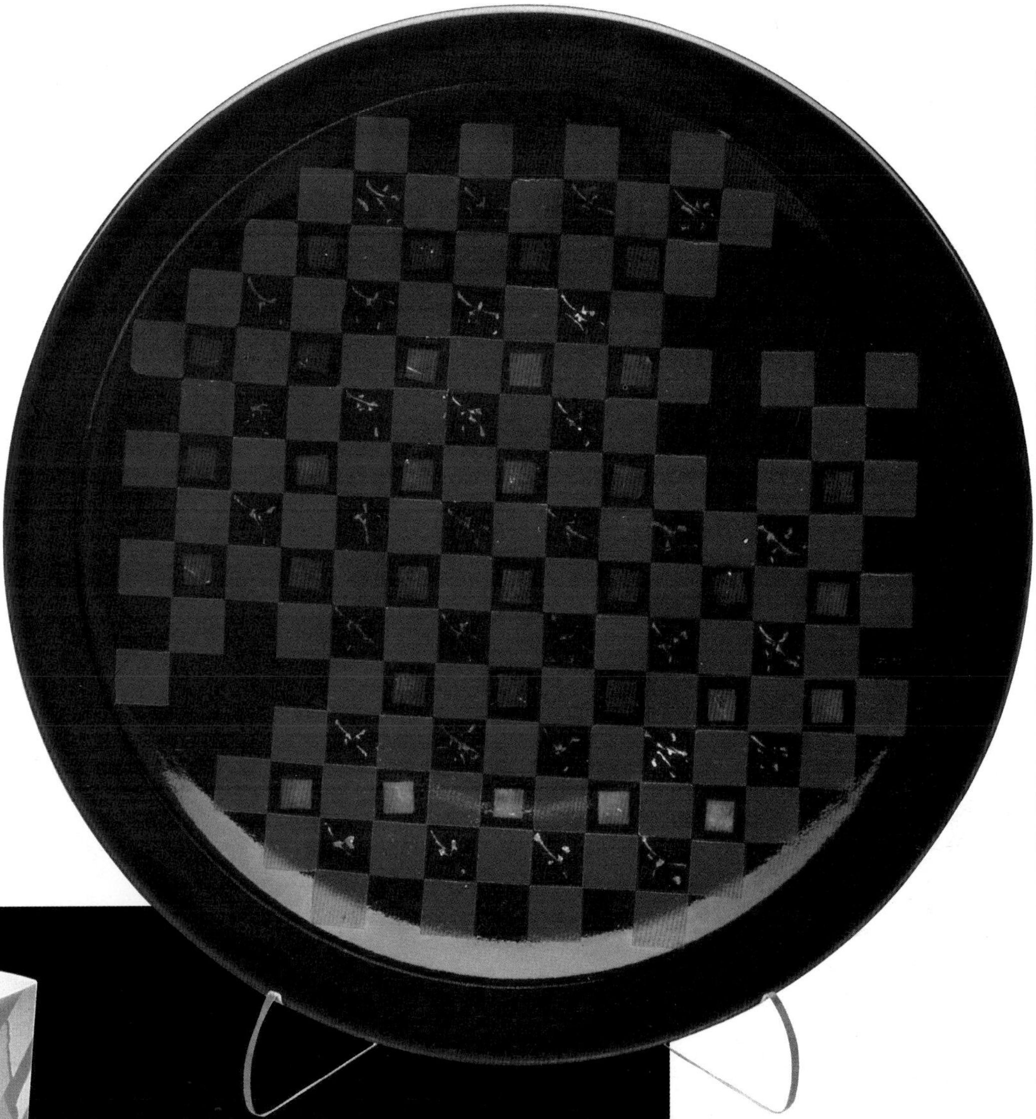

Platter by Greg Daly (Australia)
The bold color of the enamels and gold luster decoration transform a simple plate into an exciting object.

Slip-cast bone china by Sasha Wardell (UK)
Both the faceted surfaces and air-brushed color (applied through masks) add to the effectiveness of these lovely pieces.

Leeds pottery vase
An incized floral design decorated with brilliantly colored glazes.

18th century couple by Hilary Brock (UK)
On these amusing figures the decoration serves to add color and many fine details which contribute greatly to the fun of the final effect.

Handpainted ware by Andrew McGarva (UK)
In classic blue and white with peach, this tableware becomes lively and interesting through the color and subject matter chosen for decoration.

Reduced stoneware pot by Chris Myers (Australia)
A temnoku glaze has been partially removed by sand-blasting the surface.

Bowl by Jenny Beavan (UK)
Subtle color, intricate form and added details which resemble icing on a cake render this a very rich piece indeed.

Slip decoration

Platter in pink by David Pendell (USA)
Earthenware luster has been most effectively slip-trailed.

H See also pages 293 and 296 for examples of Early English slipware.

Two interesting examples of slipware by Godfrey E. Arnison (UK)

Using liquid clay of a different color to alter the appearance of a pot is one of the oldest methods of decoration. Even a coarse brown clay can be given a smooth light-colored finish with this method.

Slip can be dipped, painted, poured, sprayed, flicked, squeezed through a narrow nozzle (called a slip trailer) or used to fill a carved pattern as inlay.

Before being slip decorated, the surface of the pot should be hard enough to handle without distortion (unless it is well supported) and damp enough to absorb the slip slowly. If the pot is too wet, the slip can soak the pot and make it collapse. If the pot is bone dry, the water in the slip can soak into the surface and may crack it or, if the pot does survive to be fired and glazed, the slip decoration will often lift from the surface, taking the glaze with it. This is because the wet slip shrinks as the water content evaporates and adhesion is lost.

If strong bright colors are to be the dominant feature of your work, then a light-colored body or surface will show these colors at their best. The clay chosen could be a white earthenware, a light stoneware or a porcelain clay. Alternatively, the dark body can be coated with a light slip: this was the method used by most ancient civilizations before the use of glazes. Coating a dark body with slip will not always completely obliterate the body beneath; a dark body can burn through the coating at high temperatures and this can produce a different kind of light effect. The thickness or thinness of the slip, and whether it is poured, dipped, sprayed or brushed on to the surface, will also alter the finished result. Experiment with these methods until you achieve the best medium for expressing your design ideas.

Trailing

Slip trailing was used to its fullest effect by the seventeenth and eighteenth century potters in Staffordshire, England and most notably by the Toft family. The technique is to squeeze the slip on to the coated or uncoated surface of the pot through a fine nozzle fixed into a bulb holder made of clay, rubber or plastic.

A clay slip trailer can have several nozzles attached, depending on the thickness of the line required. The flow of slip is controlled by adjusting the thumb over the filling hole. If the trailer is made of plastic or rubber then the slip must be forced out by steady continuous pressure on the bulb to avoid air being sucked into the bulb and interrupting the flow.

The slip must be well sieved to prevent lumps blocking the trailer and at the right thickness for the design to be applied. Too thick a consistency and the slip will not flow freely from the trailer; too thin and the slip will run down the pot as soon as it is applied. The design can be built up gradually from lines and dots, using different colors contained in separate trailers.

It can either be applied directly to the body surface or to a coating of slip; the coating must be allowed to dry until the shine has dulled before trailing begins. A design can be drawn out lightly with a pencil or painted with watercolor or pricked through with a pin on to the pot.

Practising the trailing method

You need to control the trailer confidently to produce the flowing line that is the essence of slip-trail design. A flat slab of clay rolled on a board or a slab laid in a dish mold will provide suitable surfaces on which to practise. Both bases will be firmly supported: trial attempts can be washed off several times over to build up experience.

After each trial, wipe off the slip with a sponge and clean water and smooth the surface with a rubber kidney to press down the grit which has been brought to the surface by the sponging. The freedom of being able to wipe the surface clean if the design goes wrong will increase your dexterity and eventual mastery of the technique.

Various slip trailers

Supporting the pot

Decorating the outside of a wide bowl is easier if the bowl is held upside down so that the surface slopes towards you.

To trail on bands of slip, place the pot on a banding wheel. Hold the trailer against the side of the pot leading away from you, turn the pot and gently squeeze the trailer at the same time. The hand should be held steady — perhaps supported by the edge of the table, or by the other hand, or by a stick held under the arm. The end of the trailed line should then connect up with the start of the line as the pot revolves.

Some potters decorate their pots immediately after throwing while the pot is still centered and attached to the wheel so that the pot and the decoration dry together.

Slip spotting

A means of applying thick slip in spots from the end of a tool which has been dipped in slip. Decoration is built up by the variety and choice of spotting tools. You can use, for instance, the ends of a wooden dowel, or strips of wood or metal for lines, while carved pieces of wood will allow you to make a more precise design — whatever the tool, the weight and intricacy of the pattern must complement the forms. The technique is a gradual build up of dots and lines of color in a controlled pattern.

Decorating a hump mold

A slab of clay big enough to cover the mold is decorated with slip while it is flat and left until the slip is hard enough to make contact with the mold without smudging. The slab is then placed slip side down over the drape mold.

Trailing on to a thrown shape

1 Trailing on to an upright surface rather than a flat dish requires a different technique. For instance, when squeezing a liquid great care must be taken to ensure liquid does not run down.

To band lines around the pot, first center the pot on to a small banding or kick wheel. Then fill the slip trailer. (Use a rubber bulb-type trailer.) Squeeze the bulb flat, immerse the nozzle in the slip and allow the suction inside the bulb to suck up the slip until it is full.

2 Test the flow of the slip on the bench before you start working on the pot. Then spin the wheel anticlockwise with the left hand and hold the slip trailer close to the surface of the pot, with the nozzle pointing downwards. (If the nozzle is pointing upwards the slip will splatter and destroy the decoration.)

Some shapes benefit from being placed on the wheel upside down. For example, a shallow bowl will be easier to decorate this way round. Otherwise you will have to contend with the awkward angle of an upwards sloping surface. Moreover the slip trailer will run dry and may splatter — there is every chance that the design will be spoilt.

3 Squeeze gently until the slip starts to flow. As the wheel turns and the pot revolves, hold the hands steadily in one position so that the line of slip on the pot will start and finish at the same point.

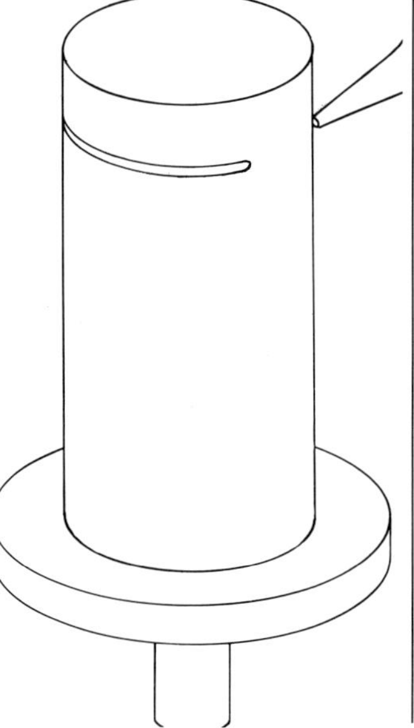

4 Further decoration can be added by placing the hand inside the pot and lifting it into a horizontal position. This will allow you to continue decorating the surface by revolving the pot on one hand as you squeeze the slip with the other. The decoration can be freely drawn while the pot is held steady.

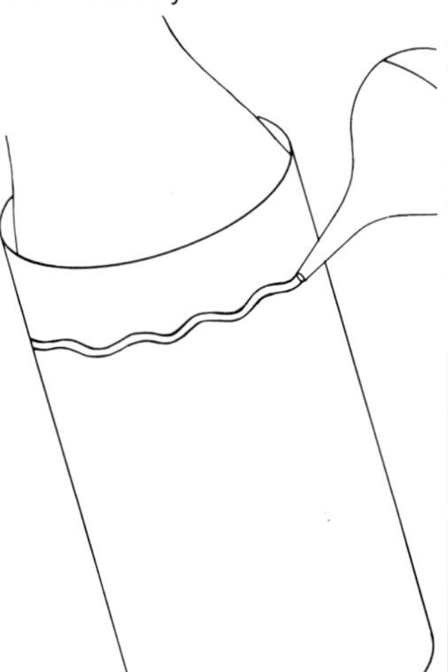

Decorating and making a slip-decorated slab

1 Flatten a slab of clay on to a board. Cover the slab with a coating of colored slip (for example, white). Allow the coating to dry until the shine goes.

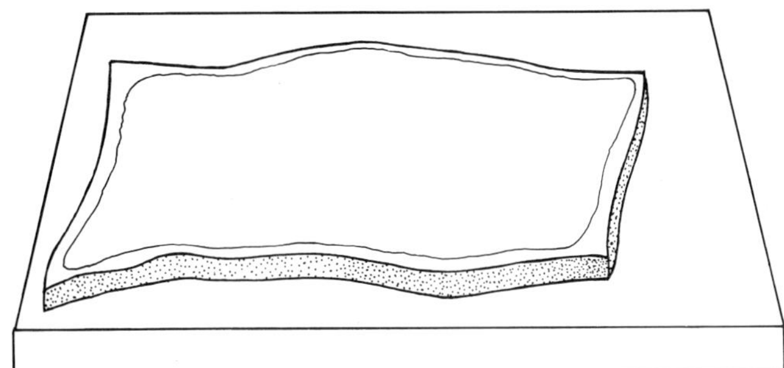

2 Trail on a pattern of lines and dots. A sharp tap on the edge of the board will flatten and broaden the lines, softening them effectively.

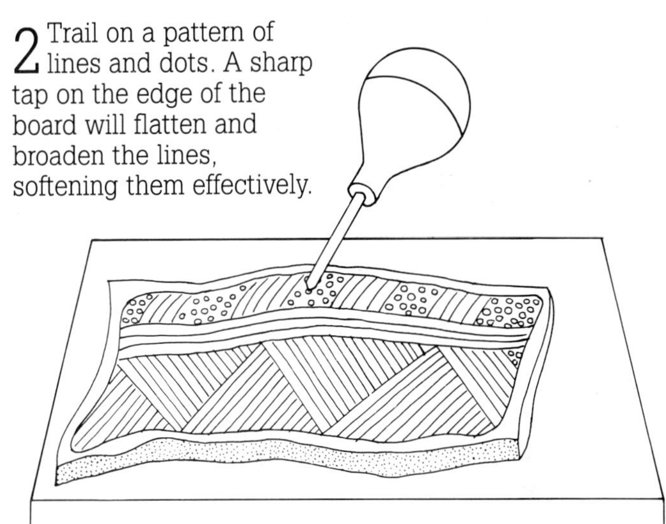

3 Let the slab dry until the slip can be touched without smudging. Place a hump mold in the middle of the slab upside down. Slide the hand and arm under the slab. Then lift both slab and mold together and turn them both over.

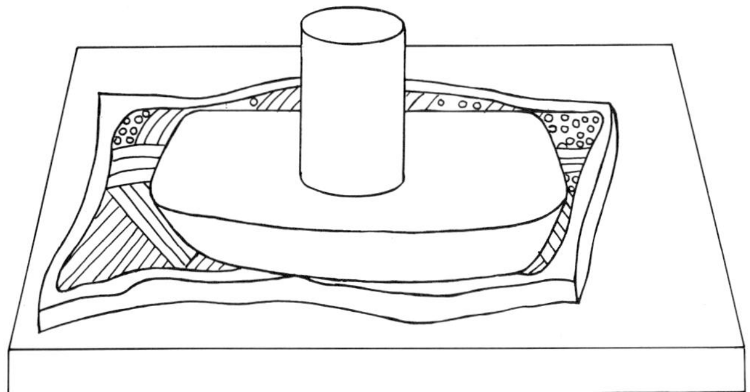

It is obviously much easier to trail patterns and lettering designs on a stable flat surface, such as a slab. Use this opportunity to create an exciting design of letters, lines or patterns.

4 Flatten the clay on to the mold with a damp sponge and rubber kidney. Trim off the excess clay around the edges with a wire, knife or wooden tool and leave to harden.

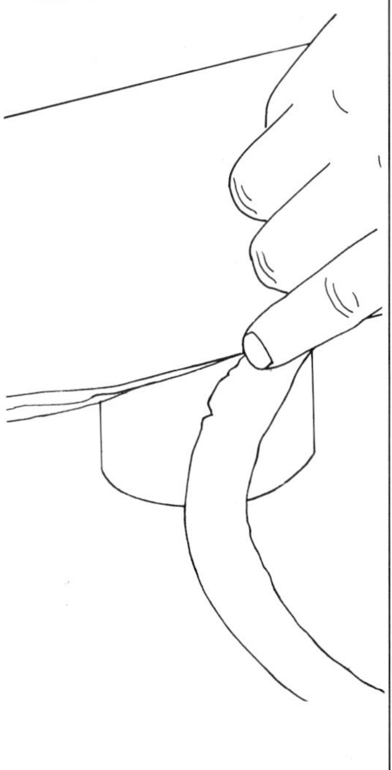

5 A foot ring can be fixed to the base of the dish while it is drying upside down. Make the foot from a strip of clay, or use a coil.

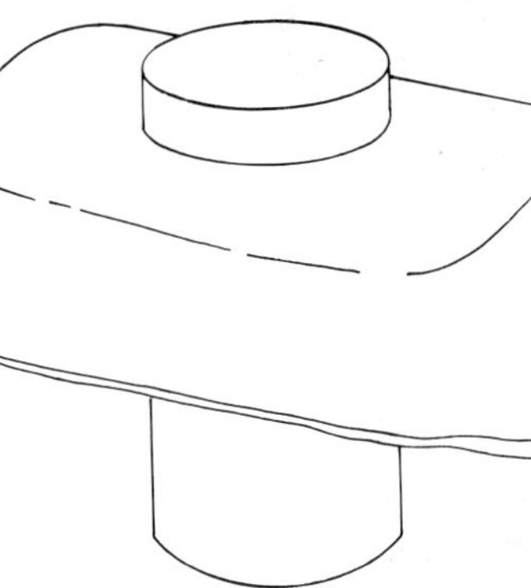

Remove the dish from the mold before it becomes too dry; the stiff leather stage is the best time to do this. Clean the edges with finger and sponge, before leaving to dry completely.

Dipping and pouring

For this method to be successful the slip must be well sieved (through at least an 80-mesh sieve) and well stirred. It will settle if left for even a short time, leaving a thin watery mixture at the top of the bucket. The slip should be the consistency of thin cream. If it is too thick it will cling to the surface in uneven patches, masking the shape of the pot; if it is too thin it will run down the surface and the body will show through.

A dipping procedure has to be devised according to the size and shape of the pot. If the pot is small, for example, it can be held with the thumb at the rim and a finger at the foot and dipped in one movement—but beware that you do not hold it too tightly and break or distort thin rims when using this method. If you dip the hand into the slip first, before picking up the pot, this will help to prevent finger-marks showing when the dipped pot is placed on the table (after it has been held for a few moments to drain). If the pot is large, there are several ways to cover the surface; these may combine dipping and pouring. If you can hold the pot easily by its foot, pour slip inside and then tip out the excess slip while turning the pot so as to coat the whole of the inside. Still holding the pot by the foot, turn it upside down and dip it straight down into the slip to cover the out-side; the air trapped in the pot will prevent the inside receiving a second coating. (Just the second stage of this method of dipping can be used if the inside is to be left unslipped.)

If the pot cannot easily be held by the foot, pour the slip inside, supporting the pot with the palms of the hands until the slip has dried a little. Then prop the pot upside down on two flat sticks across the slip bucket and pour over more slip, leaving the pot to drain and harden until it can be picked up without your making fingermarks. The rim will probably require touching up to cover the marks left by the supports. The pot can also be placed upside down over a narrower object, such as a wire or plastic mesh cakestand, which will support the pot and let the excess slip fall through to the bucket.

To dip a dish or plate, put the slip into a wide shallow bowl, hold the plate at the rim on each side and slide the plate into the slip with a rocking motion from side to side or from front to back. This should be done swiftly but smoothly.

Part-dipping creates an interesting contrast between dipped and undipped surfaces. A jug can be dipped to cover the handle only or dipped so that just the spout and the top of the shoulder are covered. Fat middle parts of the pot can be dipped separately by holding the foot and rim and pushing the pot down sideways to make a circle of slip on each side or on four sides. There are many techniques which can be improvised to suit the shape of your pot and the decorative effect you are attempting to produce.

If a molded dish is to be covered in preparation for trailing a design or if it is going to have a paper (or wax) resist pattern added, then the slip may be poured while the clay is still wet and still in the mold. The mold will give it some support while the slip dries and further work is added. Pour in a small amount of slip, lift the mold with the dish in it, tilt it to cover the whole of the inside surface and shake out the excess. Place a thumb on the edge of the dish at each side when you shake out the excess or the shaking process may also shake out the dish itself! As the slip dries it is wise to cut through the film of clay which bridges the junction of the dish and mold. This will reduce the risk of cracking or splitting when drying and shrinking.

Feathering

You can add to a basic trailed line pattern by drawing a stripped feather spine, a broom bristle or needle through the lines in alternate directions. Take care not to scratch the surface by digging too deeply into the body as you draw the needle through the slip. These soft feathery lines are totally different from the lines drawn in a hard slip. The areas around the drawn lines may be colored with oxides or stains after the slip has dried and further lines cut through the painted areas. Some of the lines can be filled with a different colored slip in the form of inlay.

Marbling

Trail or pour lines of contrasting slip into a molded dish, lift the mold and shake or twist it to mix the colors into a variegated pattern. The colors will stay separate unless the shaking continues for too long and mixes them into a muddy color. If this happens, wash the slip out and start again; don't be satisfied with the first marbled pattern that appears; keep trying until an element of control appears in the pattern of lines.

Trailed lines on a wet slip background have a soft fluid appearance and the dish can be tapped gently to flatten and spread the lines slightly. If the lines are trailed when the background is harder they will remain slightly raised, separating areas of color.

If a line is squeezed on to a wet slip surface by holding the trailer close to the surface and using a strong pressure, the slip will be forced under the surface coating. If the trailer is held away from the surface the line formed will spread into a wide panel.

Variations of trailed and poured slip

Slip can be poured from a jug across the surface of a dish in wide splashes of overlapping color. The contrast between slipped and unslipped surfaces may be all that the pot requires in terms of decoration—or more detail can be added with a fine brush.

Try covering the surface with a coating of slip and while it is still wet squeeze large pools of a different color along the edge of the dish dotted about around one half of the circumference. A second layer of contrasting colored pools can then be squeezed on top of the first. Lift the dish and shake the pools of slip across the length of the dish; the colors will not mix but will form fine line patterns across the dish. The effect of this shaking can be spectacular.

Painting

To produce a sharply defined pattern or a streaky surface the slip can be painted on to the pot. It can also be brushed over impressed decoration to create more contrast. Use a big soft mop brush, well filled with slip, and make bold simple strokes. Lines of color are best painted while the pot is turned on either a throwing or a banding wheel: support your painting hand, start the wheel and work the brush steadily against the surface. If the coating is uneven it can be smoothed out and spread with the wide end of a feather or even the fingers. The feather can also be used to merge the lines into an even coat if the slip is trailed on in circles with a slip trailer. This method of trailing and spreading works well on newly thrown bowls before they are cut from the wheel.

Sometimes one brushed coat will not be enough to cover the surface completely but always allow the slip to dry until the shine goes off before repeating the process—unless you are hoping to create a random texture with varying tonal changes. A thin coating can be the basis for a freehand design: build up layers of slip using the brush; experiment with various colors and techniques.

Painting and coating with fine slips: terra sigillata

One of the first forms of clay decoration, slip has been used for some six or seven thousand years. It was implemented and perfected by the Greeks and Romans who were able eventually to produce a form of fine slip which shone when dry almost like a glaze.

Combing

Apply a thorough coating of slip to the pot and while the slip is still wet draw your fingers through it to reveal the clay underneath and make a wavy pattern. Use card combs with different-sized and shaped teeth or try another color of slip once the first has dried slightly.

Zuni pot from New Mexico
The Pueblo Indians traditionally covered the whole surface of a pot with painted decoration on a slip background.

Dipping

1 Full dipping must be done when the clay is leather hard.

Coat the entire surface of a pot by immersing it briefly in a bucket of slip. The pot must be leather hard before dipping takes place, otherwise it will absorb too much slip and start to collapse.

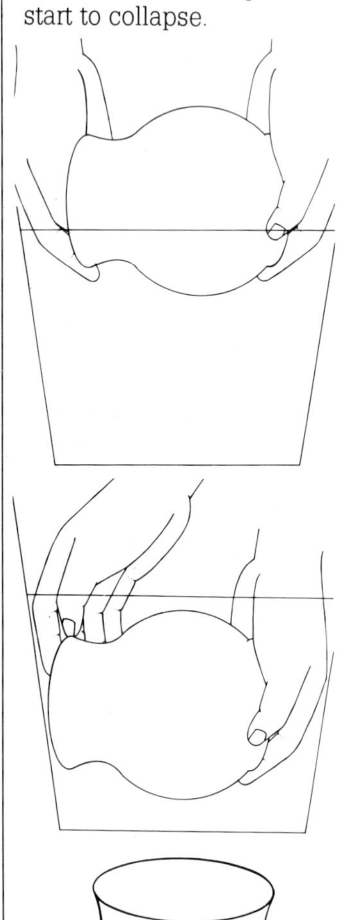

2 If outside only is to be covered, pot can be immersed upside down into the slip.

Slip does not go inside the pot if the dipping is straight down—the trapped air inside stops slip entering. Hold the pot over the bucket for a short time. Twist backwards and forwards to shake off excess slip, before turning upright.

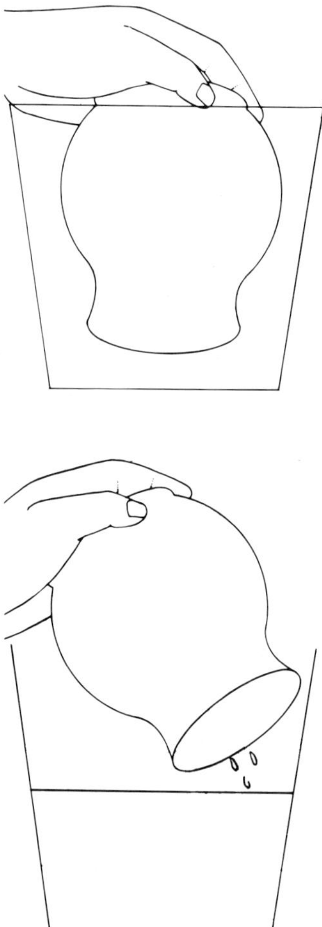

3 Patches of slip form their own curves if the sides are partially dipped at any angle.

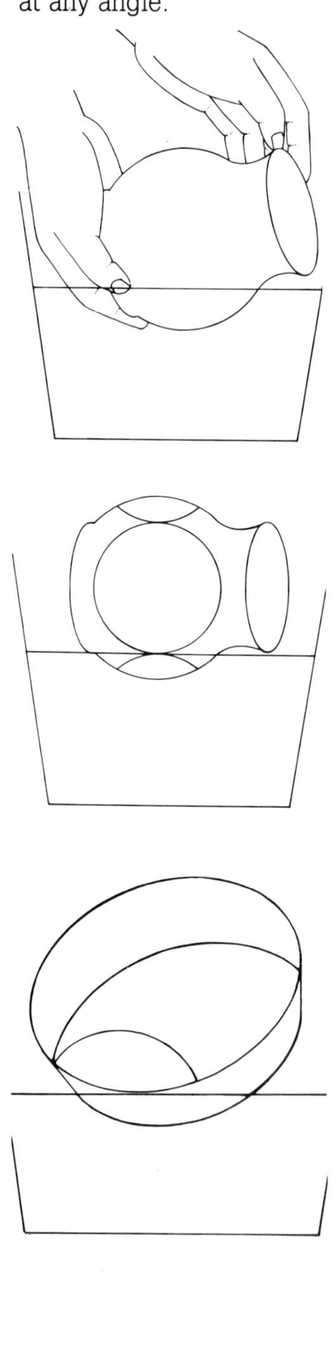

4 Combs or fingers can create a swift flowing pattern if drawn through the wet slip.

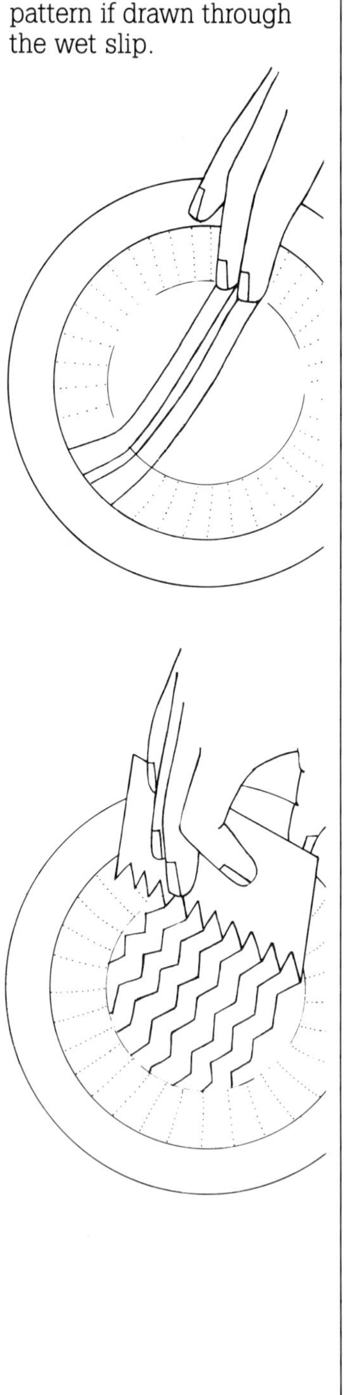

Pouring

This can be a way to decorate or cover completely the inside and outside of a pot. It is often used by potters who have access to a dark-colored clay only and need to create a light background for decoration. As a purely decorative process, pouring in lines and swirls with a spoon or ladle quickly creates a pattern across a dish or bowl. This can be added to by trailing and pouring more slip, scratching through, painting with oxides and so on.

1 To completely coat the inside of a pot, pour in a generous puddle of slip

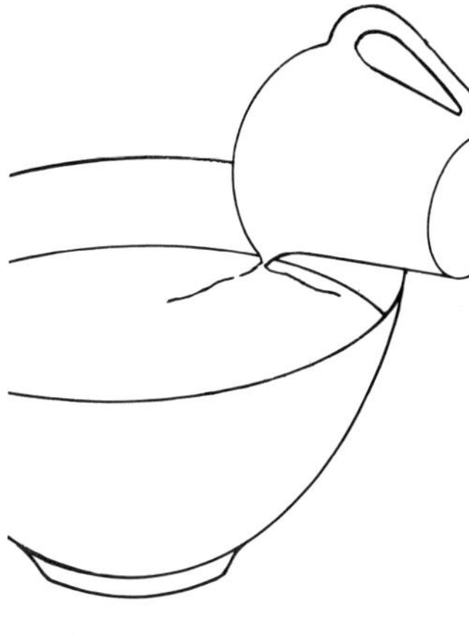

2 Revolve the pot so the inside is covered completely and then drain over a bowl with the pot supported on two sticks.

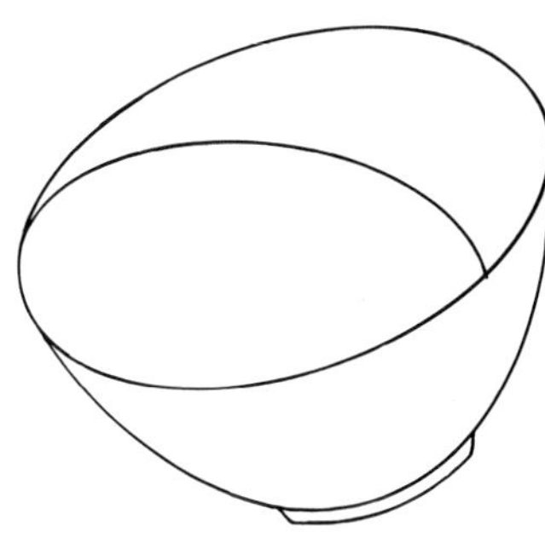

3 Make sure there are no pools of slip left in the bottom of the pot and then leave it to stiffen slightly as the slip will have softened the clay. The coating of slip can then be poured over the outside.

4 Leave the pot to drain and harden on the sticks. Do not touch it.

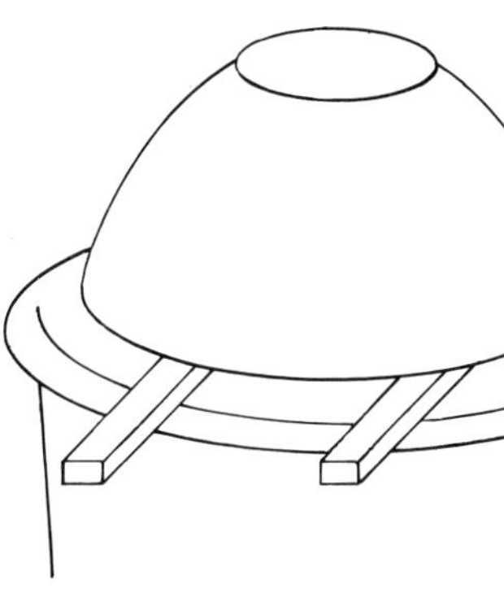

A wet-slip surface can be decorated by adding lines or blobs of slip to the wet surface and shaking or twisting the pot to create a marbled effect. This must be carefully controlled (too much shaking will result in a muddy haphazard pattern).

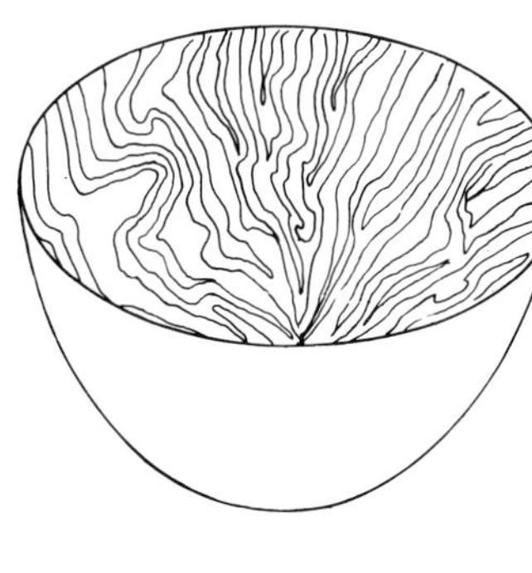

Hold the pot upside down. Pour wide ribbons of slip down the sides—straight from the jug. This needs practice, you may well have to wipe off your efforts a few times before achieving success.

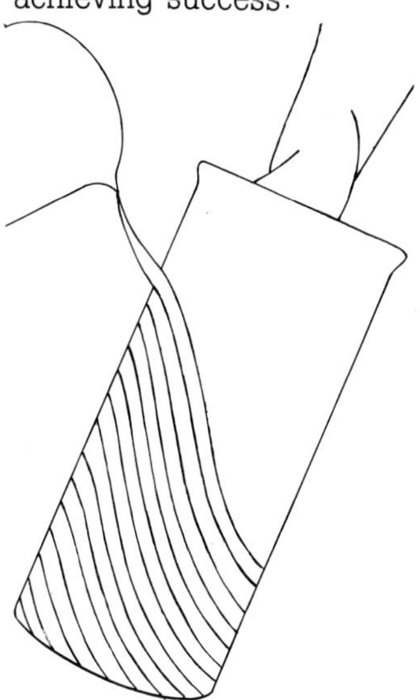

A ladle which has a single hole in it or one with a lip makes a good slip pourer for running broad bands of color across a dish. The quicker the movement, the thinner the lines.

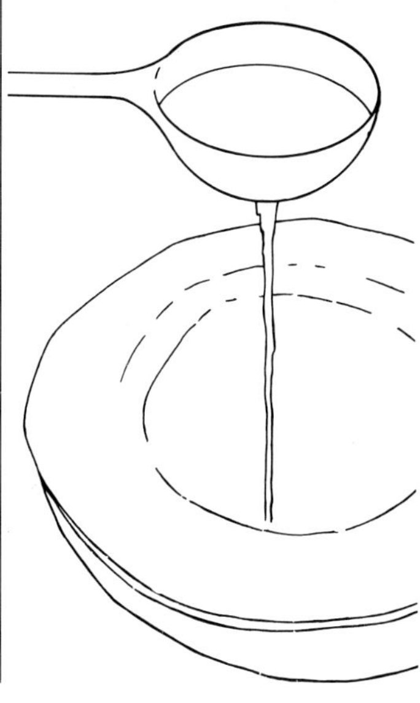

Feathering

A slip-coated surface with a pattern of close lines can be cross feathered. This is a very effective form of decoration. A stripped feather spine (or a stiff brush bristle) can be pulled through in opposite directions making this distinctive pattern. Some potters use a knife edge or piece of wire but these can be a little too stiff to manipulate easily.

Trail lines across the dish of slab and then weave one color through another by drawing the feather in alternative directions across the design to create a wave-like pattern.

1 Feathering is most readily applied to a flat sheet of rolled clay or on to a dish supported in a mold while the clay is still damp. Extra fluidity is possible if the surface has received a fresh coating of slip.

2 Carefully trail a series of parallel lines of slip across the surface of the clay. Make sure the trailer's nozzle is kept just above the wet slip surface.

3 It can be effective to trail a second series of slip lines, in a contrasting color, between the original lines of slip.

4 Tap the slab or mold, holding it up at an angle, so that the trailed lines sink into the surface.

5 Using a sharpened feather point or a fine needle mounted in a cork, lightly draw the point through the trailed lines at right angles to these. Be very careful not to cut into the clay and work fairly quickly before the slip becomes too firm to achieve a flowing effect.

6 At the end of each stroke, wipe the needle clean or the build-up of slip will reduce the fineness and definition of the lines.

The idea is a simple enough one but a great variety of patterns can be achieved.

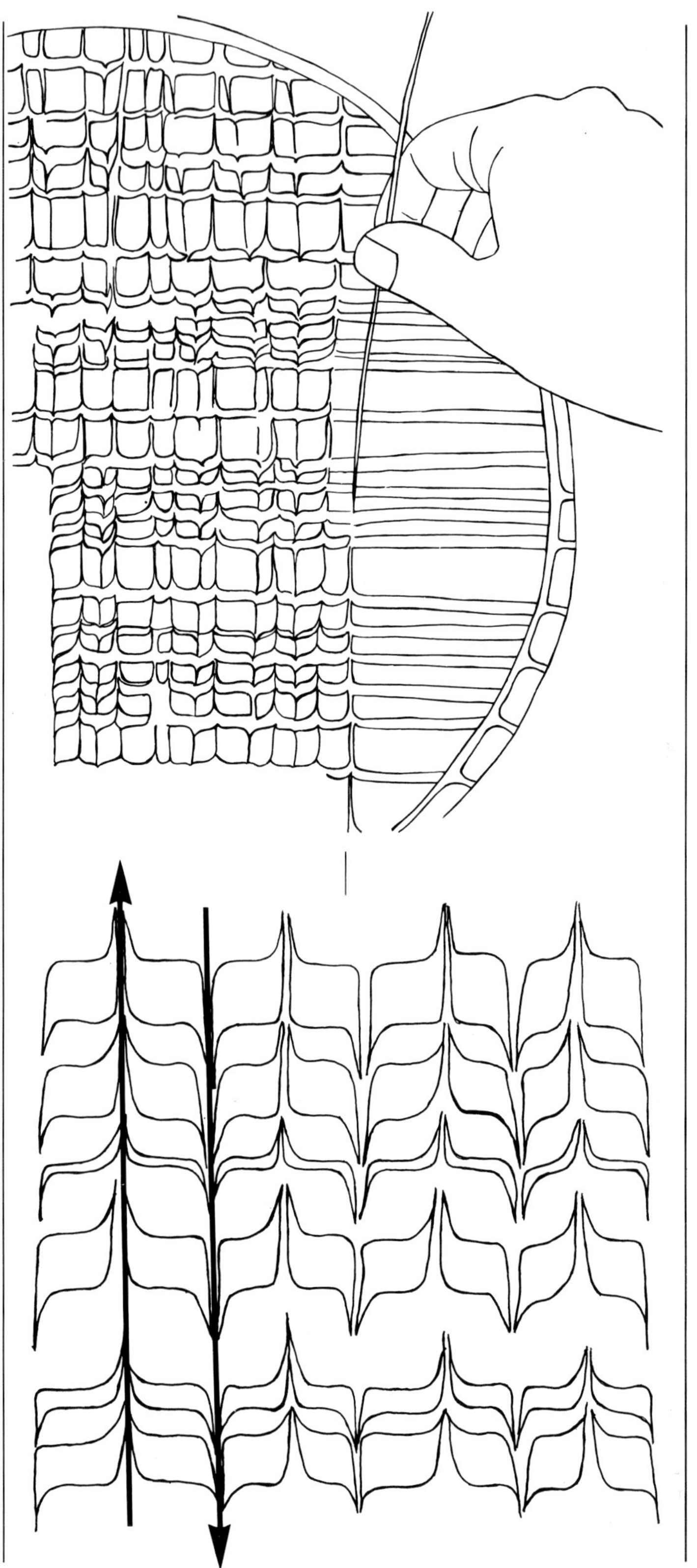

Marbling

After trailing a line pattern on to a fluid slip background, carefully shake the slips together to create a marble-like effect. Do not shake too much or the pattern will become over complicated or even muddy in appearance.

This marbling is difficult to do on a thrown dish without any support — the slip surface will soften the clay and it may collapse or start to distort. If the design has to be washed off, do this carefully with a sponge and leave the pot to harden before starting again.

Marbling is essentially a slip process, an intermingling of different colors that creates a pattern on the surface of the clay. (Similar patterns can be created with inlay and agate but the changes of color are then 'three-dimensional' and an integral part of the whole clay body, not just a surface effect.)

1 First coat the surface of the pot with a fairly thick layer of slip (thicker than would be required for normal trailing but not so thick that it later develops cracks.) Trail lines of a contrasting color across the piece in a random design.

2 Alternatively, blobs of different colored slip can be poured randomly on to the surface. These will not produce the fine lines possible from trailed slip but will create broader areas of color.

3 Tilt up the dish or slab. (If you keep the piece horizontal you will merely end up with a pool of thick slip in the base.) Shake the pot in various directions so that the slips will swirl and mix together.

4 The lines that are parallel to the direction of the tilt will become straighter and narrower; those that are at right angles will spread out into broader areas of color. Stop as soon as you have achieved a pleasing result. Too much shaking will spoil the definition of the lines and the patterns will become muddled.

Interesting patterns emerge when marbling. These depend on both the original position and contours of the trailed slip and on the movements used to intermingle the colors.

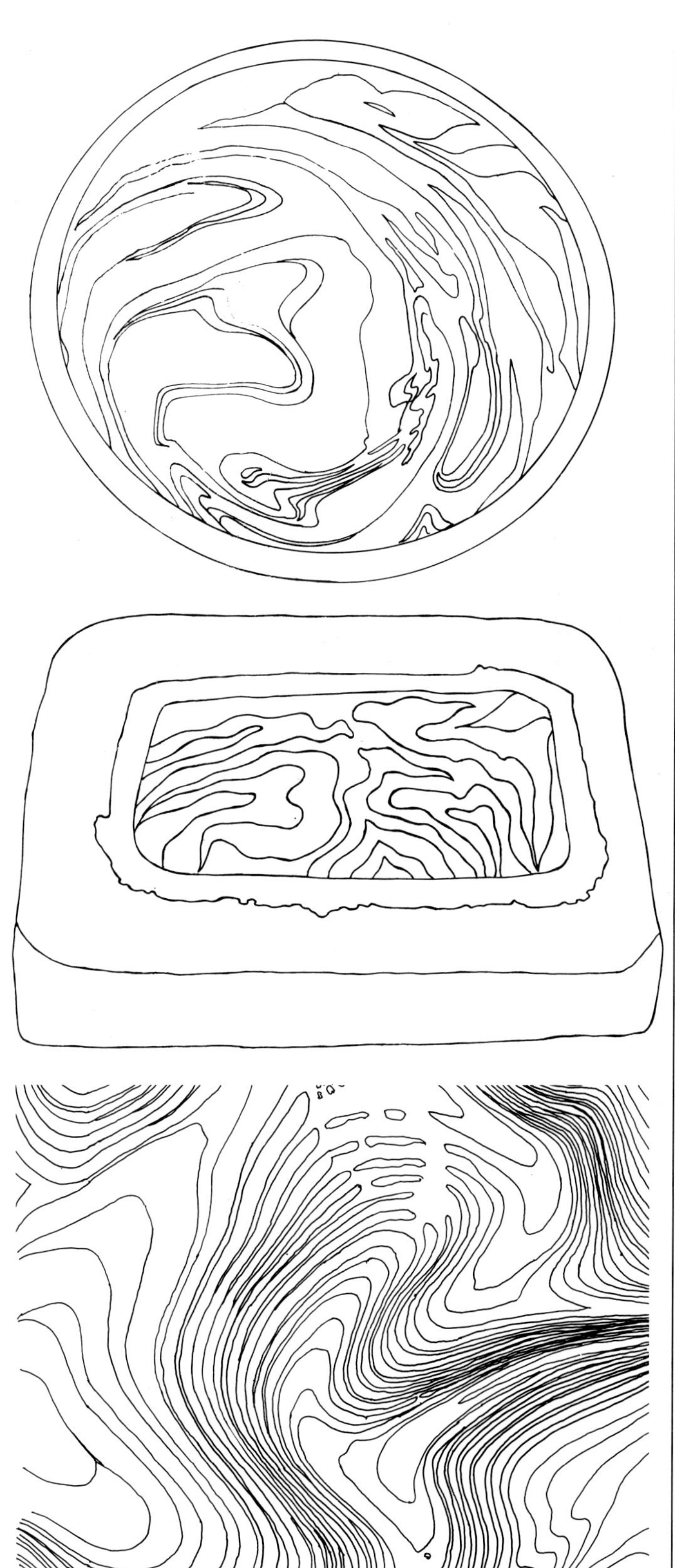

Slip trailing a dish

A dish that has been made in a mold can be supported by this while the decoration is put on to the still damp surface. The design can be trailed directly on to the clay but make sure this has been smoothed with a rubber-shaped kidney, which presses any grittiness down in to the clay and leaves a smooth surface ready for decoration.

If the decoration goes wrong, the work can be cleaned away while the clay is still supported in the mold and cannot collapse. The clay can then be smoothed again with the rubber kidney and decoration recommenced. This can be repeated many times.

The untreated damp clay surface can be trailed, cross-hatched, or filled in to create numerous designs. The slip will not run but will stay slightly raised on the surface of the dish. The design sinks into the clay as it dries but leaves a slightly bumpy surface when the pot is finished and glazed. A more fluid type of decoration can be achieved by coating the inside surface of the dish with a different coating of slip.

C see page 176-9

1 Pour the slip through a sieve into the dish — this is to ensure there are no lumps in the slip. Move the mold around until the whole inside surface is covered and then pour out the excess, leaving the mold slightly tilted to let the slip drain well. (Otherwise a pool of slip will run back into the bottom of the mold, resulting in uneven distribution.)

2 Allow the shine to disappear from the surface of the slip coating It must become slightly dull before commencing the next process. Choose two contrasting colors of slip and work around the edge of the dish, squeezing on large blobs of one color, (say, white).

3 Now, with a separate slip trailer (possibly containing black), put a blob of contrasting color into the center of the first set of blobs.

4 Gently tap the side of the mold or shake it until the two colors start to run down the surface to the center. They can be twisted or shaken in a circular motion. You will find that the two colors do not mix but will form very fine lines. (They mix only if the mold is shaken too vigorously — resulting in an unpleasing muddled effect.)

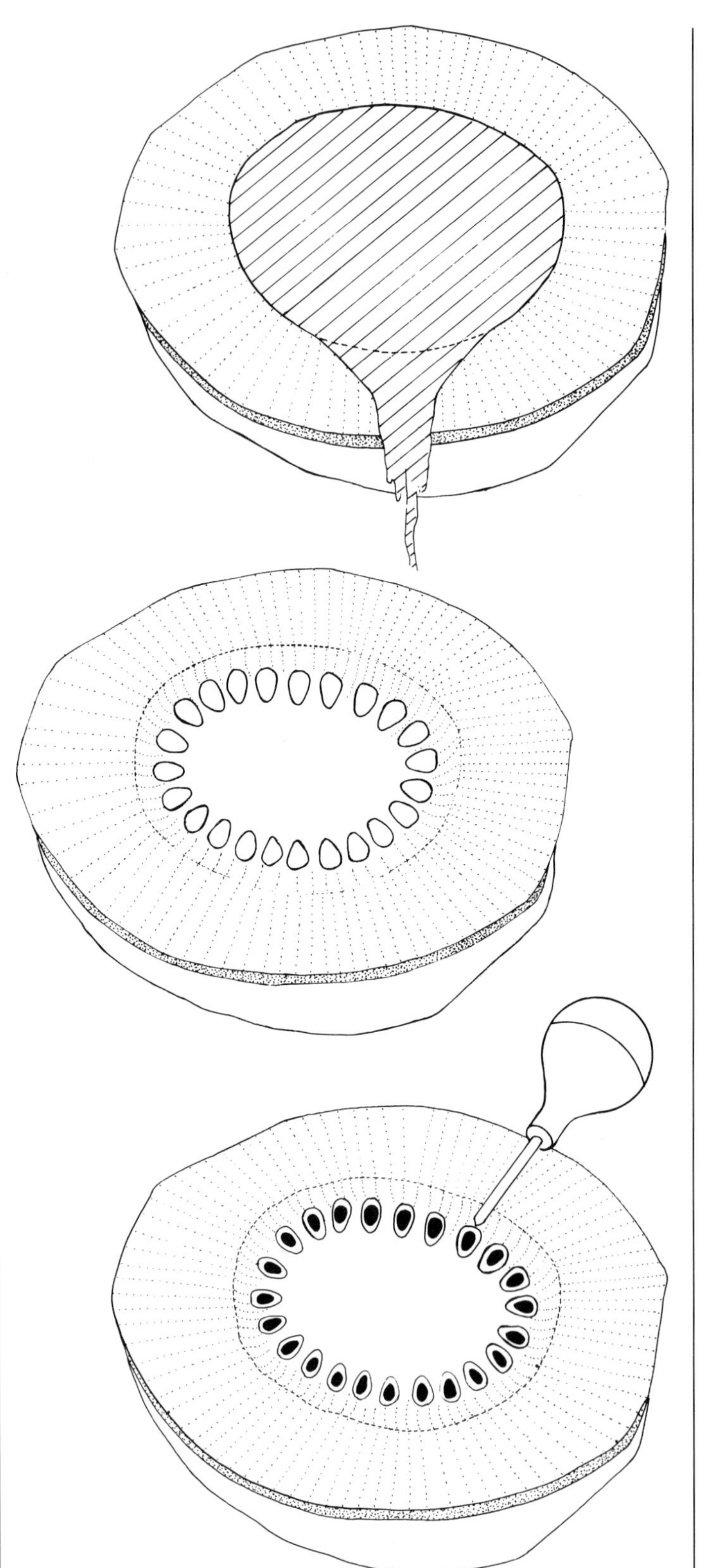

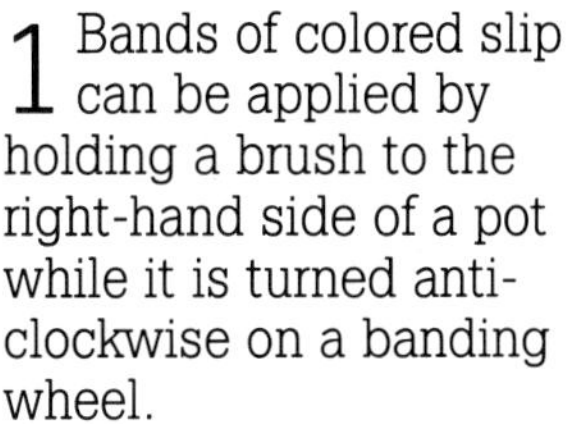

1 Bands of colored slip can be applied by holding a brush to the right-hand side of a pot while it is turned anti-clockwise on a banding wheel.

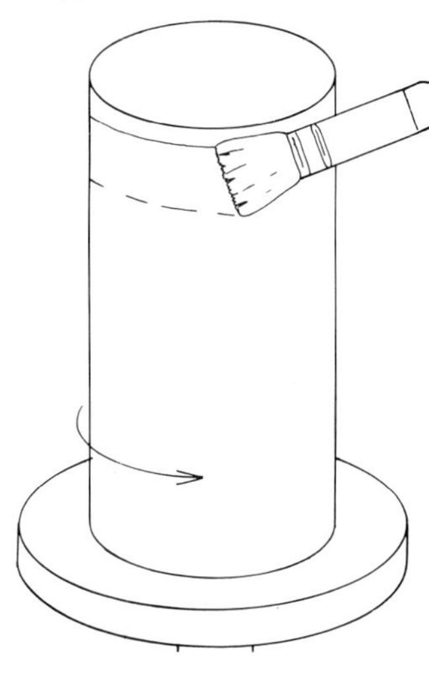

2 Free-form painting can produce a wide variety of effects and much depends on the choice of brush: a soft brush is best for large blobs of color while using a stiff bristle brush can result in a thinner layer and textured effect.

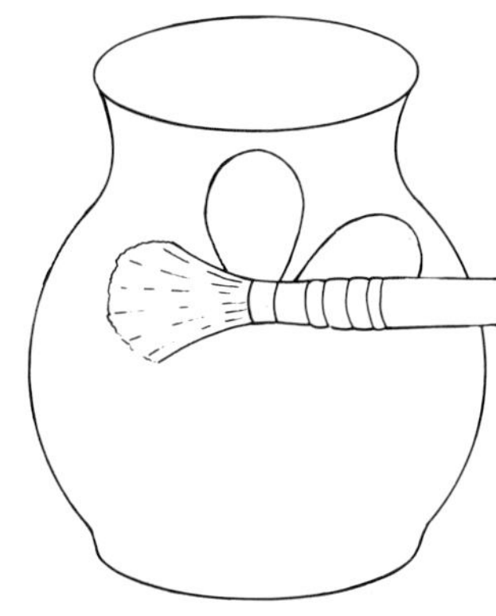

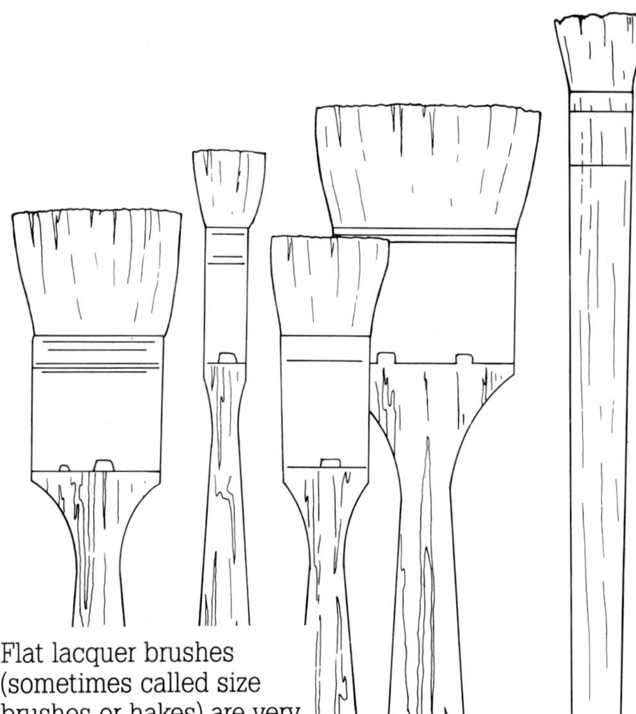

Flat lacquer brushes (sometimes called size brushes or hakes) are very soft and can be used successfully for applying slip to pots.

1 **Wet slip**
Soft edges can be created. Try drawing your fingers across the still-wet slip to create wavy lines.

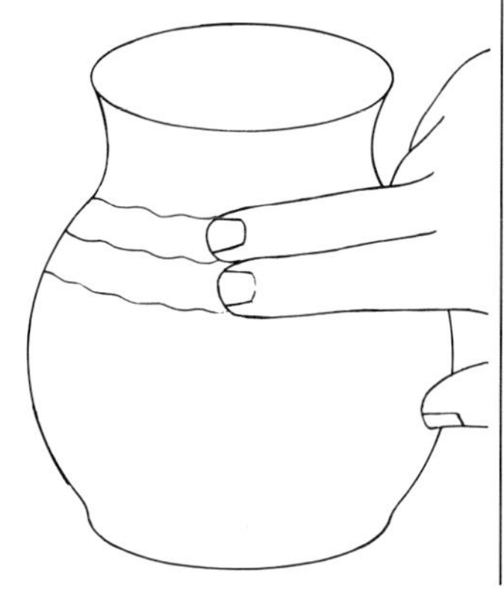

2 **Firm slip**
Sharper edges will result whern the slip is firmer. The effect will generally be more controlled and precise.

Suggested tools

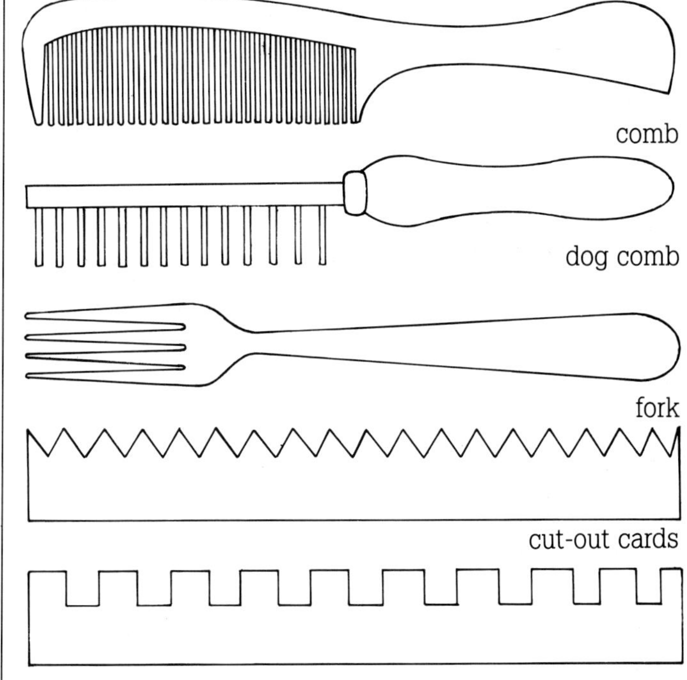

Birds of a Feather'
by G. E. Arnison (UK)
This plate has been most effectively slip-trailed in a bird design with lettering around the rim.

Tzu-chou jar
This black jar with simple white calligraphic design is from the Ming dynasty.

Casserole
by Graham Flight (UK)
Swirls of trailed slip decoration add interest to a simple form.

A collection of miniatures by Graham Flight (UK)
Bright splashes of color have been boldly applied to these miniature thrown pieces.

Molded dish by Graham Flight (UK)
Molded dish with trailed slip, painted oxides, sgraffito decoration and a thin tin glaze.

Bone china ware by Sasha Wardell (UK)
Here color has been applied by adding stains to the slip and by trailing on to the mold prior to casting.

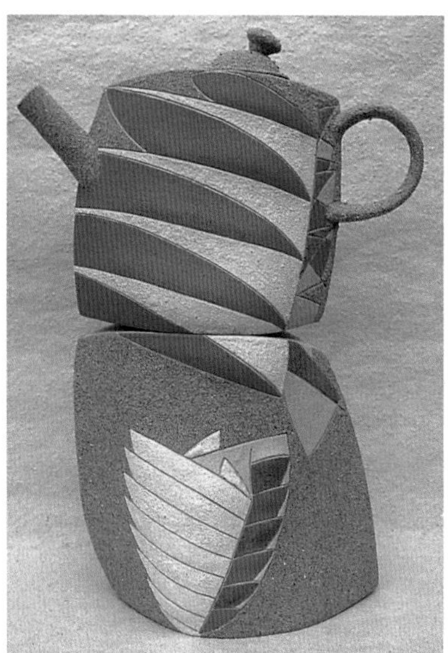

Two teapots by Jenny Beavan (UK)
These jaunty teapots have been brightly decorated with colored and textured slips to create an original and vivid effect.

Two bowls
by Graham Flight (UK)
Swirls of slip in muted browns and spiralling sgraffito complement the circular form.

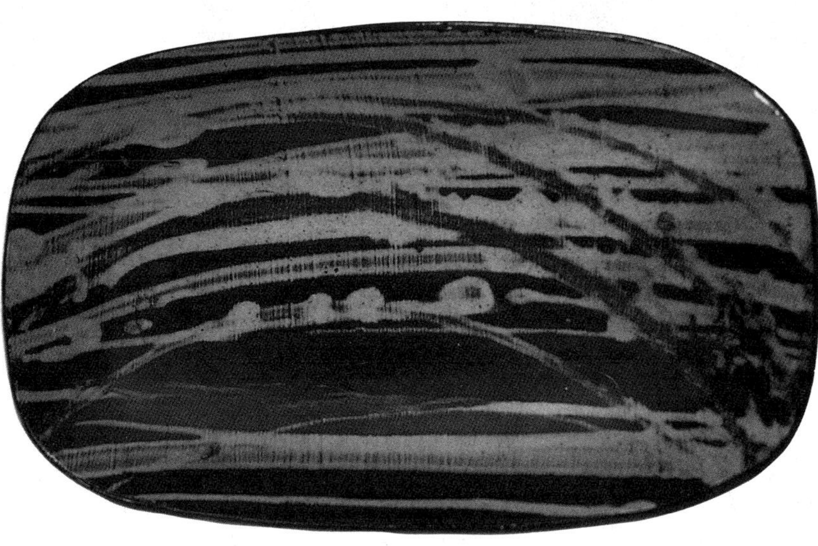

Hump-molded dish
by Graham Flight (UK)
A slab has been decorated with poured and trailed slip before forming the dish over a mold.

Dish
by Graham Flight (UK)
Freely-applied slip work in a simple bold design.

Sgraffito

This is a form of incising which involves scratching through a coating of slip, oxide or glaze to the body beneath. Using slip, various qualities of line can be made according to the dampness or dryness of the clay and the type of tool used to cut through the coating. Knives, pointed, steel modeling tools and needles are the most suitable implements. Soft slip is a lovely surface for a flowing free design cut with a looped tool. More precise marks can be made as the slip hardens to produce a satisfying contrast of soft and hard lines. Fine needlepoint patterns will show up well if the slip is left to dry completely but there might be a tendency for the lines to chip. This type of sgraffito looks at its best when applied to a burnished surface.

Try coating the dry slip with oxide underglaze pigments or body stains. Let the color flow into parts of the scratched lines and then scratch through the oxide parts as well to add another variation to the design. A dry clay surface coated in this way can be cut and scratched in a fine line design or a wider area can be removed with a nylon pad, leaving the oxide in the texture of the clay. Gently shake the dust on to the bench and then wipe away carefully with a damp sponge. ◊

Tools for sgraffito

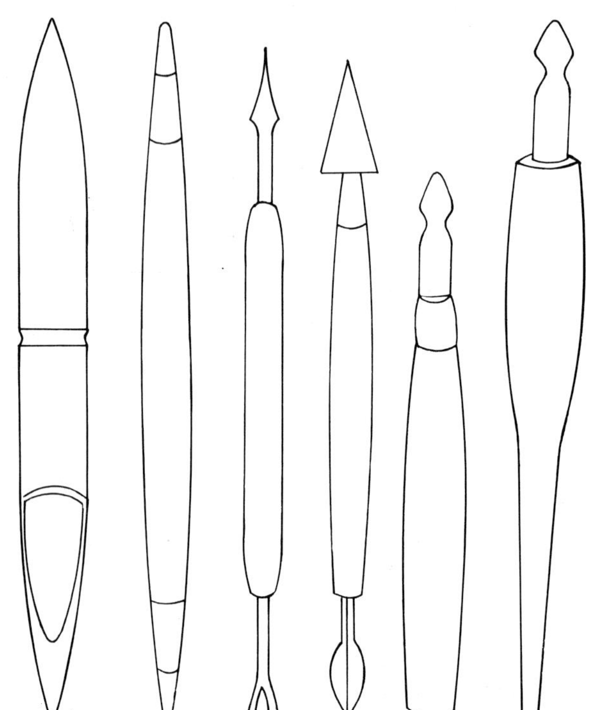

Sgraffito techniques

Sgraffito made through an oxide-painted, dry clay pot will produce a finer and more precise type of line. Coat the pot with oxide when it is bone dry and then scratch away the oxide as required.

1 Use molded dish that is still in its mold so that it is well supported while the decoration is applied. The clay should still be damp.

2 Pour in a coating of slip. Lift the mold and tilt until entire inside surface is covered. Tip out excess slip, holding the dish in both hands with the thumbs placed on the edges of the dish so it does not fall out of the mold. Make sure all excess slip is removed.

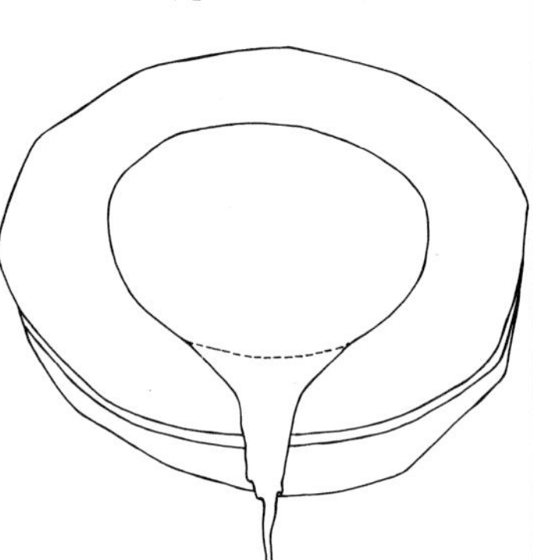

3 Leave to dry. Meanwhile you can work out a suitable design on tracing paper, using fairly bold lines that will be easy to duplicate.

4 Place your design sheet over the clay surface and trace out the design with a soft pencil. Or apply light watercolor with a thin brush around the outline. Soft slip will create a soft line. If the slip is too hard the edges of the lines will chip. Medium dry, leather-hard slip will provide a good surface for scraping.

5 Either etch out the design to leave fine lines on a background of slip or cut away the background, leaving the design as a solid area through which fine lines can be scratched.

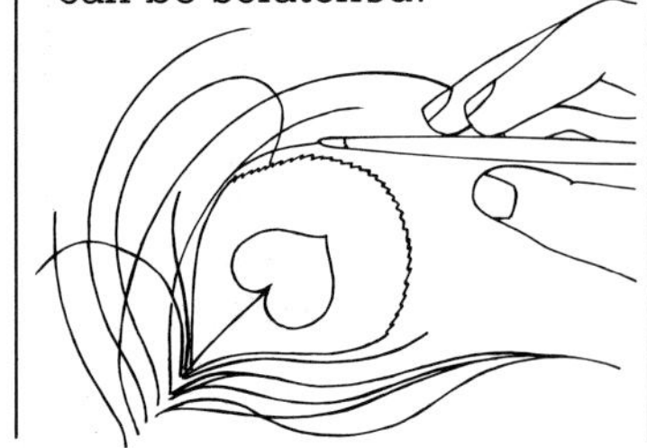

'Gaddah Cave Drawings' by Sidig El Nigoumi (UK)
The fascinating sgraffito design on this bowl was inspired by primitive cave painting.

Teapot by John Maltby (UK)
The surface of this piece has been coated with enamel with black pigment resisted by latex. Sgraffito completes the details of the design.

Slipware and resist decoration

Wax resist

Hot or cold wax painted on to a leather-hard pot will resist a coating of slip and the unwaxed areas will provide the pattern when the slip is applied. The slip may cling in small spots to the waxed areas, giving the clay a distinctly textured quality.

Masking

Another form of slip decoration is to use cut or torn paper to mask out parts of the background before pouring or spraying on slip. The paper pieces (newsprint works well) are stuck to the surface of a leather-hard pot with a damp sponge and pressed down well to prevent slip seeping under the edges. The fascination lies in revealing the design after the coating of slip has dried sufficiently for the masking paper to be removed; this should be done carefully, using a pin to lift one edge.

Other forms of resist

Found objects can also be used as resists to slip: keys, coins, card, lace pieces, plastic mesh, leaves or grasses dampened and pressed into the clay will not only resist the slip but will leave an interesting impression. Adhesive tape is excellent for creating a sharp, straight geometric pattern and stencils placed on a tile or held against a pot are ideal for a repeating pattern; the slip, oxide or glaze just needs to be sprayed on.

You can also paint wax on to a leather-hard pot and then, using a damp sponge, carefully work away at the surface to create a shallow relief in the unprotected clay.

Wax resist with slip

There are two main types of wax resist: one can be used cold and one must be applied hot. Cold wax emulsion is readily purchased and simple to use as the wax can be removed easily from the brush by immersing in hot water.

Hot wax needs to be melted in a pan immersed in water within a larger pan or glue kettle. The traditional hot wax works well but it soon dries on the brush and needs swift application. The design needs to be worked out carefully and applied quickly before the wax dries and damages the brushes. ◇

1 Painted (or poured) wax on a damp or dry surface will resist any mixture of water and pigment (or water and clay).

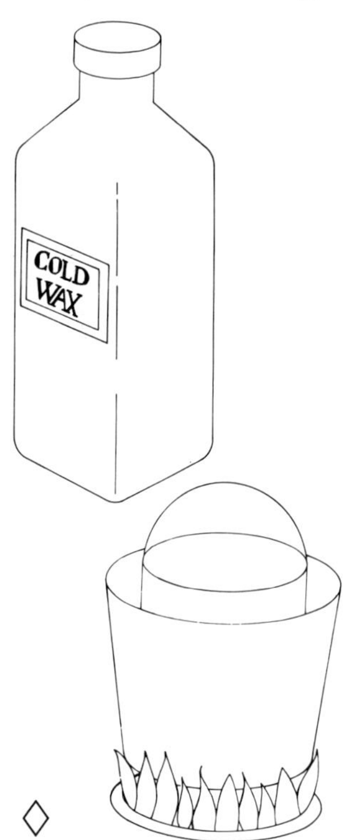

2 Paint the design on with cold wax emulsion which is ready-mixed or with wax heated in a kettle. The emulsion is easier to use and kinder to brushes.

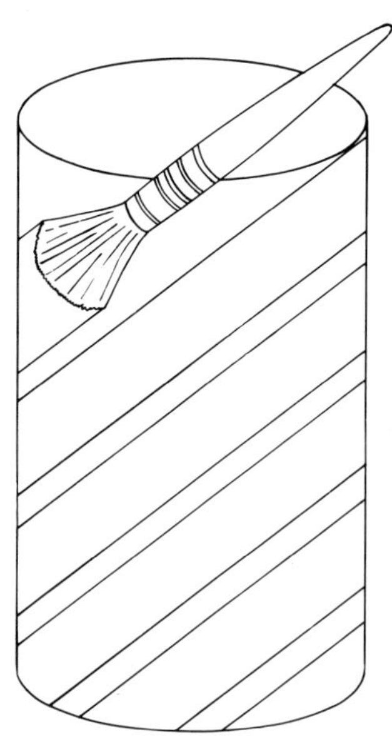

3 Allow the wax to dry. Pour or dip into a thin slip. The waxed part will not be covered and will therefore reveal the clay below when the wax has gone, while the rest of the surface will be coated with slip.

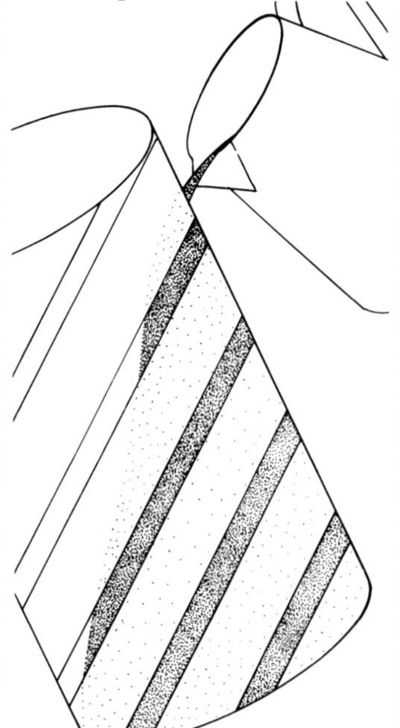

Wax resist with oxides

There are two ways of using a wax resist with oxides.

Please note that unless an underglaze medium is used with the oxides or stains the surface will remain extremely friable and dusty when dry. Pots must be handled with great care if the surface pattern is not to be spoilt and stained fingers can leave marks on other pots.

1 Paint a wax design on a bone-dry pot. Allow it to dry.

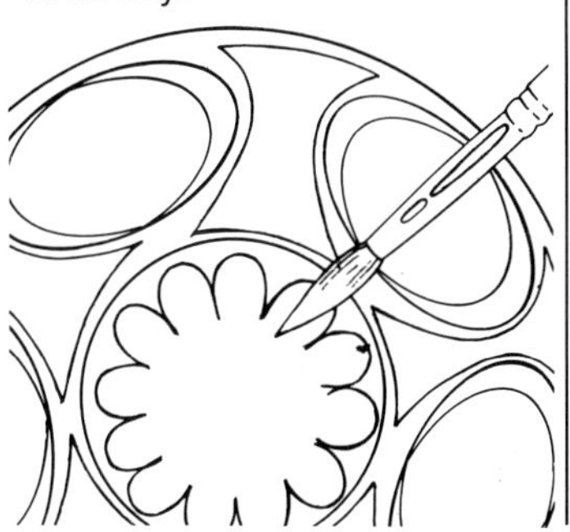

2 Paint the whole surface with an oxide and water mixture (ceramic stains can also be used). The pattern will show up as a contrast of light against the dark.

Further fine lines can be scratched through the oxide.

1 Paint the whole surface with wax. Allow this to dry and then scratch a fine pattern through the wax with a point.

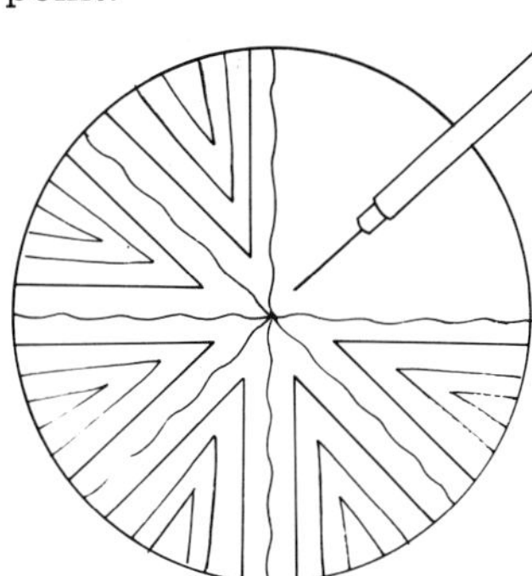

2 Paint over the whole surface with an oxide. This will stain the scratch marks only, revealing the pattern in negative.

Other resists

1 A latex-rubber solution can be painted on and will produce a fine resist. Then apply a coating of oxide, glaze or slip.

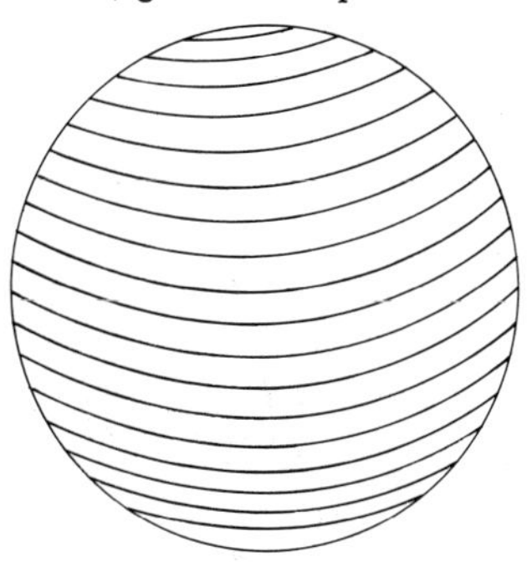

2 The latex can be peeled off before firing and the pattern will be revealed.

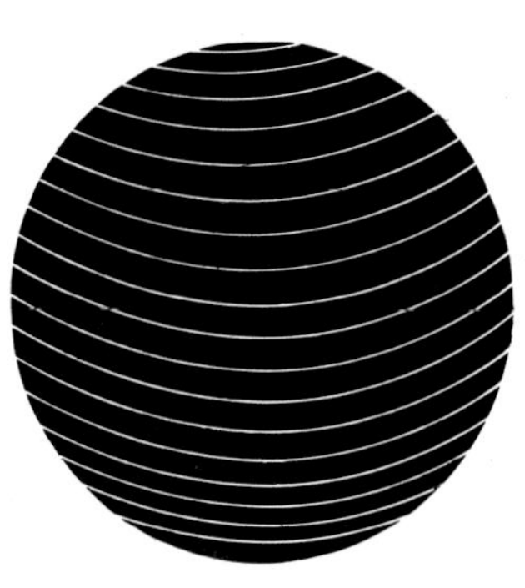

Paper resist

Paper resist works well with a slip coating. Either cut, stamp or tear paper, or use ready-patterned papers.

1 Dampen the paper shapes and stick them on to a leather-hard pot.

2 Pour a thin slip over the surface and allow the work to stand until nearly dry.

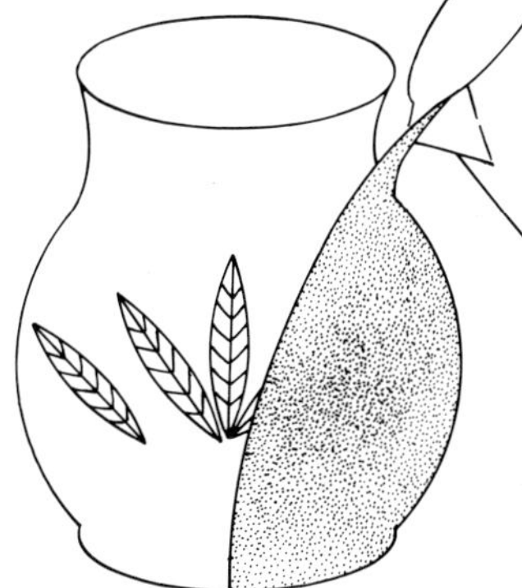

3 Remove the paper by lifting the edge with a pin and peeling it off.

Detail of work by Lana Wilson (USA)
Wax resist and a copper sulfate wash have been used to create a chequered pattern.

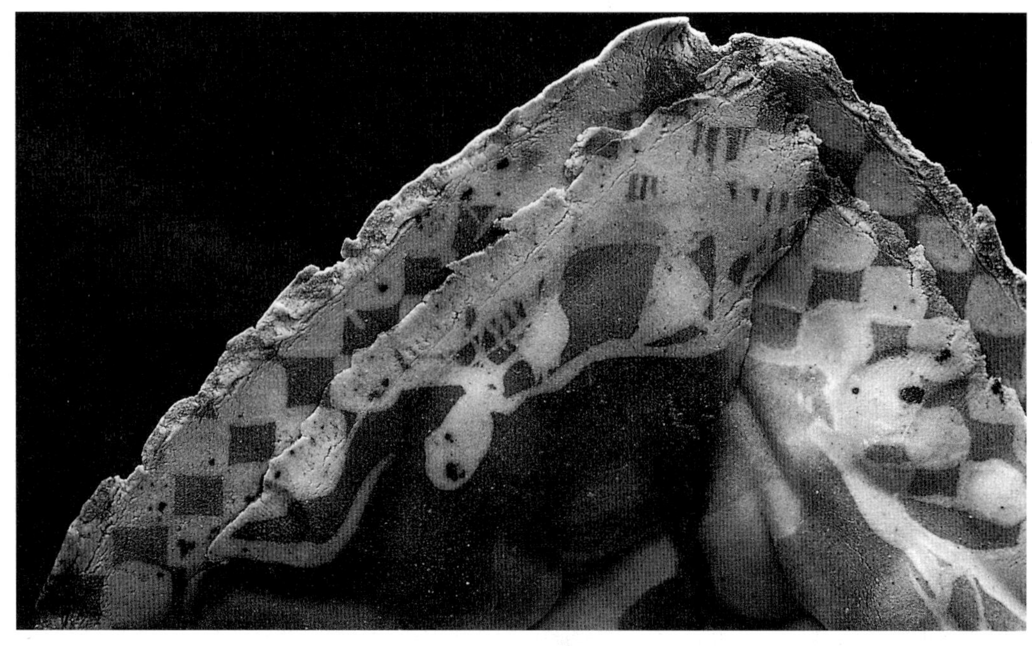

Wax-painted design by Sue Varley (UK)
Made of colored clay, this stoneware slab dish has been bisque fired and a wax-painted design and wood-ash glaze have then been applied.

Farm across the Valley by Frank Hamer (UK)
A square tile has been decorated using underglaze drawing and wax resist of colors and glazes.

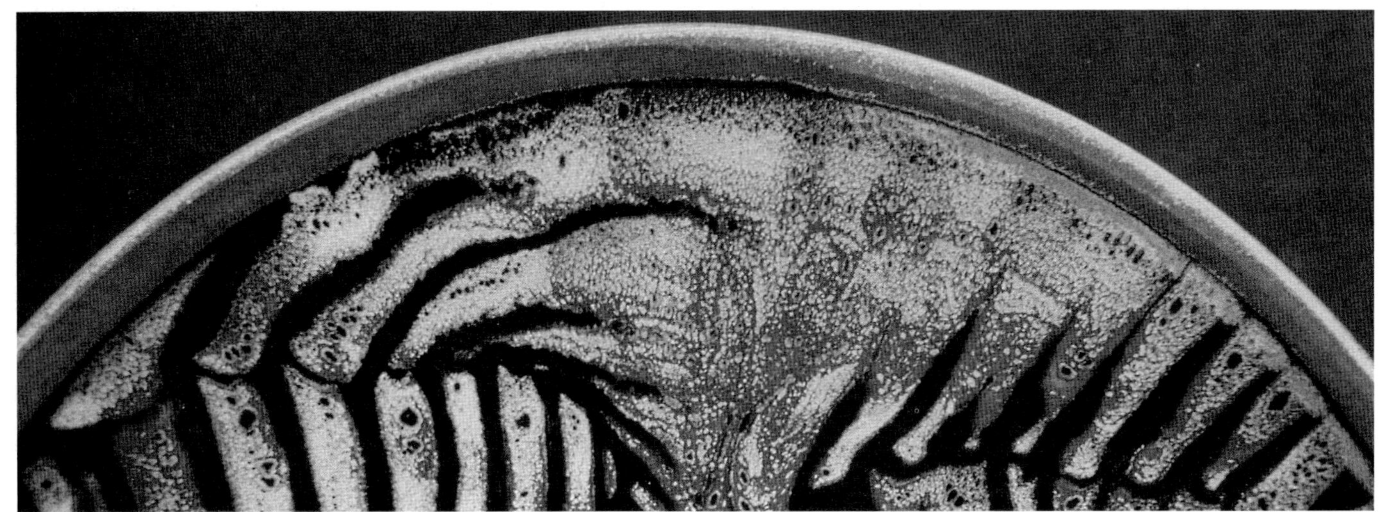

Detail of wax-resist pattern.

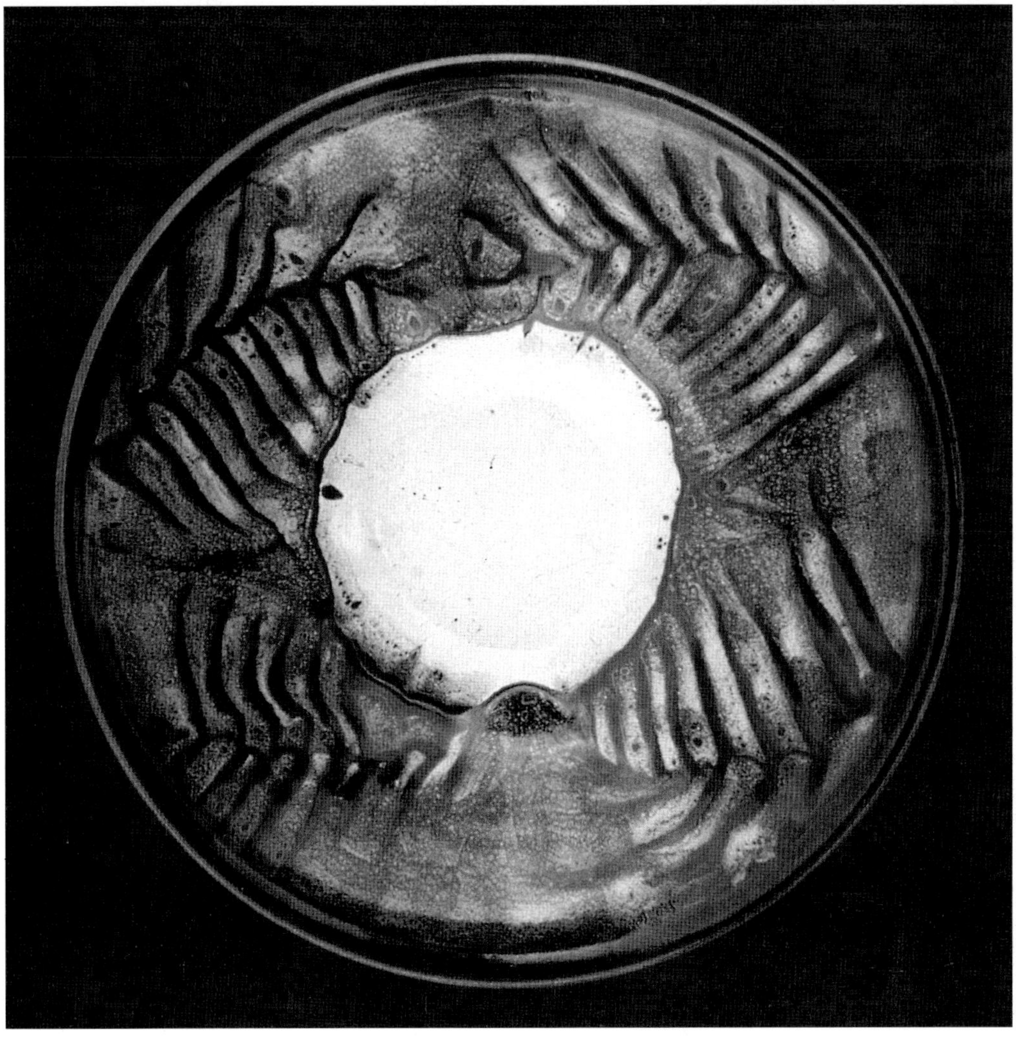

Stoneware bowl by Bryan Trueman (Australia)
Wax resist over iron glazes and a copper-red glaze have been used to decorate the bowl's inner surface.

Tiles

Using slabs to create tile patterns

Tiles have always lent themselves to the exploration of pattern and design, their flat surface providing great scope for decoration.

Using slabs in the construction of a tile panel may help towards the creation of an overall design.

First choose a suitable day. Clay for tiles is usually mixed with grog or silica sand to open up and stiffen the body and to reduce the problems of warping.

Assuming that a design has already been chosen, trace the outlines on to stiff card and cut the card into the required number of pieces of tile.

If the design for the tile requires applied, beaten or pressed decoration, then carry on with the decoration before the tiles are cut. This will help the pattern flow across the pieces of tile rather than being seen as separate entities.

Place the cards on top of the stiff slab and mark around them with a needle or pointed knife, or use a tile cutter which is stamped into the clay rather like a pastry-cutter. Apply the required decoration at leather-hard stage, perhaps adding more while the clay is damp (such as slip painting, trailing or inlay). Carefully cut out the tiles using a straight-edge and sharp knife. The tiles can then be covered over with plastic and allowed to dry slowly to prevent warping.

Tube-lining slip

This method of decoration was used extensively on Victor -ian tiles.The outline of the design is applied with a slip trailer (using the same clay as the body) when the clay is leather-hard and this then dries in a distinct raised line.

After the pot has been biscuit fired the spaces in between the raised design are then colored with oxides and glazes. The tube line shows as a white highlight when the piece is glazed. In some cases the application of a single colored glaze will highlight the tube lines while the glazes thicken and darken in the hollows.

Tube-lining glaze

This is best used when working on a flat tile or dish—after biscuit firing. It is a way of drawing with a thick glaze mixture using slip-trailing techniques. The design is drawn in outline and the contained areas are then filled with different colored glazes. It can also be used with slip trailers, using brushes for the details and highlights.

Tube lining is especially useful when distinct areas are to be filled with differently colored glazes because the raised lines act as walls to keep the areas of different color apart.

Delft tiles 18th century
Tin-glazed earthenware tiles were used to create vast decorative designs with ships, monsters and Biblical scenes as popular themes.

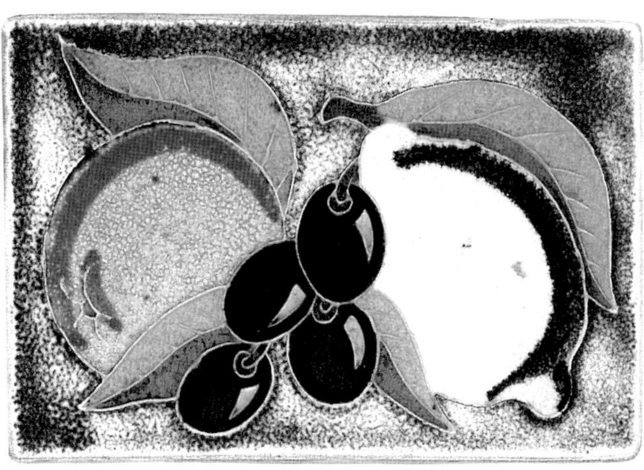

Modern Italian tile
Simple but effective decoration can be made by creating a raised outline pattern that is filled with brightly colored glazes.

A late-Victorian tile
Designs like this were printed from engravings on to a transfer and then applied to a succession of tiles to create a repeating pattern.

Patio panel by Maggie Angus Berkowitz (UK)
Poured, painted and trailed glazes have been used on buff ironstone and red quarry tiles to create a stimulating focal point in Great Ormond Street Hospital for Sick Children.

Creating color

Color in ceramics

Most color in clay work is provided by oxides, dioxides or carbonates of metals. Sometimes organic matter, such as wood ash, will provide its own color when used in a glaze recipe. The oxides can be applied to damp or dry clay, to one-fired biscuit ware or to the glaze before the second firing. They can be mixed into the clay to change the color or be rubbed dry into it before firing. Metallic oxides also provide color when added in percentages to any glaze. They react in different ways and have a wide variety of shades according to the glaze composition and the firing technique employed.

There is also an extensive palette of colors in the form of commercially-prepared body stains that are reliable and effective. These can be bought in powder form or as paint, crayons or felt-tip pens which have been specifically developed for use under, over or in a glaze.

Oxides

Oxides are used in all stages of manufacture. A combination of manganese, iron and cobalt, for example, will make a slip black; they can be painted over a dry slip to provide added color. It is possible to sgraffito through an oxide or to paint it on over a wax resist. They can be sprayed on (or mixed with a thick medium) to run through a silk screen on to tiles or transfers and, of course, they provide the vital color when mixed with a glaze. A wide range of colors can be produced from combinations of these metallic oxides and the results will vary according to their use, the composition of the glaze and the type of firing to which they are subjected.

Manganese dioxide

This is a most useful oxide giving colors from rich brown to black and, under some glazes, a purple brown.

Cobalt

Cobalt oxide produces a very strong blue. Only small amounts need to be added to any mixture of glaze or slips or when mixing underglaze pigments, as too much can ruin the effect and make it very dark. Small percentages in slip will give a rich blue; on a white glaze it can dominate the design.

The best stage at which to add cobalt carbonate to the surface of green ware is when the clay is dry, before the pot and the oxide are fired together. This allows the oxide to fire into the clay and any loose particles of oxide can be washed from the surface before it is glazed.

Cobalt is perhaps best used in its carbonate form. This is cheaper and produces results that are still very strong in color. It is usually tempered with additions of other oxides, such as iron, to reduce the harshness of the blue when used as a pigment for painting. Be careful not to pick up any traces of cobalt on your fingers or the merest touch elsewhere on the pot will blemish the surface when fired.

Copper

Copper oxides or carbonate provides a range of greens to turquoise when in oxidation, or reds when fired in a reducing atmosphere. Loose particles of copper oxide can fly about in the kiln, speckling other pots nearby. (This can be used to advantage if used in a controlled environment. Clay containers holding small amounts of powdered oxide can be placed nearby selected pots in the kiln.) It will appear brown or gray under or in certain glazes containing zinc.

Iron

This most useful colorant is mainly used in the form of red iron oxide and is present naturally in many clays when dug—showing as black, terracotta red, mustard or green-yellow ocher.

When present in quantity it will lower the melting point of clay making some iron-bearing clays useful as slip glazes. It can produce an enormous range of colors from yellow to dark brown, rust-red to black, or even subtle grays, greens and blues in some reduction-fired glazes, depending on the amount used and the other ingredients present.

Other oxides commonly used by studio potters include chromium oxide which gives yellows or red at low temperatures, and in a tin glaze can produce a delicate pink. Nickel oxide will provide a more muted muddy green than copper or perhaps a gray-blue. In some oxidized glazes rich blues and purples can be obtained from this oxide.

Apart from these basic oxides there is a wide range of prepared stains available for coloring bodies by painting under or over glazes. They are mixed with water or gum, or perhaps a little of the chosen glaze, to make them easier to apply to a dry surface, or they can be mixed with clay bodies or slips (5 to 15 per cent) as for oxides.

Onglaze enamels are low-firing colors which are painted on to the glazed surface and fired into the glaze. They are applied mixed with an oil medium and ground while mixing. They have the advantage of showing the color as it will actually appear. They are fired up to 800°C, and more can be added and fired again — the temperature being regulated according to the color, finishing with precious metals, gold and silver and lusters. These are expensive to buy but are delightful in their rich finish.

Lusters

Lusters are fine metal coatings which have been fired in a reduction atmosphere. The Persian lusters made at Rayy in the 12th and 13th centuries and the Spanish of the 14th and 15th centuries are some of the finest seen. More recently, William De Morgan used it to great effect in England around 1900 and his tiles are particularly fine examples.

They are painted on to the glazed and fired surface and refired in a smoky atmosphere at 1200-1470°F (650-800°C). This will melt the luster into the surface of the glaze. The surface will be iridescent and may need to be polished with cotton balls and pumice powder to reveal the color.

H see page 297

Applying lusters

Commercially produced lusters are readily available. These can be painted, sponged or air brushed on to the surface but if several colors are used, their initial appearance will be a rather sticky brown. This can be confusing unless you are quite clear which colors form the various parts of the design. The true colors are revealed only after burning away the oil medium (which provides the necessary local reduction). It may be wise to fire each color before adding the next.

Interesting effects can be obtained if you break the surface tension of these lusters (before they are completely dry) by applying tiny amounts of solvent in selected places. Detergent solutions or paraffin both work well, causing the luster medium to retreat from the point of contact. This can be controlled with practice and is very effective. This technique has been exploited imaginatively by many potters including Geoffrey Swindell (UK).

C see page 207

Enamels

These are prepared colors made from various fluxes and alumina silicates colored with metallic oxides which, when applied to a glazed surface, will melt into this surface at 1200-1472°F (650-800°C). The range of colors made from these ground metal salts is both wide and bright. They can be painted on to a glazed surface by mixing with a fat oil of turpentine; this facilitates the flow and holds the colors on to the glaze until they are fired in. Practise on flat glazed tiles that can be easily wiped clean if a mistake is made; try thinning with turpentine for a wash effect, or perhaps spraying and masking with stencils, or printing.

Firing must be slow at first with plenty of ventilation in the kiln and kiln room in order to burn off the oil medium without any bubbling. If the firing is too fast the oil will boil and bubble marks will be left in the color. Some colors will require a lower firing than others and these will then need a second firing at slightly lower temperatures. The firing temperature should be just high enough to seal the colors into the surface and regulated according to the particular color—finishing with precious metals, gold and silver, and lusters. These are expensive to buy but are delightful in their rich finish— and have the advantage of showing the color as it will actually appear.

French faience plate 17th century

Majolica and underglaze

Majolica

Majolica is sometimes called onglaze because the color is applied over an unfired, dry glazed surface. (It may sometimes be referred to as inglaze when the color fuses with the glaze.) It was originally painted over a clear glaze with a coating of white slip over the body underneath to show up the colors—until the discovery in the ninth century that using 5 to 10 per cent tin oxide in an earthenware glaze made it white and opaque. This development quickly spread from Mesopotamia to North Africa and thence to Spain, Italy and Northern Europe. In Spain and Italy it became known as 'Majolica' ware (after its transportation by Majorcan ships), in France as 'Faience' (after Faenza in Italy) and as 'Delftware' in Holland and England—although basically it was the same technique. **H**

The pigment is best used finely ground with water and a little of the base glaze. As the strokes need to be confident and even, practise on a spare pot before you attempt the designs on your best pieces. Experiment with a variety of brushes in a calligraphic manner.

Mark out the design with a fine brush. First make some guidelines, using water with a little vegetable dye (or mix a little dye with wax resist if that is being used). Too much oxide or stain will act as a flux with the glaze and will run down the pot, so be careful. Fine lines scratched through the oxide but not the glaze can be added. Mistakes cannot be readily rectified because the glaze surface is friable and easily disturbed. Clumsy brushwork may also remove large amounts of the powdery glaze. For these reasons Majolica is a technique which may prove discouraging to the inexperienced potter. Wax resist on the tin glaze also works well prior to brushing oxides.

Underglaze

This is the application of oxides and pigments on to a biscuit-fired pot with a transparent or semi-opaque glaze covering.

The brilliance of Chinese Ming porcelain of the 15th century shows this technique superbly. With cobalt painted decoration on a thrown porcelain body covered by a clear glaze, the effect is rich and splendid. **H**

Oxides can be painted, sponged, poured or sprayed. The application needs practice; some colors, such as cobalt, need only small amounts to give a strong color. If the application is too thin it will disappear in the gloss firing; if too thick it may bubble and erupt or go metallic black. This volcanic bubbling can be used to great effect under a thick opaque glaze — manganese chloride applied thickly works very well, for example.

Try to develop the design using one brush stroke only to make a clear even line; touching up strokes or overlaying will show and perhaps turn the color metallic.

Parts to be left clear may be masked with wax resist before painting.

When painting with oxides and commercial underglaze colors, a little gum arabic mixture with the oxide and water will harden the pigment so it is less likely to wash away when the glaze is applied. Thin water-color paint or a light pencil are useful for sketching out designs on the pot before painting with oxide. These guidelines fire away in the kiln. Sprayed oxides need a water or oil-based medium to help application. Stencils or masking tape cut and stuck on provide interesting resist patterns. If oil is used as a medium it needs to be slowly burnt off in a low firing to prevent the oil bubbling and boiling under the glaze. Underglaze crayons make a useful variation in underglaze color. Used like pastels, they work best if the body of the pot is light-colored and smooth.

Dry unfired clay is a good surface to paint on, especially if some form of sgraffito is being done. (See page 180.) The basic oxide colors work best with further applications of lighter colors after the first firing. Using manganese, cobalt, copper and iron, the thickness of the coating is not quite so critical as when painted on a biscuit surface. A wash under running water after biscuit firing will remove any excess oxide on the surface, leaving the fired-in stain intact and ready for glazing. Scratched designs are most successful when made through the oxide to the dry clay. If mistakes are made the oxide can be rubbed away carefully with wire wool or a nylon pad (wear a mask and wipe away dust with a damp sponge) and then another coat of oxide can be applied. ◊

A coating of oxide partially rubbed away will leave the color staining in the grain of the clay. This is sometimes all the decoration required for a stonelike form, fired to 2280°F (1250°C) without glaze. Wax resist on dry clay works well. If a coat of oxide is painted on when the resist is dry, the pattern of spaces around the resist appear magically. Paper resists, masking tape and stencils can be experimented with — try using a mixture of oxide and water sprayed on with a hand-operated garden spray.

H see pages 268 278 282 285 288

Earthenware pot by Gordon Cooke (UK)
Gordon Cooke uses simple ash and dolomite glazes with metal oxides and body stains painted on to the raw clay.

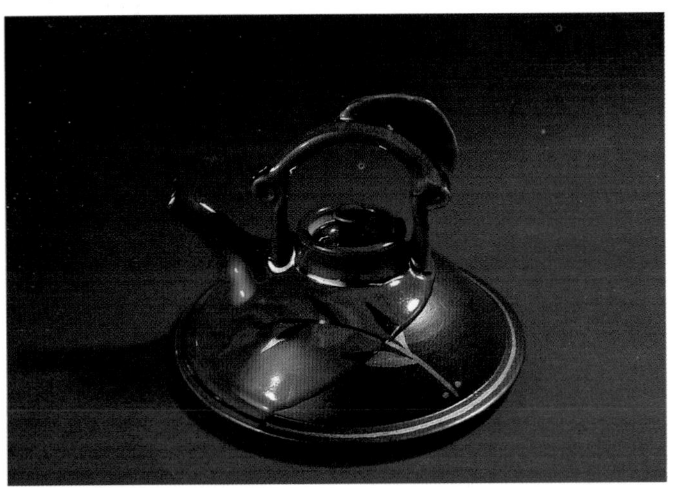

Raku teapot on stand by Bryan Trueman (Australia)
The deep-blue color on this teapot has been created by an alkaline glaze and various oxides.

Dish with luster decoration by Marianne Cole (Australia)

Color in ceramics

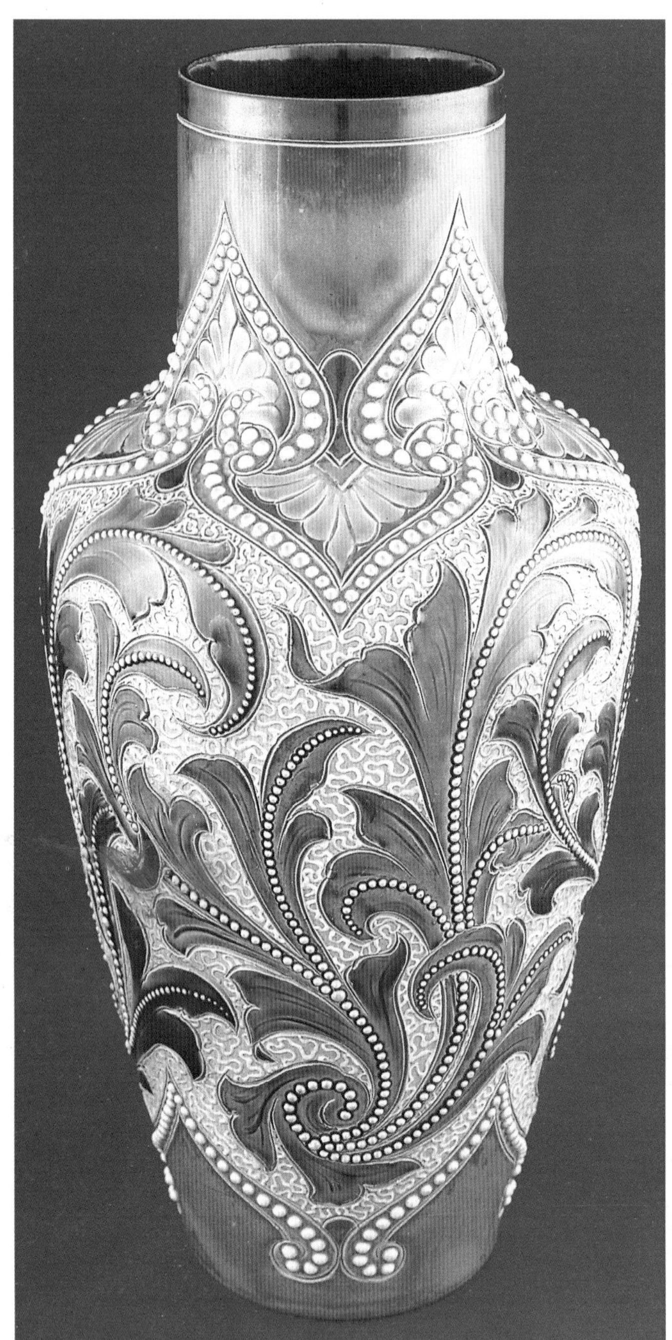

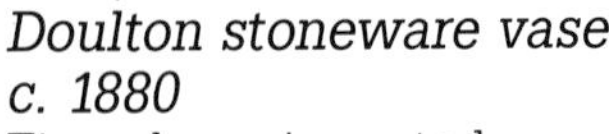

Doulton stoneware vase c. 1880
Fine glazes in muted colors, floral scrolls and applied decoration all contribute to the exquisite work on this elegant vase.

Raku bottle by Walter Dexter (Canada)

Pit-fired ware by Ray Rogers (Australia)
Astronomical bodies seen through a telescope inspired the cloudy swirls of color on this piece.

Tree
by Tessa Fuchs (UK)
The golden gleam of the leaves — even those fallen from the tree — add to the atmosphere of an autumn day.

Luster glaze pot
by Greg Daly (Australia)
Copper and silver have been used to create a superb lustrous surface speckled with red.

Poison-arrow frog
by Tony Bennett (UK)
Color has been built up over three or four firings, applied by a combination of spraying and hand-painting.

Color and effect of decoration

Chinese porcelain 1662-1722
Figure of laughing man with an unusual mix of textured, geometric, floral and landscape designs on his robes.

Victorian ribbon plate
The expansion of the railways and travel in Victorian times opened up the market for holiday souvenirs.

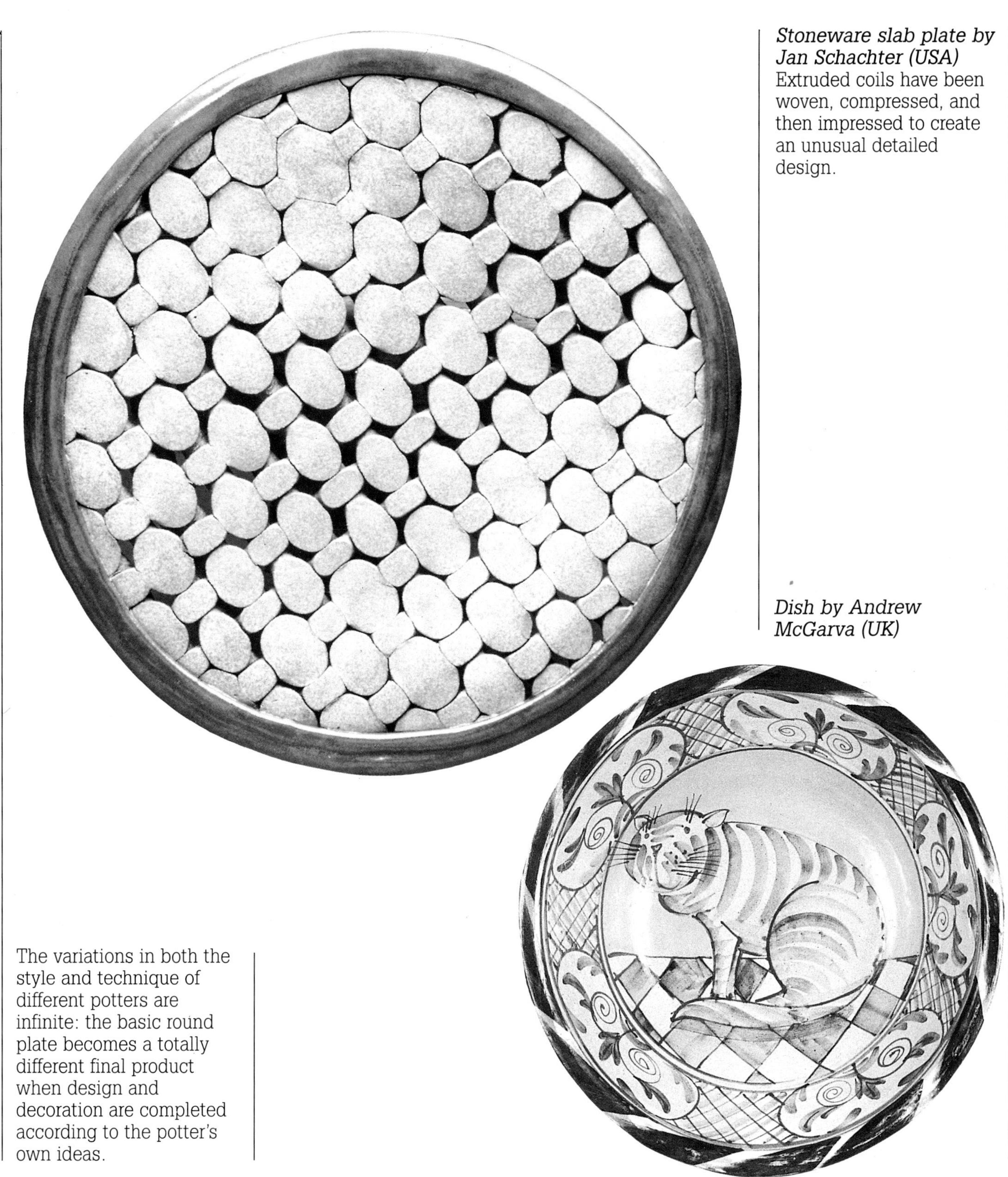

Stoneware slab plate by Jan Schachter (USA)
Extruded coils have been woven, compressed, and then impressed to create an unusual detailed design.

Dish by Andrew McGarva (UK)

The variations in both the style and technique of different potters are infinite: the basic round plate becomes a totally different final product when design and decoration are completed according to the potter's own ideas.

6 Glazing

Thrown casserole by Jane Hamlyn (UK)
The ware has been salt glazed with polychrome splashes and the detailed decoration created with roulette impressions.

Introduction to glazing

Glazing pottery presents the beginner with a bewildering variety of materials whose properties have to be appreciated and understood if disillusion and disappointment are to be avoided when compounding your own glazes. It is possible, of course, and perfectly acceptable to purchase reliable prepared glazes of almost every conceivable color and texture. It is also true that some professional potters have little or no interest in mixing their own glazes and prefer to direct all their creative energies towards other aspects. But, if they use glazes in any part of their work, mixing their own will have given them an intimate knowledge of their selected materials and the way in which they behave under different firing conditions.

Much of the mystique attached to glazes arises through a misunderstanding of the constituent materials or the principles involved. There exists also a certain secretiveness among those potters who do not wish to reveal the composition of their glazes and this has helped to perpetuate and magnify the mystery of glazes. However, it cannot be denied that even the most experienced potters do occasionally meet problems in this area of ceramics so that they sometimes obtain results which may not always be easy to explain. There are those too who search endlessly for some kind of ideal glaze. Above all, you should remember that glaze technology, like all the other techniques described in this book, can provide only the means to an end and, therefore, it should not be allowed to become an impediment to the expression of your ideas.

Glazing composition

In simple terms, glaze is normally composed of three basic ingredients. Silica, the glass-forming element, is the first of these; secondly, some kind of fluxing agent(s) is necessary to reduce its melting point so that it becomes compatible with the particular clay body beneath; and, finally, alumina which acts as a stabilizer and helps to bind the glaze particles together when still in the raw state. While the first two are essential elements in glaze composition, alumina (as clay) is not always necessary (as in salt-glaze and certain crystalline glazes, for example). However, it should be noted that when little or no clay is present the applied mixture tends to be powdery and easily removed from the surface of a pot prior to firing. Extra care is required when handling and packing such glazes in the kiln.

You can learn more about the potential of the various minerals used in glazes by mixing small amounts of each singly, and in combination, with water to make little pellets which can be fired individually on bisque tiles or shallow dishes. Keep a thorough record of the way they behave at different temperatures. Some of your samples will be seen to have melted and flowed while others may appear unaffected by the heat. It will be logical to experiment initially with a further mixture from those materials that melt easily and those that do not. This is a fairly crude but often an effective method of learning — especially if it is supported by reading widely about both the theory and practise of glaze technology.

The aim in these rudimentary exercises is to arrive at a recipe which will ensure the production of a glaze that will melt sufficiently to fit the pot but not so much that it will run down and ruin the kiln shelves!

Melting points

At earthenware glaze temperatures, around 1100°C (2010°F) fluxes with low melting points are required and these usually contain some form of either soda, potassium, lead oxide or boric oxide. In practice, various natural or man-made compounds of these are likely to be used together with small amounts of other fluxes more commonly associated with stoneware glazes.

Far more materials will be found to melt at temperatures in excess of 1200°C (2190°F) so that the choice of fluxes is much wider. (Lead compounds become volatile at high temperatures and their use is, therefore, confined to earthenwares.) This makes stoneware and porcelain glazes an especially exciting and fruitful area to explore because the range of potential components can offer a greater richness and subtlety in pure glaze quality.

Experimentation

If you discover that you have a particular interest in formulating glazes then you will enjoy controlled experimentation with unusual materials and mixtures. You are advised to read more widely around the subject while doing so. Many thousands of glaze recipes have been published during recent years and there are a number of books devoted entirely to the study of glaze composition and analysis. Some of these are listed in the bibliography at the back of this book.

Glaze chemistry can be daunting to many people and, rather than become confused or discouraged by the enormity of the subject, you should try to approach it as a practical and fascinating adventure. Indeed, most potters have just two or three basic glazes which they know well and that they can adjust for color or texture as needs arise. Tests of different combinations of materials and glaze mixtures can be carried out easily while firing more familiar glazes. Beginners often search for spectacular glazes in the misguided hope that they will compensate for any weakness of form — they will never do that!

It is wise not to rely too heavily on any recipe, however obtained, without thorough testing. Materials are not always consistent in quality and considerable variations can occur, especially if sources differ. There can be no substitute for direct involvement with glaze preparation and firing. Equally, the assessment and recording of results will give you confidence so that you will be able to concentrate upon the aesthetics of pottery-making rather than be hamstrung by technical inadequacies. Try to learn the names and properties of the glaze ingredients described in this chapter before engaging in your practical experiments.

You will also discover that glazes do not always react in the same way. Their behaviour under reducing conditions in a flame-burning kiln may be radically altered when the same recipe is fired in the oxidizing atmosphere of an electric kiln. Both firing procedures have their advocates among potters who design their glazes accordingly.

If you discover a preference for ash glazes you will find them particularly rewarding when you explore the potential, not only of mixed bonfire ash but also ashes from a single identifiable source. If this should be from a particular tree, for example, the results may vary according to that part of the tree which is burnt or even the time of year when it is cut down. The same species grown on different soils may also provide noticeable differences, perhaps in color, even though the recipe proportions are unchanged.

Mixing glazes

Overlaying glazes of differing viscosity usually produces more interesting results than if they are mixed together prior to application. For example, if a 'dry' glaze containing a high proportion of clay (e.g. 50 parts kaolin and 50 parts mixed wood ash) is applied over a more fluid type of glaze (e.g. 85 parts by weight of Cornwall stone, 15 parts whiting, and 10 parts red iron oxide) and fired to cone 9 a broken texture or a mottled surface is likely to be obtained. This is because the underlying glossy glaze moves as it melts, fracturing the surface of the drier and more refractory glaze above. By reversing these layers you will be able to produce quite different results.

Similarly, the use of two or more glazes of contrasting colors or tones (light over dark or dark over light) also offers interesting visual possibilities. As always in any such experiments however, you would be wise to safeguard against the risk of creating too fluid a combination that will run down the pot and glaze your kiln shelves! Remove glaze from well up the side of your test pieces and place the pot on a piece of old kiln furniture or a thick layer of silica sand. Remember that glazes which perform satisfactorily on a horizontal surface may be at risk and, in any case will probably look quite different, when they are fired instead on to an upright form.

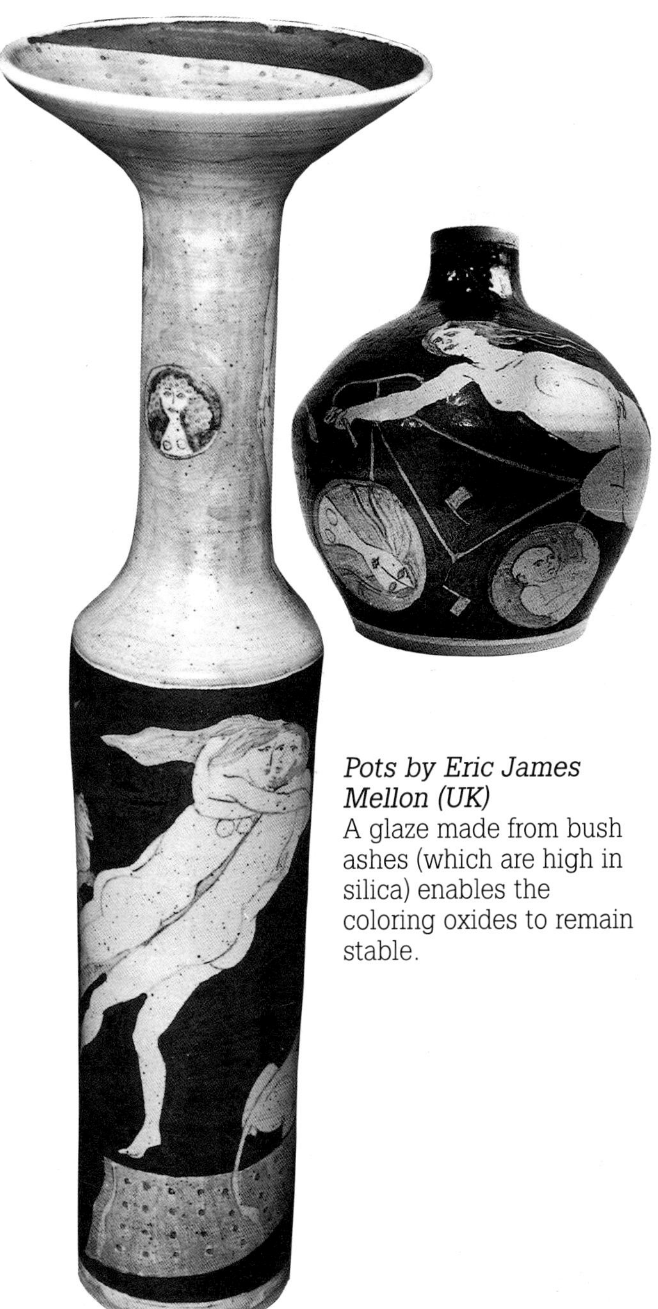

Pots by Eric James Mellon (UK)
A glaze made from bush ashes (which are high in silica) enables the coloring oxides to remain stable.

Different glazes

Low-firing glazes

Lead, soda and borax are the main fluxes used in low-firing glazes, (below 2010°F [1100°C]). They help to develop a wide range of colors and surface textures when mixed with coloring oxides but each can have certain disadvantages. ◊

Lead in its raw state is highly poisonous; soda and borax are soluble in water and are difficult to control. Therefore these substances are used as a frit. Fritting is a way of making these materials safe and usable. The flux is combined with silica and heated to melting point. The molten liquid is tapped off into cold water where it shatters and is then ground to a fine powder ready for use. Most potters use lead, soda or borax in frit form when making a low-temperature glaze. Many suppliers stock a variety of frits ready for use.

Low-temperature glazes using soda or potassium oxide as a flux are called alkaline glazes.

Alkaline glazes

Alkaline glazes were first used by the Egyptians to make small objects with bright colors. The pieces were not glazed in the usual sense of the word; that is, they were not covered with a glaze coating by pouring or painting. Instead the glaze was incorporated into the body in the form of soda mixed with silica sand and flint — known as Egyptian paste or frit body. During drying the soda migrates to the surface and forms a deposit which melts to form a glaze when fired. Coloring oxide, usually copper and cobalt, was also mixed in and created the familiar blue and turquoise of early Egyptian and Persian pots.

The paste was a crumbly mixture and because of this was confined at first to making small pieces such as beads or objects pressed into molds. These pieces date from at least 5000 BC and exhibit the first known use of glazes.

The colors derived from alkaline glazes are clean and brilliant. Copper and cobalt make turquoise and vibrant blues; manganese gives violet and purple hues; iron provides straw-yellow to rich browns. They can be glossy or matte, but are soft and have a tendency to craze over most bodies. However, they can be used as reliable alternatives to a lead glaze when fired to around 2010°F (1100°C).

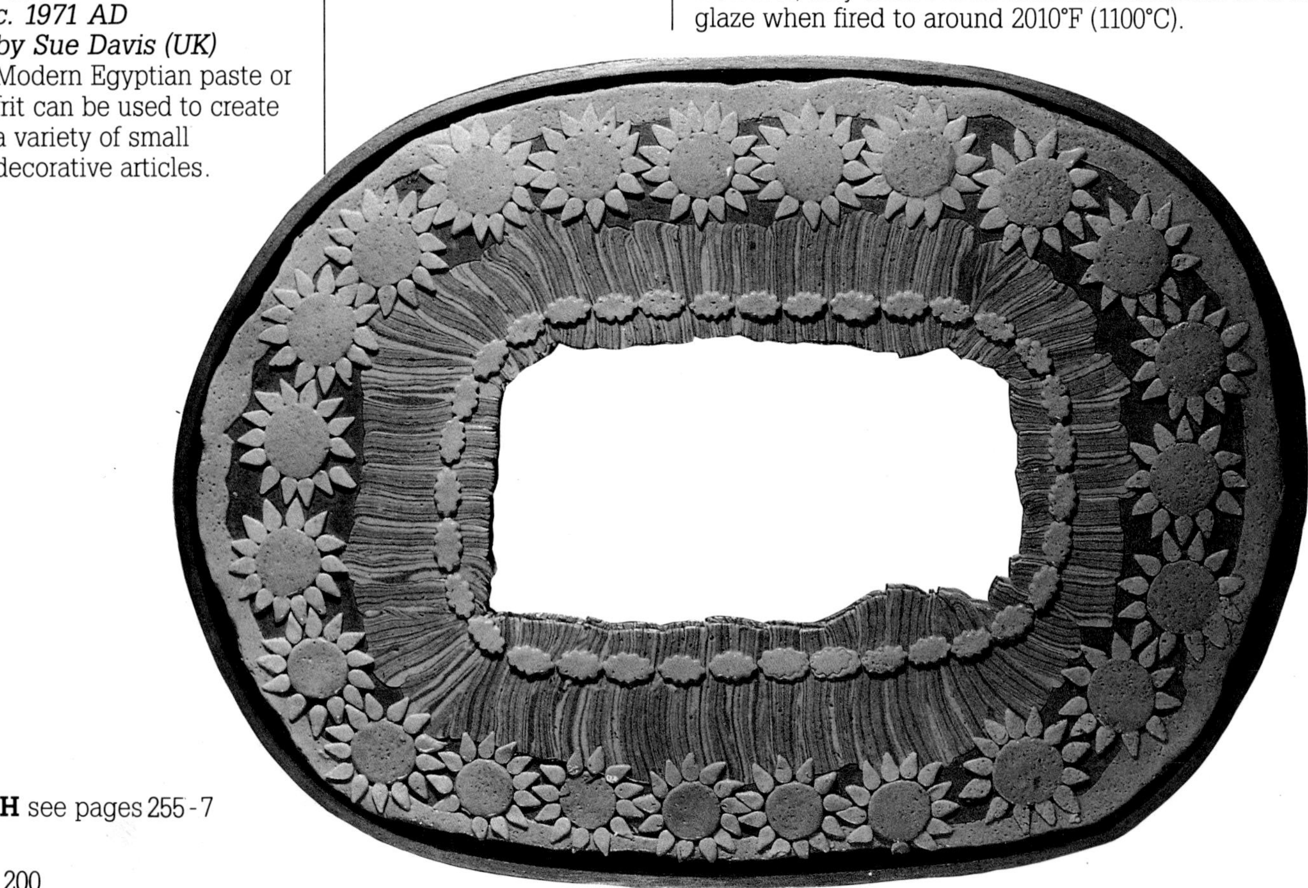

Egyptian paste
c. 1971 AD
by Sue Davis (UK)
Modern Egyptian paste or frit can be used to create a variety of small decorative articles.

H see pages 255-7

Egyptian paste

Sue Davis is one potter who advocates the case of Egyptian frit as described below:

'Because of its consistency it is difficult to form into much else other than beads and small sculptures, as the Egyptians did. However, with research there are recipes which manage to keep the same exciting colors and enable you to manipulate the paste even to the point of throwing.

As the paste relies on ingredients migrating to the surface, they should be weighed including the water. The clay content of the paste should be kept below 40 per cent as higher would inhibit the glaze formation. At least 10 per cent of some soluble soda compound is necessary. The color is also added to the body and I have found both oxides and body stains, up to 10 per cent, interesting. Experiment!'

First weigh and mix all dry ingredients except the soda and coloring pigment. Now weigh the hot water. Add the soda and color to half the water; it will need a good stir. Pour this mixture into dry ingredients and mix as well as possible.

You will find you have a residue of soda left. Boil the other half-measure of water and try to dissolve the leftovers. Add to the potion.

Tip out on to a board and wedge. DO NOT use a plaster bat or disaster will strike. You can leave it on the board to dry a little if needed. Now the frit is ready for you to mold.

Egyptian paste (5000 BC)

1705°F (930°C)

Ingredient	Amount
Feldspar	40
Flint	20
Kaolin	15
Ball clay	5
Sodium bicarbonate	6
Soda ash	6
Whiting	5
Fine white sand	8
Water	20

Egyptian paste (circa 1971 AD)

1705-1830°F (930-1000°C)

Ingredient	Amount
Ball clay	35
J Frit (P2245).	12
Soda ash	6
Fine white sand	10
Water	20

Color

Color	Oxide
The famous turquoise	2% copper
Rich blue	2% cobalt
Pink	1% manganese
Brown	5% manganese
Lime green	5% chrome

Points to remember:

Egyptian paste is porous: keep it covered with plastic and never let it dry out.

The paste responds well to agating and marbling.

Mixing slightly wet and sieving evens out the color but make sure the paste is very dry before firing.

The soda may continue to migrate after firing; washing with a little vinegar in the water will overcome this problem.

Protect your kiln bats with silver sand and be prepared for a little gentle after-fire grinding. Surfaces not exposed to the air will not glaze, and consequently can be fired on that part.

Modern Egyptian paste can be used for mirror frames, lidded pots, brooches, bowls, sculptures and beads (which should be fired on Nichrome wire).

Lead glazes

Most glazes firing between 1742°F (950°C) and 2010°F (1100°C) have a lead base. Lead was first used as a glaze about 1000 BC when, in the highly poisonious form of galena, it was dusted on to wet pots and combined with the clay of the pot to form a clear glaze. Lead glazes were simple to mix and have a relatively low melting point — they are reasonably hard, glossy and easily cleaned. They were used over all body colors and slips, or colored with oxides — copper, green and iron-browns being the most common.

Glazes containing lead in a safe fritted form (low solubility frit) are ideal over underglaze colors which have been painted on to a light body or white slip coating, or over trailed or painted slip decoration. When opacified with the addition of 8-10 per cent tin oxide they also make an ideal majolica glaze for over-painting with different pigments.

Raw lead is poisonous and, should never be used. Great care must always be taken when mixing any glazes. Wear a mask and always wipe any spilt powder away with a damp sponge. Unless properly formulated, lead release can still take place when the pot is used to prepare food with an acidic juice. In particular copper should not be used as a colorant in any form with lead-based glazes intended for food because this will increase the risk of lead release. The glaze should be tested, in case of doubt, to ensure it meets the requirements concerning lead release.

There are many good lead frits produced by manufacturers. Frits are almost glazes in themselves, needing only the addition of 15-17 per cent of ball clay to make a good clear glaze, or red clay to make a 'honey' glaze. As already mentioned, the addition of up to 10 per cent of tin oxide can make a fine majolica glaze, letting the color of the body shine through where the glaze becomes thinner on rims, edges and handles.

High-fired glazes

Covering a dark body with slip to lighten the color and painting slip decoration using different colored clays has been used since pots were first made. With the development of Chinese high-firing kilns as early as 500 BC, certain slips would melt when temperatures of 2190°F (1200°C) or above were reached — and thus glazes were created. It was also discovered that wood ash blown through the kiln from the fire mouth created a glaze. Abundant materials such as feldspar (a decomposed granite found in many parts of the world) were the bases of simply formulated but beautiful glazes made and used by the Chinese potters for centuries before the high-firing, down-draught kilns were developed in Europe.

At temperatures of 2190°F (1200°C) and above, a simple mixture of flux, silica and alumina will produce a glaze more readily than glazes fired to 2010°F (1100°C). Feldspar, containing fluxes such as potash, soda and lime is almost a natural glaze on its own and will melt to an uneven milky white at 2280°F (1250°C); some of the early Chinese wares were glazed in this way.

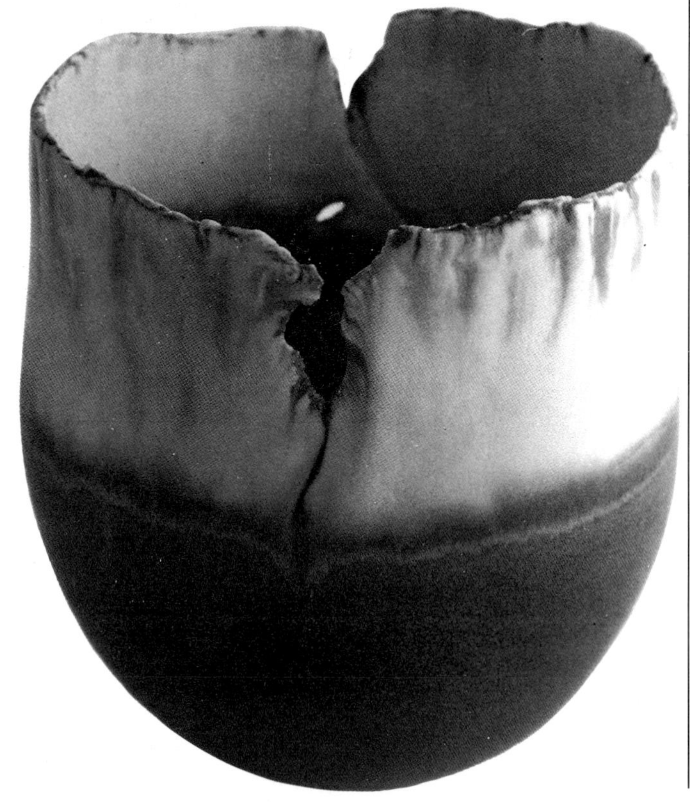

Deep bowl by Mary White (Germany)
A dark-brown mottled glaze and a dolomite glaze have been combined on this porcelain thrown form.

Celebration glaze by Sally Bowen Prange (USA)
A porcelain treasure chalice has been given a gleaming glaze surface with gold luster on a silicon-carbide slip glaze.

John S. Cummings (USA)
Porcelain with a fine celadon glaze has been complemented by the simple piercing.

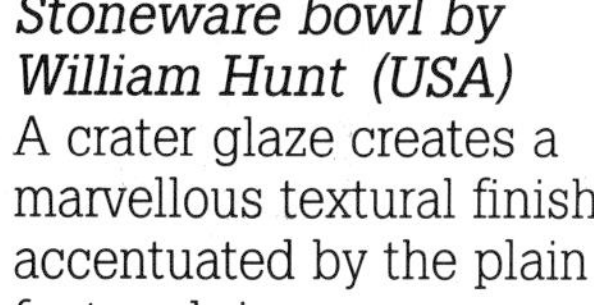

Stoneware bowl by William Hunt (USA)
A crater glaze creates a marvellous textural finish accentuated by the plain foot and rim.

H see pages 266 - 7

Ash glazes

The discovery that partial glazing occurred when ash was blown through a high-firing kiln to settle on the rims and shoulders led to the intentional use of ash as a flux in glazes.

Ash is derived from any organic material which is burnt. The commonest ash used for glazes is wood ash, but ash from grass, straw, vegetables, coal and even wool can be used. Finding a constant supply that will give consistent results may be difficult but any wood ash from burnt felled trees or wood that has been treated or painted — like old furniture or floor joists burned on a demolition site — can be utilized.

Ash contains silica, alumina and flux and many trace elements, the proportions of which can vary enormously. Some ash, if it contains a predominance of flux, will melt alone to become a glaze around 2280°F (1250°C); some with more silica than anything else will be dry and rough. If the ash melts easily it may require only a percentage of alumina in the form of clay, to make an acceptable glaze. If too dry, additional fluxes can be used together with percentages of feldspar and/or clay to make a good base glaze to which colorants may be added.

Preparing ash

The ash should be a light gray color when properly burnt. Sieve out any large pieces of charcoal and then soak the ash in water, stirring with a piece of wood. Take due care over this operation; the ash and water mix can be very caustic and will burn the skin. Rubber gloves are advisable to protect your hands. ◊

Allow the mixture to settle and then remove the water off the top, and wash again. This repeated washing removes the soluble salts from the ash. Some potters prefer to wash their ash many times, while others leave the ash unwashed. It depends on the source and the fired results they have obtained. I have used a fine gray ash from a wood-firing stove which burned off-cuts from a wood yard. By sieving the larger pieces out, I was able to mix a fine glaze using the basic 40 parts wood ash, 40 parts feldspar potash and 20 parts clay for a semi-matte stoneware. 50/50 of ash and clay makes a dry sculptural finish which can produce shades of yellow or red-brown depending on the clay used. The Rhodes recipe for a wood-ash and talc glaze makes a smooth buttery glaze which is very good for oxide painting. (See *Clay and Glazes for the Potter* by Daniel Rhodes.) It is wise to sieve the ash and water mixture through an 80-mesh sieve (or even finer) before drying and storing the material for subsequent use.

Reduced and oxidized glazes

The choice of oxidized or reduction firing of glazes is usually dictated by the type of kiln that you have.

Reduction firing involves introducing a smoky atmosphere through the reduction of the oxygen supply, controlled by the closing or partially closing of dampers and flues. This is best tried on a fuel-burning kiln. Occasionally electric kilns can be reduction fired by introducing through the spyhole a burning substance such as moth balls or slivers of wood but this can damage the elements; and is not to be recommended. (Several oxidized firings will be required in between to counteract the effect of reduction on the elements.)

Electric kilns are normally used for oxidized firing although it is possible to buy kilns in which the electric wire elements are enclosed in ceramic tubes in order to overcome the problems posed by reduction atmospheres. Even so, the results are scarcely comparable to those obtained in a live-flame kiln.

Wood-firing kilns which have periods of oxidization and reduction naturally as the kiln is stoked produce ideal conditions for a reduction firing. Other fuels such as oil, gas, coal and coke may be used, oil and gas being the most readily controlled of these.

Each type of firing offers its own attractions and each potter has his or her own particular favorites; one is not necessarily better than the other. While oxidization produces a wide range of colors, reduction produces subtle, often less predictable variations which cannot be attained in any other way.

Iron and copper are perhaps the most versatile of the coloring oxides in reduction. A small percentage of iron in a simple feldspar, whiting and clay stoneware glaze (about 0.5-1%) will make a pale-blue or green celadon glaze over a light-colored body like porcelain. It is especially effective over a modeled or cut pattern. The color of an iron glaze may be gray, blue-gray or olive-green through to black. Copper oxide will change from its usual oxidizing color (green), to a rich copper red — but this can be an elusive color unless the firing conditions are carefully controlled.

Raku

The clay used for Raku must have a heavily grogged, open texture to withstand the thermal shock of being placed into the red-hot kiln, taken out again when the glaze is melting, and immersed in water or various combustible materials (such as sawdust, shavings, peat and leaves) to induce partial or total reduction.

Raku glazes are usually fired between 1380°F (750°C) and 1560°F (850°C) after first having a biscuit firing to about 1830°F (1000°C). The glazes can be used with additions of oxide applied under or over this, or with painted slips on the body. Leaving parts unglazed allows the additional contrast of smoke-blackened clay against either a white ground or a brilliant color.

The appearance of reduced copper oxide is often spectacular when used in a Raku firing. The glazes used for Raku are often high in alkali. This, combined with the copper and a post-firing reduction, induced by burying the red-hot pot in sawdust or some other combustible substance, often makes the copper appear lustrous green to copper-bronze or a deep red — depending on the thickness of the application and the degree of reduction produced.

Raku-fired pots can be cooled rapidly by immersion in water immediately after post-firing reduction has taken place. This will prevent the glazes being subjected to further change due to re-oxidization.

Raku is exciting and spectacular to watch and is no longer mere experimentation. It has been developed to a high art by many potters who produce beautiful individual pieces of work. (For more details on Raku firing see pages 243-4.)

Using enamels

Onglaze enamels are prepared colors mixed with flux. They have the property of melting at low temperatures and are used over a fired glaze. They are then refired to around 1380°F (750°C) — this will fuse them into the glazed surface without melting the glaze itself. Practice in the application is needed so experiment first on a glazed tile that can be easily wiped clean if mistakes are made. The colors are mixed with an oil medium to adhere them to the glazed surface and allow for ease of application. They can be applied by brush or sponge or spray.

Try thinning with turpentine for a wash effect or mask the surface with stencils and spray.

The firing should be slow at first, with plenty of ventilation in the kiln and kiln room to allow the oil medium to burn off. If the firing is too fast the oil will bubble and blemish the surface color. Some colors benefit from a low firing followed by a second firing at slightly lower temperatures again.

Two shallow bowls by Jan Schachter (USA)
An ash glaze has been used to create a beautifully rippled surface on the first bowl. The tenmoku glaze on the second has been resisted by pieces of palm fronds; when these burn out in the kiln distinctive gold markings remain.

Luster plate 17th century
From Valencia or Manises this earthenware enamel plate depicts a pelican pecking at its breast.

Bowls and jug by Lisa Katzenstein (UK)
Yellow ocher glaze has been sponged on to produce a speckled ground with on-glaze transfers superimposed.

Porcelain pots by Geoffrey Swindell (UK)
A basic dolomite-type glaze has been used with varying quantities of copper carbonate to achieve color.

Boats by Tessa Fuchs (UK)
The gleaming glaze finish adds to the sense of light and movement in this delightful piece.

Black basalt dishes (detail) by John Wheeldon (UK)
Luster decoration has been applied over a matte stoneware glaze. The attractive decoration has been accomplished with resists, soft rubber stamps and brush work.

Tureen by Sandy Brown (UK)
Sandy Brown uses oxides and commercial underglaze stains in powder form which she adds to the transparent glaze as a medium. She then uses wallpaper and paint brushes, glaze trailers or hands and fingers to apply the color.

Lidded pot by Bryan Trueman (Australia)
Deep cobalt-blue glaze complements an Eastern-inspired pot.

Apple by Derek Davis (UK)
Tin glaze over copper has been used to produce the gleaming apple surface.

Joan Campbell (Australia)
Low-fired porcelain with pattern produced from seaweed fired over.

Carved porcelain by Karl Scheid (Germany)

Plate, bowl and storage jars by Jane Hamlyn (UK)
Salt-glazed stoneware has been raw-glazed and once-fired and has the typical 'orange-peel' texture.

Porcelain bowl by Ann Kenny (UK)
The piece revels in a rich glowing lustrous finish, with gold enhancing the other colors.

Glaze materials and methods

Porcelain bowl
by Greg Daly (Australia)
A gold luster glaze has been complemented by gold and silver decoration.

H see page 271 and 276-7

The importance of fluxes

Glazes are often referred to by the kind of flux used in their composition — for example a lead glaze, an alkaline glaze, a dolomite glaze and so on.

Often more than one flux is used in a glaze. Lead, soda and borax are the main low-temperature, very active fluxes. Feldspar is naturally occurring mineral composed of silica, alumina and a flux.

The most common feldspar used by potters is potash feldspar — so called because potassium provides the major flux content. Feldspar contributes a proportion of flux in high-firing glazes (that is glazes fired over 2190°F or 1200°C) but it is usually aided by others such as calcium oxide and magnesium oxide.

To make a good basic glaze, feldspar (together with a small percentage of whiting) will form a useful transparent glaze to which may be added coloring oxides or opacifiers.

The fusion point of a glaze can be controlled by increasing or decreasing the amount of flux which is present in the composition.

In an earthenware glaze, that is a glaze firing at around 2010°F (1100°C) the combination of fluxes might be borax or lead, whiting and potash feldspar, with flint as the main source of silica and with clay to contribute mainly silica and alumina.

Achieving the correct mix and conditions

The right balance of flux, alumina and silica has to be achieved to make a particular glaze work. Too much silica in its composition will make the glaze too refractory, or heat resistant; not enough will produce a glaze that is soft and liable to craze. Too much alumina will make the glaze matte, or dry and opaque. Too little or no alumina will effect the adhesion of the glaze so that it is more likely to run off the pot.

The range of glaze materials available from suppliers today can be confusing to the beginner but, once the basic chemistry is understood, these ready-prepared constituents certainly ease the burden of the glazemaker. Ancient potters had to compound their glazes from minerals dug and ground and pulverized before mixing. There must have been many variations in the end result, causing delight or disappointment. Most potters still experience that element of surprise and anticipation when opening a glost kiln. The action of very high temperatures on clays and glazes, especially the color change of oxides and clay bodies in a reduction firing, the effect of smoke and flame and the transformation these bring about, often make the long process of preparation and forming all the more worthwhile.

Using lusters

Lusters are metallic salts mixed with a fat oil medium and applied to a glazed surface (see page 188).

One technique of luster painting uses salts of metals painted on a fired glazed surface before refiring in a reduced atmosphere to about 1380°F (750°C) — that is, in a smoky kiln where the temperature is just high enough to fix the color to the surface. This method probably originated in Egypt from glass-making processes and was used in Mesopotamia from the 9th century onwards. The glittering effect obtained by the reduction of copper oxide was also used with great effect on both the Persian and Hispano-Moresque wares.

The iridescent coating of metal is usually polished after firing to bring out the color. A more common technique is the use of commercially prepared metallic pigments. These are ready-mixed with a reducing medium which burns away to create a local reduction. Silver, gold and bronze produce a bright metallic surface but many other colors are also available.

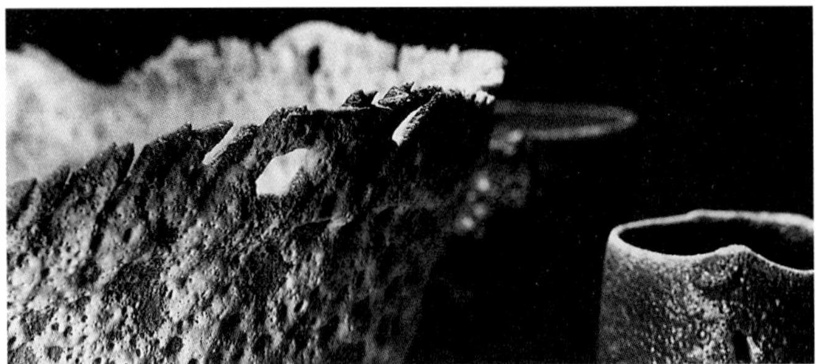

Vessel by Sally Bowen Prange (USA)
A matte glaze helps suggest a barnacled or coral surface.

Using glaze at stoneware and porcelain temperatures

Clay fired to stoneware and porcelain temperatures is almost vitrified and will not need a glaze all over to make it watertight. If high-fired pots are placed on stilts the shape will be distorted, the stilt will embed itself in the foot of the pot or, if the pot is very heavy, it may even collapse over the stilt, making this irremovable.

Most stoneware and porcelain is placed directly on the kiln shelf on a layer of silica sand or an alumina bath wash, with even support across the base. The foot ring or the whole base must be either resisted with wax or latex or wiped clean before it is placed in the kiln. If the glaze tends to run or is thickly applied, it should be removed a little way up the side of the pot to prevent it sticking to the shelf.

Glazing biscuit ware

Biscuited pots are very absorbent and this is used to advantage when glazing. When the glaze mixture is put on the pot, the water content sinks into the porous surface leaving a smooth layer of glaze over the pot. The absorption is immediate; this means that if the pot is held for too long in the glaze a thick coating will build up on the surface. Glazing must be accomplished smoothly and quickly. Your confidence will increase with practice.

Tools required when mixing a glaze

A good set of scales with fine adjustment (digital if you can afford it) which is capable of weighing fairly large amounts: 11 lbs (5 kgs)

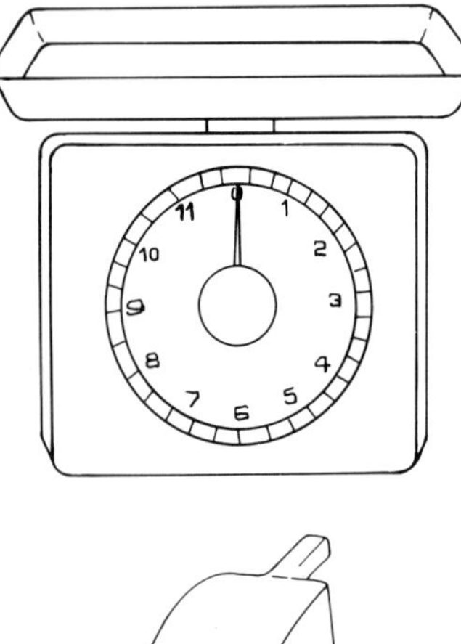

A scoop of some sort.

Various sieves: 60, 80, 100, 120-mesh

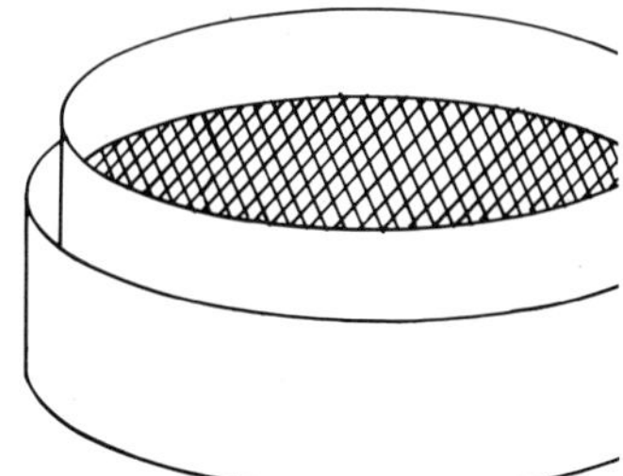

Pair of wooden battens for supporting sieves over buckets.

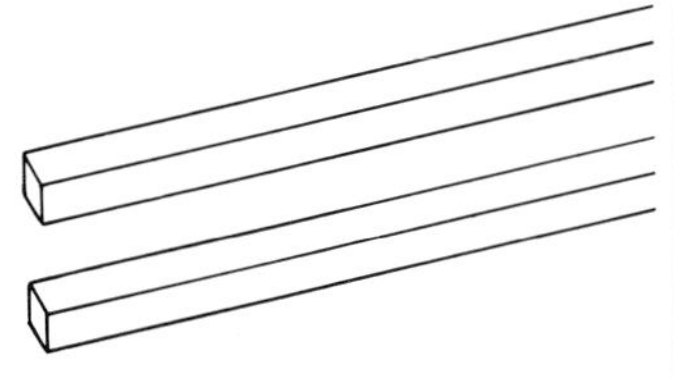

Lawn brush: a strong bristled brush will be needed for sieving. Washing-up brushes intended for saucepan scrubbing are good!

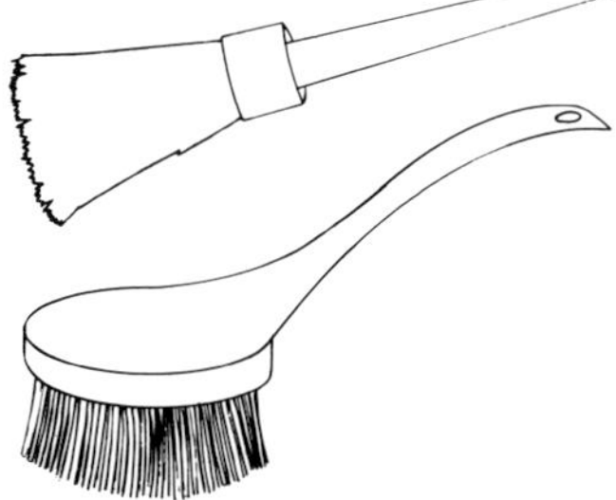

Jugs: both large and small, plastic measuring type.

Plastic buckets with handles and lids. These can be quite expensive so look out for 'throw-outs' from food shops.

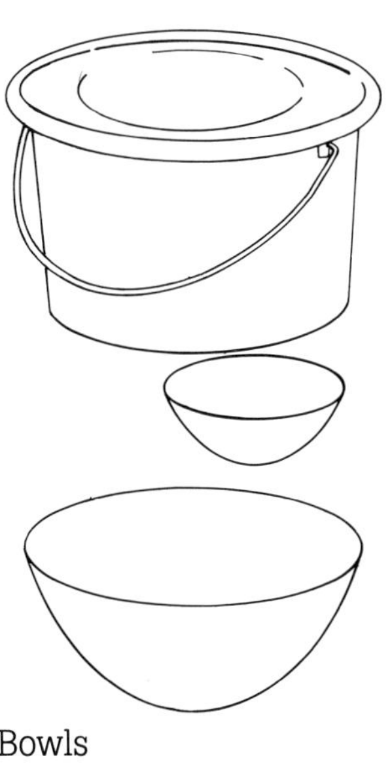

Bowls

Mixing dry powders ◇

Remember, before glazing your pots you must remove all dust particles from the surface with a damp sponge or 'crawled' glazes will almost certainly be the disappointing result.

Wear a protective mask

1 Take a clean dry bowl. Measure out the first ingredient.

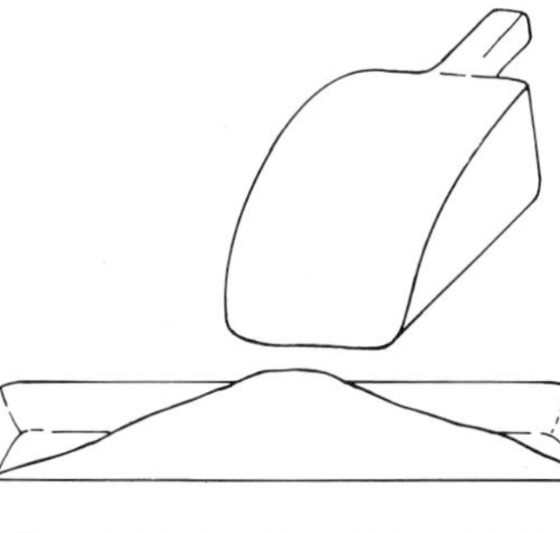

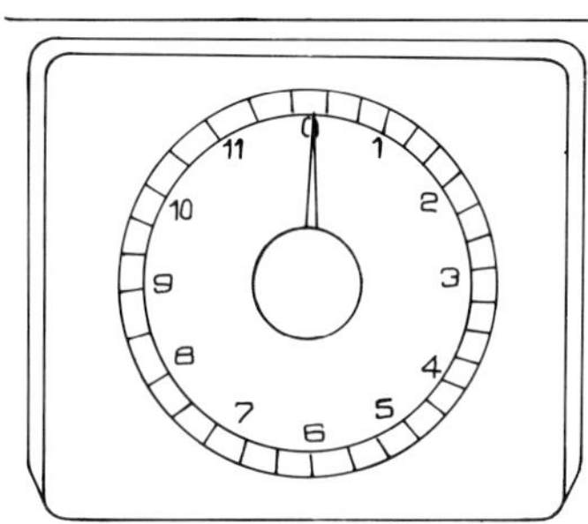

2 Place this in a confined area on the edge of the bowl. Before mixing, the other ingredients should also be isolated in piles at different points around the bowl. In this way you can make a final check and ensure no ingredients have been overlooked.

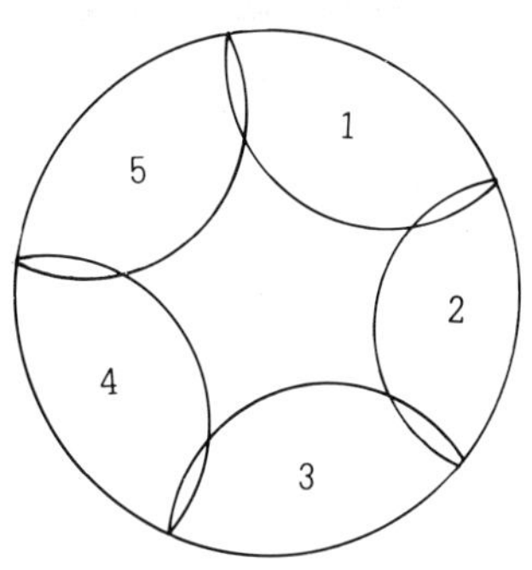

3 Cover the ingredients with water and stir together; then leave to soak for 30-60 minutes. Stir again.

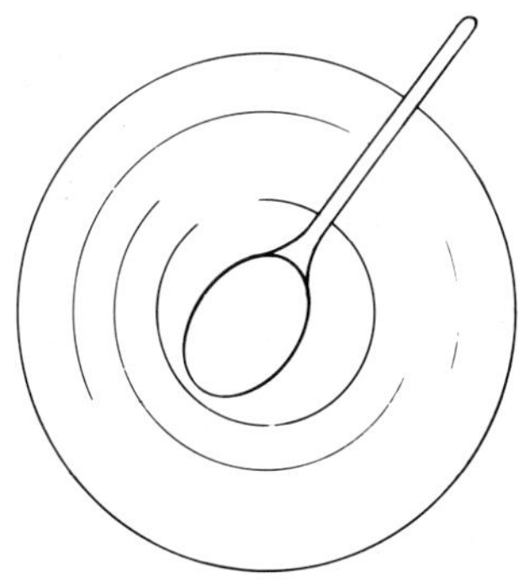

4 Pour the mix through a 100-mesh sieve. If any deposit is left, push it through the sieve with a brush or rubber kidney.

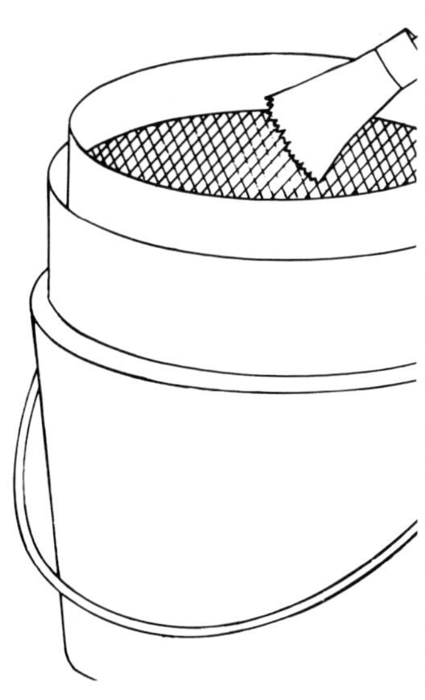

5 If you have used the glaze before, then you will know how thick it needs to be. If not, keep the consistency like thick cream and test it on a trial piece. Then thin another sample down and try that on the same piece — to determine the correct thickness. Apply some underglaze and onglaze colors too and test their reactions with the glaze.

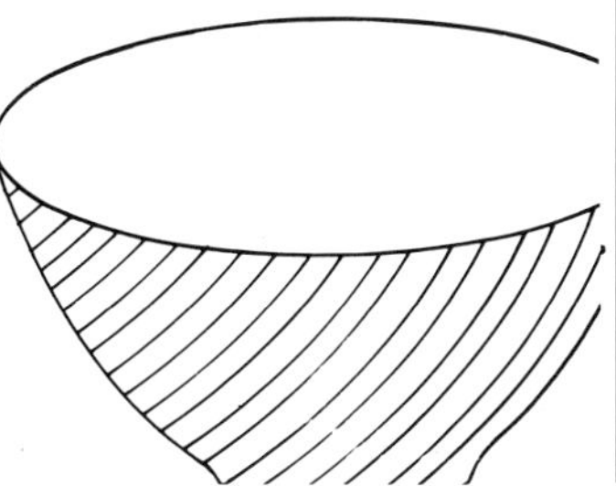

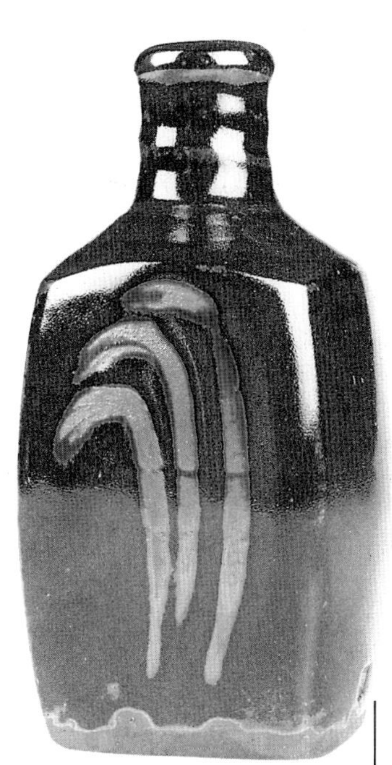

Vase or bottle by Jim Malone (UK)
Made from a thrown and beaten square, this piece can be used as a bottle or vase. Thick tenmoku has been applied over ochre slip and then finger wiped.

Applying the glaze

Glaze can be applied by painting, pouring, dipping or spraying — or any combination of these methods. Usually the choice depends on the size and shape of the pot and, most important of all, the effect you are aiming to achieve in the final firing. A few things have to be considered before commencing. If the pot is to be fired to earthenware temperatures (about 2010°F 1100°C or below) then the clay under the glaze coating will remain porous. Earthenware pots, if they are to be used for food preparation, need to be glazed all over, including the base. If the foot is left unglazed gradual absorption of liquid and food particles will make the pot unhygienic. Moreover, dark stains will appear under the glaze and cause it to craze.

This means that the pot cannot stand directly on the kiln shelf but must be placed in the kiln on a stilt or stand. Stilts come in many shapes and sizes to suit all types of pots and dishes. The most commonly used are three pronged and made of high-fired clay. The pot rests on the stilts during firing. Afterwards the stilt is removed, leaving only three small marks which can be smoothed with a grindstone. Do not run your fingers over the base of the pot until you have cleaned the stilt marks; small slivers of the stilt sometimes stick to the glaze and are razor sharp.

Dipping

Small pots can be dipped fully into a bucket of glaze. Stir the glaze well, hold the pot with the thumb at the rim and the finger at the base, having covered the hand in glaze (the glaze on the fingers will reduce the marks made by them when the pot is removed). Immerse the pot in the glaze, easing your hold slightly to let the glaze reach the parts under your fingers, and remove the pot, shaking off the excess glaze. If the pot has a waxed foot and is clean of glaze (or is going to be cleaned later), then the pot can be placed directly on to the table to dry. If it is to be fired on a stilt then place a few old used stilts on the table and place the pots on them to dry.

Large bowls can be dipped if you have a large enough container to take the bowl and hold a fair depth of glaze. Hold the bowl at either side with both hands and dip first vertically and then in a curving movement to cover the bowl, lifting out and letting the excess drain away.

Pouring and dipping can be combined if the pot is just too big to hold between finger and thumb. Take a jug full of glaze and fill the pot to the rim. Then pour it out immediately. Hold the pot by the foot and immerse it rim-down to the point where your fingers are holding. Keep it vertical and the air inside will prevent any glaze entering the coated inside.

Pouring

If the pot is larger and more awkward than can be held in one hand, coat the inside first and allow to dry. Then place the pot upside down over the hand, hold it over the bucket of glaze and pour the glaze over the outside. The outside can also be coated while supporting the pot on two sticks or rods placed across the top of a clean bowl that is large enough to catch the excess glaze. To make a large pot more accessible, place the bowl, supports and pot on to a banding wheel so that the whole thing may be turned while you pour the glaze.

Painting

Use a large soft brush to hold plenty of glaze and work across the pot in short strokes. At least two or three coats are necessary to make a good covering and even then the brush strokes may still show when the pot is fired. However, the thinness of the glaze and the brushstroke may become part of the effect you are trying to achieve. Watercolor-like images can be built up with different colored glazes. The glaze can also be banded on; the pot is placed on a wheel and a brush of glaze held against it while the wheel is turned.

Spraying

Glaze can be sprayed with a hand-pumped spray or with a spray running on compressed air for continuous power. Spraying must be done in a spray booth with a built-in extractor fan. ◇

This is an ideal method if you wish to blend thin coats of glaze into graded colors or thicknesses or for glazing very thin pieces of porcelain which would become saturated if dipped or poured. By spraying colors over resist patterns or through stencils and handheld masks, a pattern of overlapping shapes can be built up.

Dipping or pouring inevitably involve more direct handling of the object being glazed. Therefore large or awkward shapes which need to be raw glazed (that is, have the glaze applied to the unfired clay) are in less danger of being damaged if the glaze is sprayed.

C see page 86

Applying the glaze

Always stir the glaze with the hand first, making sure there is no deposit left on the bottom of the bucket.

Dipping

1 Place the fingers underneath and the thumb on the rim.

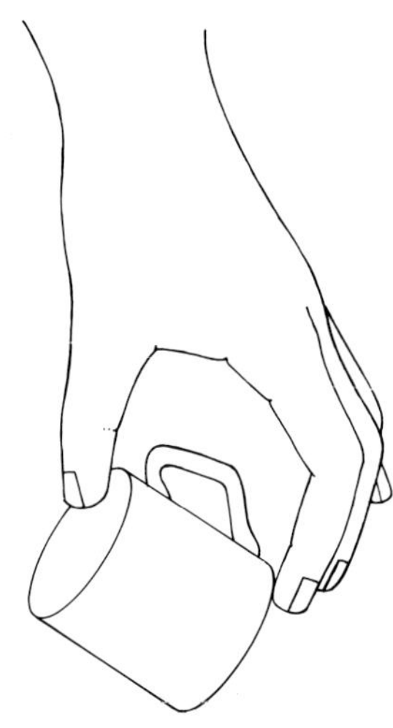

2 If the pot can be comfortably held between finger and thumb, then it may be fully immersed — provided the glaze is deep enough.

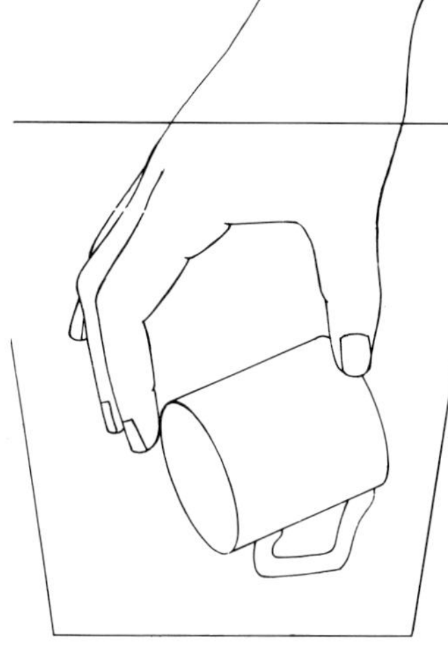

3 Dip the pot in and out of the glaze smoothly and quickly. Drain it upside down for a few seconds and then place it on a stilt on the table to dry. Touch up any finger-marks before the glaze dries; it will dry almost immediately as the water sinks into the porous clay.

Dipping and pouring

Use this method if the pot cannot be held in one hand to immerse or if there is insufficient glaze for dipping.

1 Fill a good sized jug with glaze. Hold the pot upright and fill the inside with glaze halfway and then tilt and twist the pot to ensure even coverage.

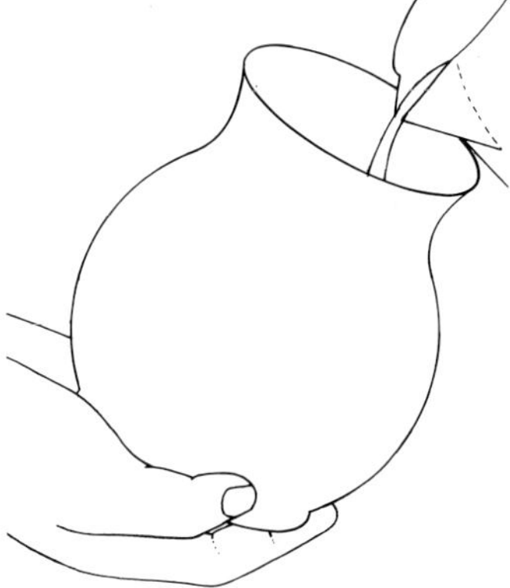

2 Carefully pour out the glaze into the bucket, making sure the glaze has covered every part of the interior surface. Allow it to dry for a few seconds.

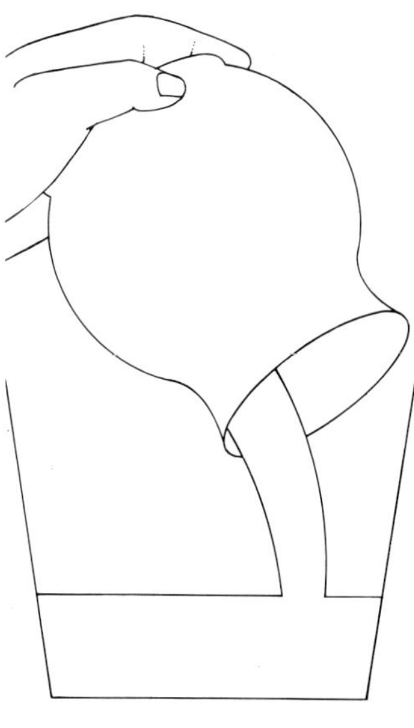

3 Then, holding the pot upside down by the foot, lower it down into the glaze.

The air trapped inside the pot will stop the inside receiving a second coat, while the glaze displaced by the bulk of the pot will make the level of glaze rise sufficiently to cover the outside. Smooth out any marks made by your fingers

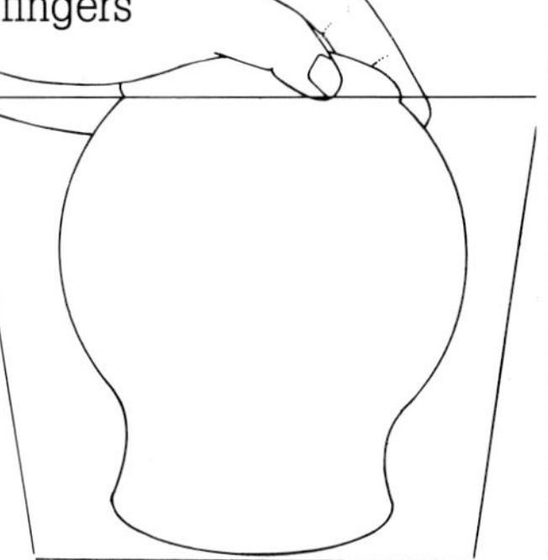

Alternatively the outside can be glazed by holding the pot upright, with the fist inside, and lowering it to the rim in the glaze.

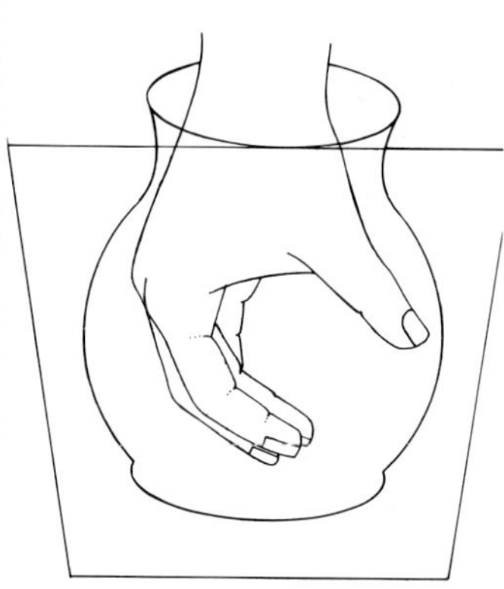

Pouring

If the pot is large it may require a different method of glazing.

1 Place the pot on the table and pour in a generous amount of glaze.

2 Immediately pick up the pot in both hands and pour the glaze back into the bucket, turning the pot as you do so to ensure that all the inside is coated.

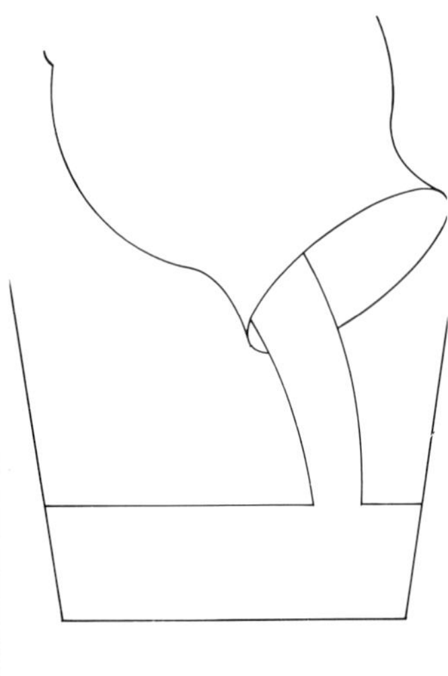

3 Leave pot to stand until the inside is dry.

4 The outside can be glazed by holding the pot over a support—such as your hand, a funnel, a banding wheel, or on two sticks placed across the top of a bucket.

First make sure you have a large bowl underneath to catch the drips and then pour the glaze over the pot.

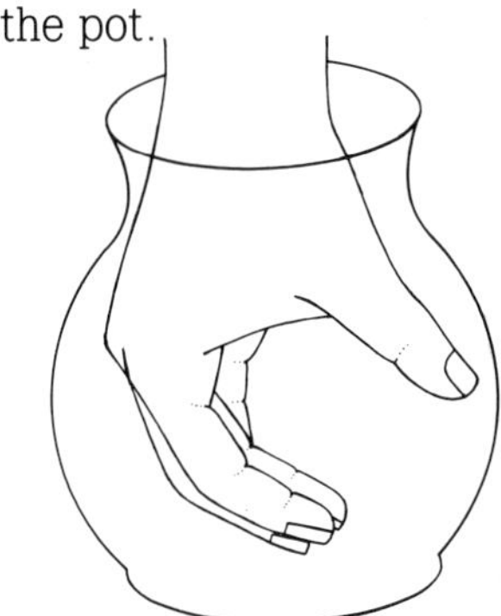

Supporting a pot upside down

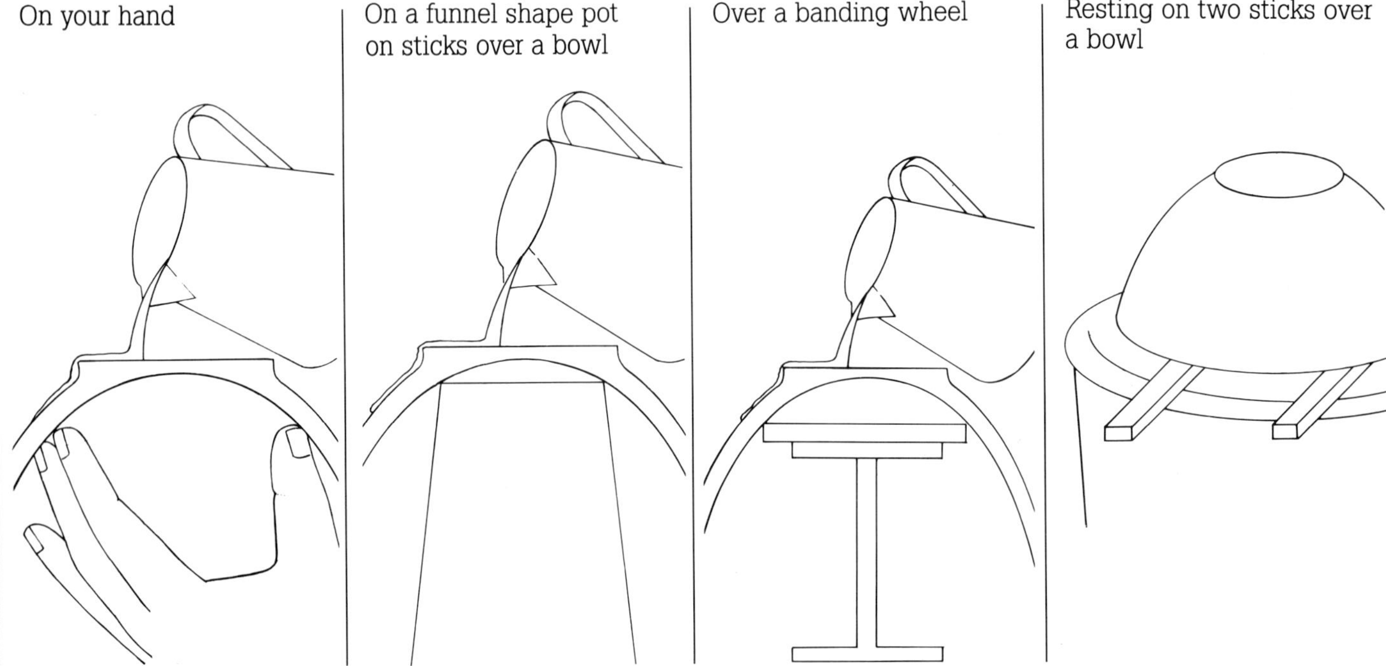

Painting

1 Place the pot on top of another shape on a banding wheel. Spin the pot slowly, holding a large full brush of glaze against the side. Several coats will be required.

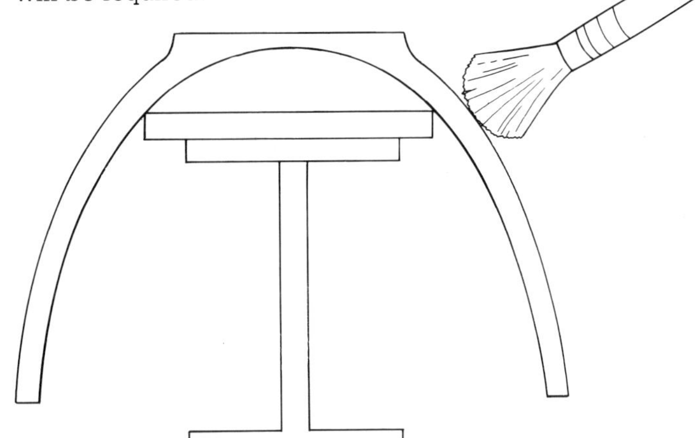

2 Paint the outside first. Allow this to dry and then paint the inside in the same way.

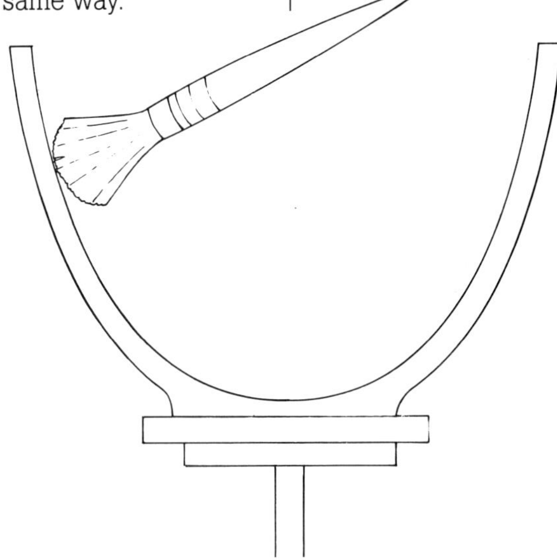

Spraying

The spray gun is powered by a compressor. The glaze is atomized by compressed air. Several effects can be successfully achieved by spraying: these include even coating, fading edges, overlapping, gradation of color and masking.

It is useful to stand the pot on a banding wheel so that it can be turned while spraying. Raw glazing can also be accomplished quite safely by this method.

Spraying must be done in a spray booth that has a means of extracting the glaze particles from the air and conveying them outside. ◊

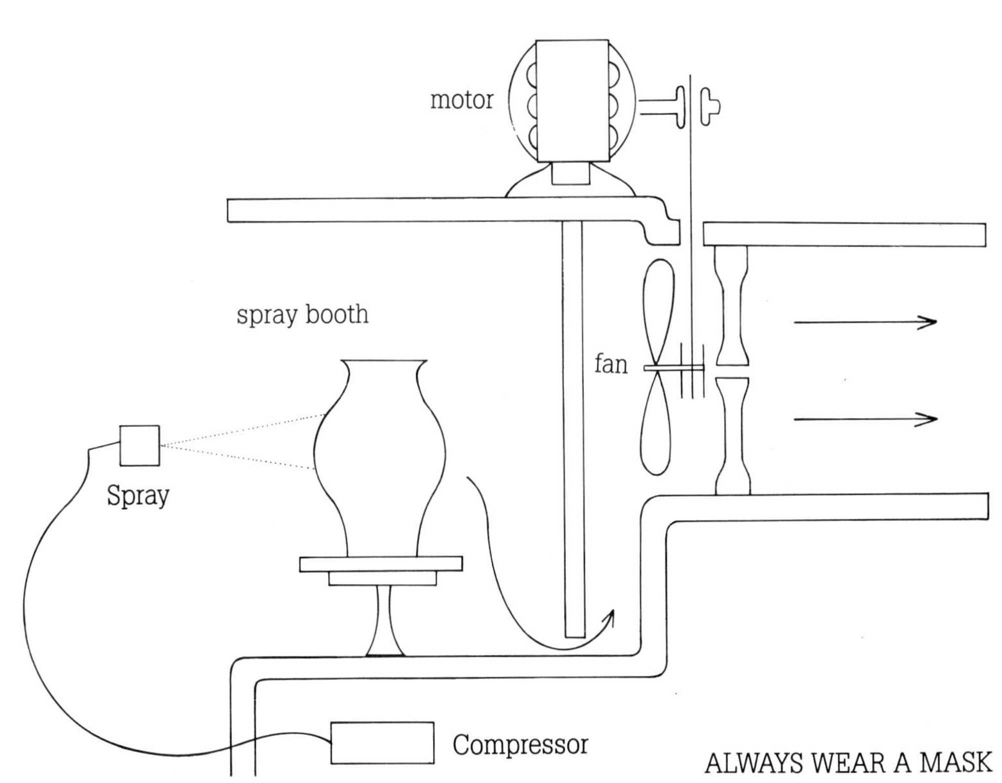

ALWAYS WEAR A MASK

A reference list of glaze materials.

This is a simplified list of the materials most commonly used in the composition of pottery glazes. It is not necessary to become involved with complex chemical formulae, atomic weights or molecular equivalents in order to produce satisfactory glazes, unless you develop a special interest in chemistry and glaze technology. Many potters prefer to work empirically and intuitively once they have learnt the basic essentials. There can be no realistic substitute for practical experience. The sense of achievement and satisfaction experienced when you first discover how to compound your own glazes from raw materials will probably encourage you to read more widely around this intriguing aspect of ceramics.

Remember that the composition of the various glaze materials will not always be the same because individual sources differ. Published recipes, and any descriptions of their behaviour under particular conditions, should be taken only as a guide; let them act as a starting point for your own experience.

Silica

This is the principal glass former that must be present in all glazes. It is one of the most common elements and often occurs naturally in combination with other minerals but the major source of silica is in a crystalline quartz which is finely ground and generally refered to as flint. Flint has a very high melting point around 3119°F (1700°C) but, when heated it combines readily with a variety of fluxes which lower the fusing temperature. Adjusting the proportion of flint in a glaze recipe will raise or lower the point at which that particular mixture melts. The higher the melting point of a glaze the harder it becomes and its resistance to acid attack is similarly increased.

Alumina and clay

Alumina plays an important part in most glazes because its extremely refactory nature modifies the fusibility of the glaze mixture. It is rarely used in its pure form and is more usually incorporated in the form of clay.

Certain clays containing large amounts of iron and other impurities when dug and will, if fired high enough, be transformed into acceptable slip glazes with no other additions being necessary. Most clays can be used as part of a glaze recipe and they will contribute some silica as well as alumina to the mixture.

China clay (kaolin) is normally preferred because it is relatively pure and is less likely to affect color adversely. The presence of clay in a glaze also helps to keep it in suspension while in slop form. (Up to 2 per cent of bentonite, an extremely plastic colloidal clay, is a good suspension agent for stored glaze mixtures.) When large amounts of china clay are used the glaze surface tends to become matte and opacity develops.

Feldspars

Feldspars that consist of silica and alumina combined with alkalis — notably *potassium* (potash feldspar) and *sodium* (soda feldspar) — possess all the ingredients for a natural glaze and these two forms of the true feldspars are those most commonly used by studio potters. Other feldspar materials, known as feldspathoids, include *Cornish stone* (sometimes called Cornwall stone, China stone or Carolina stone) and *nepheline syenite* — both of which can be used as the major ingredient and principal flux in many glazes, especially those intended for stoneware and porcelain bodies.

Fluxes

The essential component of any glaze, and that which determines much of its character, is the flux or fluxes which are necessary to lower the melting point of silica and alumina. Of course certain amounts of flux will be present in any feldspathic glaze but these will rarely be sufficient to produce an acceptable surface. Further amounts of flux will usually be required and there are various natural sources of these elements available in insoluble form. Try to learn the part that each of these can play in composing your own glazes. The following materials are the main fluxes favoured by studio potters:

Ashes

Vegetable ashes of all kinds, properly prepared with all organic matter removed, contain proportions of silica and alumina with various alkalis such as potassium, calcium and magnesium — together with traces of metals such as iron. They can provide a powerful primary or secondary flux, according to the amounts used, at stoneware temperatures. Ash produced from trees will give varying results according to the kind of tree, the season in which it was cut, and the soil in which it grew. Experimental work with ash glazes can be an exciting and rewarding adventure, especially if you have access to a kiln capable of being fired in reducing conditions. Ashes can form thin glazes on their own but the addition of clay will help to bind the particles together for easier application as well as 'fleshing' out the glaze. In some recipes they are included to perform as secondary fluxes, giving the glaze a distinctive character.

Barium carbonate

This a very useful flux for stoneware glazes because it encourages the formation of bright colors from metallic oxides like copper, manganese and nickel. It is, however, poisonous in its raw state and should always be handled with great care and discretion. It is rarely used in quantities above 15 to 20 per cent because it has a strong matting effect, and large amounts can lead to problems such as 'crawling'. When used with lead frits at earthenware temperatures a smooth matte surface can be achieved. ◊

Borax

Borax is a good flux at all temperatures but it is soluble and is best introduced to the glaze batch in the form of a borosilicate frit. An alternative source of boron is *colemanite* which contributes both boric oxide and calcia to a glaze. Colemanite is a powerful flux for both low and middle temperatures — it is usually supported by other fluxes with the glaze recipe.

Calcium carbonate

In the form of whiting this is the most useful source of calcium oxide in glazes and it is one of the cheapest and most versatile of fluxes available to the potter. In stoneware glazes fired in reduction, whiting is often used because, amongst its other attributes, it encourages the development of celadon colors when traces of iron oxide are present and rich reds with small amounts of copper oxide. *Wollastonite* (calcium metasilicate) is an alternative source of calcium in high-temperature glazes.

Dolomite

Dolomite consists of almost equal amounts of calcium carbonate and magnesium carbonate. It is a popular flux for stoneware and porcelain glazes. Dolomite glazes often have a smooth buttery appearance. A standard glaze based on the well-known Rhodes recipe, for example, is potash feldspar 49 parts by weight, dolomite 22.5, China clay 25, whiting 3.5. This glaze has a firing range between cones 8-10 and responds to colorants remarkably well under both oxidizing and reducing conditions. Mauves and violets can be obtained in combination with small amounts of cobalt oxide while copper produces a range of colors from pale grays to black; sandy to gingery browns; and pinks — according to the proportions used and the prevalent firing conditions. This same glaze can be snowy white in oxidation but even the smallest amounts of iron in the underlying body will tint it a cream color in a reduction firing.

Lead

Lead is extremely toxic and can be safely used only in the form of a frit (*lead bisilicate, lead sesquisilicate*, or combined with boron as *lead borosilicate*). Lead sesquisilicate glazes carry a greater risk of solubility when subjected to acid attack and the increased awareness and public concern for health and safety has reduced the popularity of such glazes on domestic wares despite the valuable qualities offered by lead in many earthenware and low-temperature glazes. It can produce rich, bright clear colors and glossy surfaces. ◊

Lithium carbonate

This is an extremely powerful flux so that only small amounts are normally used. It acts in a similar way to sodium and potassium but it also brightens colors obtained from metallic oxides while at the same increasing the firing range of glazes. Lithium is also available as *lepidolite, petalite* and *spodumene*.

Magnesium carbonate

Rarely used alone, magnesium carbonate is normally preferred in a natural combination with other elements such as dolomite and talc because it is slightly soluble. The qualities and color response are similar to that already described under dolomite. Magnesia is refactory at lower temperatures where it often produces sugary matte glaze textures.

Talc

Talc is a magnesium silicate, used as a secondary flux in stoneware glazes, which has the added advantage of improving the fit of glaze and body due to its low thermal expansion.

Zinc oxide

This is often used in crystalline glazes. As a secondary flux it is normally included with other stronger fluxes. It acts best as a flux when only small amounts are present in a glaze. Higher proportions (above 10 per cent) produce mattness and opacity.

Coloring oxides and commercially prepared coloring agents

Many different colors and effects can be achieved when the following oxides are mixed in varying quantities. Results will depend on the glaze composition, firing procedures and temperatures. Experiment and carefully record all your discoveries for future reference.

Iron oxide

This is a most versatile and widely used oxide which is naturally present in most clay bodies. Apart from the very pure white china clays, varying percentages of iron will be exhibited in clays as buff, ginger yellow, ocher, green, brown, red or black coloration. It is also used to modify other stronger colorants such as cobalt or copper oxides but it can be mixed in various proportions with any others, of course, for different tonal or color contrasts.

When using iron oxide in or on a glaze the most notable difference in results will be apparent in the high-fired stoneware or porcelain glazes, especially between the high-fired oxidized and reducing atmospheres.

Earthenware clear glazes which have iron added produce yellow, tan or red-brown to black. They are rich, warm and affected by the color of the clay they cover (becoming darker on terracotta bodies). If the glaze used is made opaque by additions of the oxide, iron oxides will produce a pale gray or yellow to brown. The color range at this temperature is much wider if the iron oxide is painted over the glaze before firing. The white glaze will make the iron appear as a pale yellow if a small percentage is painted thinly. The range of colors and effects can be further developed as the potter's painting experience improves.

Iron alone added to oxidized stoneware glazes will give variations of pale yellow to brown-black according to the amounts added. The amounts can vary between 0.5 to 10 per cent.

When the kiln is fired in reduction the range of color produced with additions of iron oxide is widened considerably: gray, pale to dark green, pale blue-gray or blue-green, olive to browns and blacks. The pale greens are known as *celadon* glazes and are made by adding small amounts of iron to a clear or semimatte glaze. Over a dark body the result will be a gray-brown perhaps changing where the color of the body breaks through. The subtleness of the various shades of green shows best when applied over a white body, even more so if the surface is cut or incised. The glaze will pool in the uneven surface and becomes darker green where it is thickest.

Japanese potters introduced a heavy iron-content glaze to produce a glaze known as *tenmoku*, a rich black-brown glaze which breaks to orange-red in parts. This is still a popular glaze among studio potters today.

Manganese oxide

Used mainly as manganese dioxide, manganese introduces purple browns in or on to a glaze. It requires a fairly high percentage (at least 3 - 4%) to produce a color. Alkaline glazes intensify the purple tint; in lead-based glazes the color tends to be more brown than purple. Combinations of manganese, iron and cobalt produce black or dark brown; if applied thickly under a tin glaze they bubble through the glaze and produce an interesting, volcanic blistered effect. In combination with copper a lustrous bronze to gold can be obtained at 2280°F (1250°C) in electric kilns.

Copper oxide

Copper has been used by potters as a colorant to produce greens and blues since ancient Egyptian times. In an earthenware lead-based glaze it appears green, ranging from pale to dark green according to the amounts used (0.5 to 5 per cent). Add too much and the surface becomes gray or black and metallic. An alkaline glaze produces Egyptian or Persian blue, or turquoise with the addition of some copper. Stoneware glazes fluxed with calcium also produce green in oxidation. Used in a reduced glaze, copper oxide develops hues ranging from pink to a vivid blood-red color. Alkaline glazes with small amounts of tin and copper oxide create the best results. Interesting color variations, often lustrous, can emerge if heavy copper-content alkaline glazes are used in a Raku firing with a combination of greens and reduced reds. Copper produces oranges, pinks, grays and blacks in oxidation when combined with (or sprayed under or over) dolomite glazes (see also page 240).

Cobalt

Cobalt carbonate is a pinkish-purple powder. Cobalt oxide is a strong black powder and acts as a powerful intense blue colorant; only a small percentage is required to produce a colour (0.5 per cent gives a strong blue while 1 per cent will make a dark blue). As the oxide is a very expensive item this is just as well!

Alkaline glazes with cobalt make an intense, rich bright blue. Using lead glazes will subdue this brilliance slightly. Applied alone, it can be almost too stark but additions of china clay will subdue the rawness — as will combinations of iron or nickel — to make a grayish blue. Cobalt is usually modified by additions of other oxides in higher proportions. A wide range of hues can be obtained for all temperatures (e.g. 0.5 per cent cobalt carbonate plus 3 per cent rutile in the dolomite glaze recipe given on page 319 produces an onyx-like green.)

Nickel oxide

Produces a gray-green or black (1 to 4 per cent). It is often combined with other oxides such as cobalt and iron or copper to make gray tones of that added oxide. In some alkaline glazes rich blues and purples can also be achieved when fired in oxidation.

Chromium oxide

Bright to rather dirty khaki greens are the basic colors produced from chromium oxide, but variations in glaze compositions can produce pink, yellow or brown. In a low-fired lead-based glaze 1 to 2 per cent of chromium oxide in the form of potassium dichromate will produce a red glaze. In a tin glaze chromium will make pink. Combined with cobalt oxide it will give blue/green results.

Tin oxide

Tin oxide is used as an opacifier when up to 10 per cent is added to a clear glaze. Tin oxide (ashes of tin) was first used in the Middle-Eastern countries, the whiteness changing the whole appearance of the wares and thus allowing the development of onglaze color painting and lusters which were refined to a high degree. These spread to all countries of the Middle East and Europe, to become known as majolica decoration.

Rutile

This can be obtained in different strengths; light or dark. Rutile is an ore containing iron oxide and titanium dioxide which is useful in modifying the colors of other oxides and for producing broken or mottled textures. It is a good source of titania and is often preferred to the pure dioxide for use in certain crystalline glazes.

Glaze stains

Oxides can be used to color a clay body, added in percentage to dry clay before mixing. For a wider range of colors most manufacturers produce a good range of readymade body stains which are clear in color and reliable in the firing. Glaze stains can be added directly to a glaze and generally are very reliable provided that the manufacturers' instructions are followed correctly.

Underglaze colors

These are available in a wide range of colors and are mixed with a gum, both to aid adhesion and to make the mixture more fluid before the glaze is applied. Colors in the form of underglaze pastels and pencils can also be bought.

Tureen from Marseilles
In the 18th century the tin-glazing of earthenware produced a white glazed surface suitable for fine decoration.

Stem fruit bowl by Frank Hamer (UK)
A standard plum tenmoku glaze creates glowing color.

Tile by Frank Hamer (UK)
Natural landscapes have been drawn in underglaze pigment with overlapping glazes.

Teapots by Lisa Katzenstein (UK)
Shaded glazes have been used to produce gentle color tones. Glaze enamel transfers and an airbrushed enamel trim have been added.

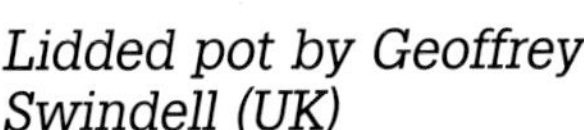

Lidded pot by Geoffrey Swindell (UK)
A resist technique has been used to create streaky effects with crazing of the glaze emphasized by rubbing color into the crazes.

Porcelain beads by Pauline Lurie (USA)

7 Firing

The firing transformation

Clay objects when dried may appear to be strong and durable. In reality they will break easily and will quickly return to the plastic state when immersed in water.

Of all the skills the potter has to learn, the manipulation of the fire to produce physical and chemical changes in the clay is perhaps the most exciting, the culmination of a long process which can result in a pot of great beauty.

However great their ability and however wide their range of experience, all potters share the same sense of joy and trepidation when a kiln is opened after a successful firing and warm pots are handled for the first time; the heat of the fire will have transformed the clay into a personal permanent statement.

Firing your pots will provide the ultimate test of your new-found skills. Whether you are merely packing and firing bisque in an electric kiln or are embarking on a more ambitious glaze firing with wood, oil or gas, you will need to understand something about the process involved before you begin. The basic principles will remain constant, although potters may develop their own individual firing patterns — best suited to the kind of work they do.

Electric kilns offer the most straightforward and generally trouble-free method, but in most cases the atmosphere within the chamber is unchanged throughout the firing cycle. This means that their very reliability (many kilns are fitted with fully automatic controllers) offers less excitement because the potter is only minimally involved, compared with the requirements of living-flame kilns. That is not to imply that pots fired in an electric kiln are likely to be inferior, just different and with more predictable results.

Clearly the potter must learn to understand the effects of heat on clays, glazes and other ceramic materials at particular temperatures. He or she must also appreciate that the work of that heat in transforming earthy matter into ceramic is also conditioned by the length of time taken to complete the process. The other important factor, which creates physical and chemical changes affecting the appearance of clay bodies, and especially the color and textural interest of glazes, is the kind of kiln atmosphere induced during the firing. A smoky, or reducing, atmosphere for instance is normally confined to flame kilns and it is usually introduced in the later stages of a stoneware or porcelain glaze firing but there are a number of exceptions to this rough guideline, notably encountered when pit firing, or working in Raku, or in producing certain lusterwares at lower temperatures.

Space-age technology has made considerable use of ceramic materials and one fairly recent development proving of great benefit to potters is that of ceramic fiber in its various forms. This material enables kilns to be so well insulated that costs and firing times can be cut enormously. Later on in this chapter I have described the extremely rapid and successful firing of a small kiln insulated with ceramic fiber blanket. The speed of firing appears to have little, if any, detrimental effect on the quality of the ware and it has the great advantage of allowing the potter early access to the fired pieces. Pots can be reglazed and refired many times in any kiln but the whole process, including any evaluation and adjustments needed, can be so much faster through the use of ceramic fiber.

Always keep careful records, not only of the clay bodies and glazes used, but also note down how tightly the kiln has been packed; the rate of temperature rise; the points at which alterations were made to the input of energy; the relationship of air and fuel throughout the firing (not necessary with electricity, of course); and any other factors which will help you to learn more about the process and which may be useful to remember for future reference. Colors and surfaces will have changed, often dramatically, after a glaze firing and it is all too easy to forget precisely what you did and, who knows, you may have achieved some wonderful qualities that you will wish to repeat.

Ceramic fiber kiln
The metal casing has an interior lining of thick ceramic fiber for improved insulation. This kiln is lightweight and tiny and temperatures can be accurately controlled.

Development of kilns

Bonfire and open-pit firing.

This is the most primitive method used to fire pottery but it does offer its own unique qualities and possibilities for the imaginative potter. It is still practised in some parts of the world and, in fact, techniques of potters as far apart as Nigeria, New Mexico and Ecuador turn out to be very similar, but it is also used most successfully by a number of contemporary studio potters around the world.

With great care the pots are stacked in heaps on a bed of twigs and perhaps a layer of dried dung or dampened ash to contain the heat and raise the temperature of the fire. Dry grass, which burns quickly with intense heat, is piled on near the end of the firing time to lift the temperature to its highest point, usually around 1472° to 1652°F (800-900°C). In Mexico large pots are first fired individually on a bed of embers to dry the pot. The wood fuel is placed around the pot in a conical pattern and gradually increased until the pot is seen to be fired, the temperature being judged by experience. The remaining burning wood is knocked away and the pot is left to cool on a stand of stones. These large water holders can be 4 to 5 feet (1-1.5m) high. They are beautifully made, symmetrical in style, and burnished — perhaps taking seven to eight days to complete from coiling to finished firing.

If you try this method of firing you will quickly discover its limitations. After a time, no matter how much fuel is added, the final temperature will reach a plateau. Many of the pots will break, being unable to withstand the thermal shock of the raw flame no matter how careful the initial warm-up.

However, the advantage of this type of firing is that it can be carried out anywhere there is a supply of clay and fuel.

The low temperature of the pit preserves the burnished effect that often completes the pots. This is created by rubbing the surface with a shiny pebble which makes it harder and slightly less porous, so providing a good surface for decoration. The effect of the smoke and flame on the clay darkens the body and blackens the surface, some part of which may be fired to a reddish/pink where a constant flame has licked the surface of the pot; oxidization and smoky reduction firing will both be present in these methods of firing

The lower temperatures of these firings demand less expertise from the potters. However, the scale and symmetry of some magnificent African handbuilt pots and the strong

Bonfire firing

The first primitive firings of pots and figurines took place in bonfires and produced a low-fired porous clay.

H 251-2 264 306

decoration are the work of highly talented craftspeople whose knowledge of their art in handbuilding is unsurpassed.

The tradition of making smoked black pots in the manner of the North American Indians of New Mexico has been rediscovered and perpetuated by Maria Martinez and her family since the early 1900s. The clay is blackened by the carbon produced when a damp fuel or smoke-producing substance is placed in the fire.

A refinement of bonfire firing is to use a pit dug to perhaps 18 inches (45cm) to 2 feet (60cm) deep, lining it with brush and twigs which become a bed of hot embers as the firing takes place, so retaining the heat directly around the pots.

Pots were piled on to a bed of brush and twigs; they were covered with wood grass and broken pieces of pot.

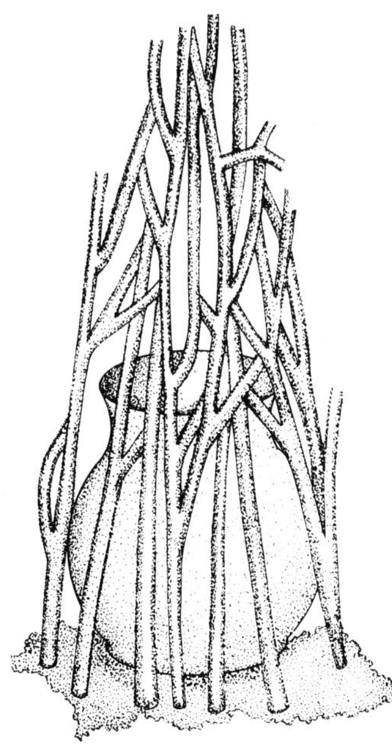

Large single pots were sometimes fired individually. They were preheated on smoldering embers, and then the fire was gradually built up to a high heat.

The updraft kiln

A seemingly simple modification in the construction and firing of the pit kiln produced a great difference in its heat retention and efficiency. By building a low wall more pots could be stacked in the kiln and the heat retained could be introduced at different points around the wall. The top was covered with broken pots and wet clay, and had holes to encourage the flow of air; the heat from the fire filtered into the kiln's perforated floor. Such kilns were virtually rebuilt after each firing. This type of updraft kiln was used by the Romans and the remains of many such kilns have been found in Britain. The Greeks built more permanent kilns around 1000 AD with a domed roof and a chimney which enabled them to have more control over the heat and atmosphere inside. The updraft kiln was used in the Middle East and in Europe. It is possible to reach temperatures of 1652° to 1922°F (900°-1050°C) which was a sufficiently high temperature to mature most earthenware clays and low fluxing glazes.

The potters of Korea, China and Japan had been using a superior type of updraft kiln since at least 1000 BC, called the bank or cave kiln. A cave was cut into the side of a hill where there was a high percentage of clay and sand in the soil. A flue or chimney was dug down into the cave from the top of the bank, creating a narrow fire mouth where the heat was concentrated, a wide space to set the pots and a flue to encourage and control the flow of air. The pots were set on wedges of clay to keep them standing straight and were piled together, touching each other. As the firings took place, the walls of the cave, cut from clay and sandy soil, would harden and fire as the pots did. The fired surface reflected the heat and made the kiln permanent and the potter was able to reach temperatures of 2012°F (1100°C) — higher than any reached in the world at that time.

The downdraft kiln

From the cave kiln the Chinese developed the downdraft kiln method. This included a modified chimney which drew the heat up through the pots, down again to the base of the chimney and then out. This was a more efficient way of using the heat as the temperature increased because the pot was allowed more time to absorb the heat before this was led out of the chimney.

The shape of the interior and narrowing of the space leading to the chimney increased the temperature. Kilns built of the abundant fireclay increased their ability to withstand the higher temperatures to 2372°F (1300°C) which were required to mature more refractory materials (such as kaolin/feldspar and quartz). Thus stoneware and porcelain of great beauty could be produced. Variations of this style of kiln are still used in China, and elsewhere, today. The kilns are composed of several chambers built into a sloping hillside. The heat travels upwards through each successive chamber before reaching the chimney. The temperature is increased by stoking individual chambers with fuel when the previous or lower one(s) have completed firing. Each successive chamber is thus preheated by the exhaust from lower ones — thereby increasing the temperature in a very economical manner.

Other ways were found to divert the passage of heated air so that it circulated around the kiln rather than straight up through the pots. Baffle walls built inside the firemouth deflected the heat. Instead of being built into the roof, the flue for the escaping gases was joined to the side of the kiln, with the opening leading from the base of the kiln. This had the effect of drawing the gases around the interior of the kiln, the heat naturally rising from the fire to the top of the stack of pots and then being drawn down again through the pots to the base of the chimney. In this way the heat could reach all the pots, exposing them to greater heat. The improvement in construction, using refractory materials and better fireboxes to control the atmosphere, led to a controllable increase in the range of temperatures. Although the kilns were capable of producing high-fired wares in red or buff clays, it was not until the properties of kaolin were explored in Germany in the early 18th century that Europeans discovered, and were soon exploiting the secret of porcelain making.

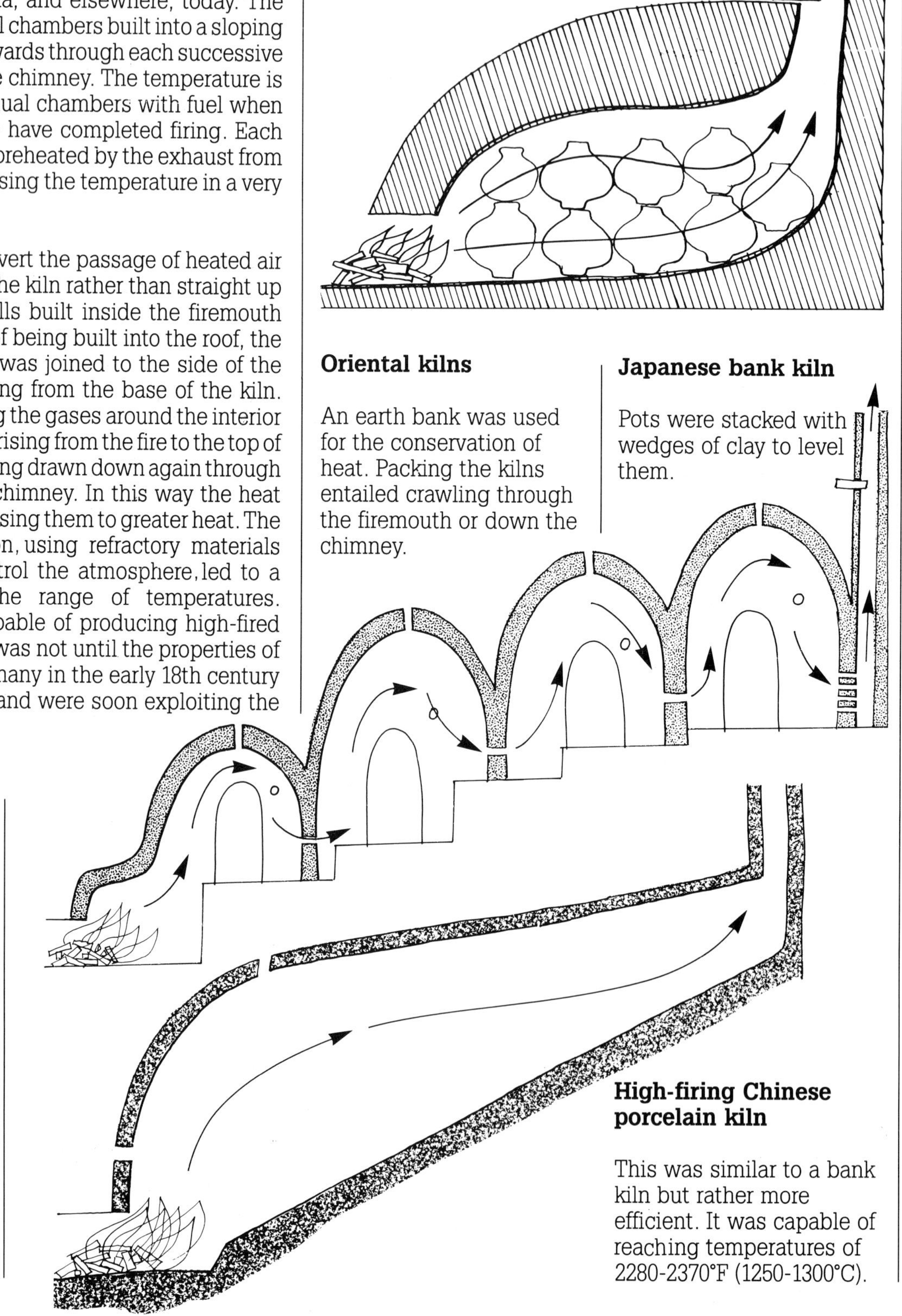

Oriental kilns

An earth bank was used for the conservation of heat. Packing the kilns entailed crawling through the firemouth or down the chimney.

Japanese bank kiln

Pots were stacked with wedges of clay to level them.

Japanese climbing kiln (built on a slope)

Heat rose from the firehole through each chamber.
Fuel was fed into each chamber to increase the heat.
Pots were packed into each chamber through a doorway which had to be bricked up for each firing.
Japanese kiln covered by a roof with space at the sides for both storing and drying.

High-firing Chinese porcelain kiln

This was similar to a bank kiln but rather more efficient. It was capable of reaching temperatures of 2280-2370°F (1250-1300°C).

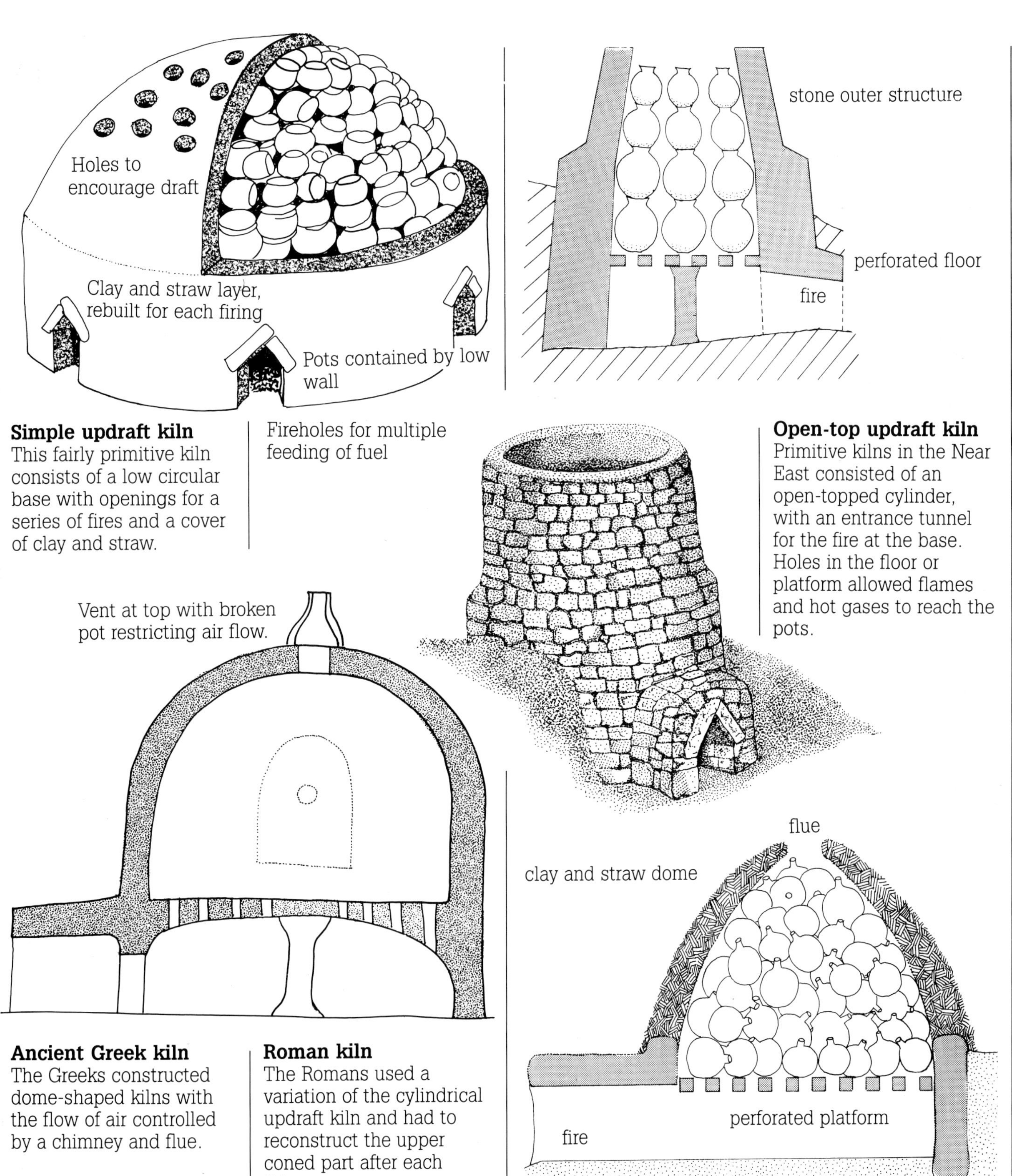

Simple updraft kiln
This fairly primitive kiln consists of a low circular base with openings for a series of fires and a cover of clay and straw.

Open-top updraft kiln
Primitive kilns in the Near East consisted of an open-topped cylinder, with an entrance tunnel for the fire at the base. Holes in the floor or platform allowed flames and hot gases to reach the pots.

Ancient Greek kiln
The Greeks constructed dome-shaped kilns with the flow of air controlled by a chimney and flue.

Roman kiln
The Romans used a variation of the cylindrical updraft kiln and had to reconstruct the upper coned part after each firing.

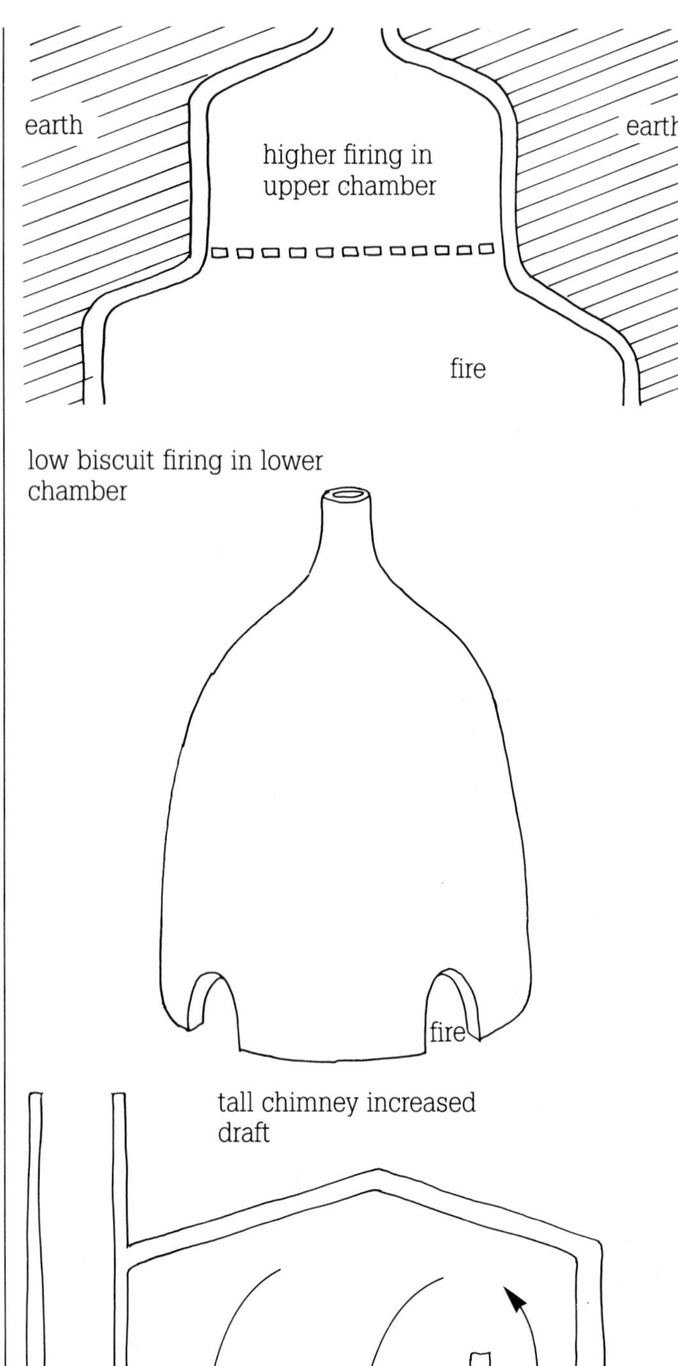

Variations on the Western updraft kiln

Most of these kilns were built into the earth to provide insulation. Biscuit ware was fired in a lower chamber; higher-glazed pots were in the hotter upper chamber (the heat was drawn here by the flue).

European kiln

This utilized several fire holes and the increased draft provided by its tall chimney. Kilns fired with coal attained higher temperatures.

Downdraft kiln

Heat is drawn into the kiln and rises up when it meets a baffle wall.It is drawn down through the pots, through the floor and out through the chimney. This movement of hot air is controlled by flues and vents.

Saggars

Pots were fired in 'saggars'. These fireclay containers, which were stacked in the kiln, protected the fine ware from the flames. They were stacked in 'bungs' one on top of the other in a kiln.

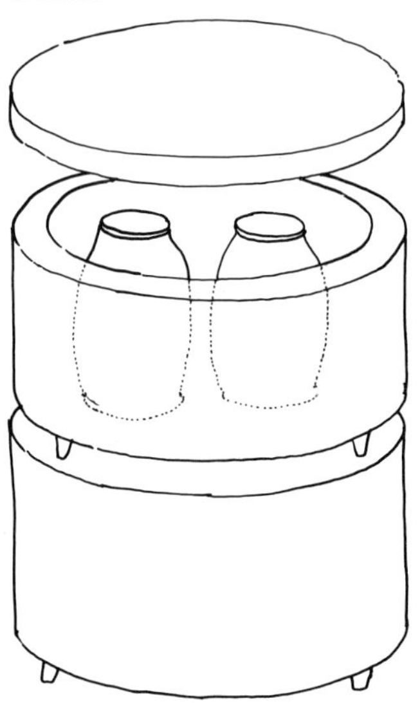

Saggars could be used singly or stacked in a "bung".

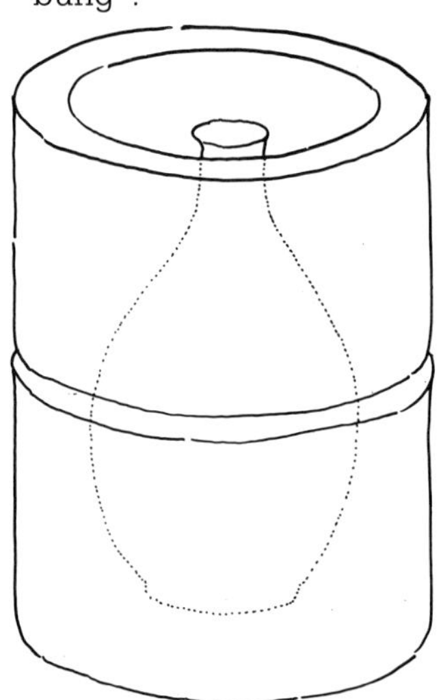

Multiple saggars were used to accommodate larger pots.

Modern kilns

Choosing a fuel

Ultimately the choice of fuel depends on availability, cost and the effects the potter wishes to produce.

Wood firing is reasonably cheap if you happen to have a forest on your doorstep but is labor intensive as the wood must be cut into suitable lengths and dried well. Sawdust firing is slow and not always practical as the results are unpredictable; none the less, sudden bursts of flame can create very interesting effects, while surrounding the pots with dry leaves will slow the burning and produce a blackened surface.

Electricity, gas and oil kilns

Electricity is generally expensive; gas and oil are cheaper, more controllable and can accommodate a reduction firing. If possible, experiment with as many of the following methods as you can in order to discover the various effects possible with different firing methods.

Coal-fired kilns

Used from the start of the Industrial Revolution in Europe, coal was plentiful and cheap and, with careful tending, produced high temperatures for firing. The kilns were mainly simple updraft types; that is, the fires were lit at the bottom at various points around the kiln and the heat was drawn up through the wares by the suction of the high chimney. Some kilns were built with the fire holes being tended from outside, with the open air coming directly into the fire. Later ones had the kiln enclosed by a building in the form of a giant flue pipe which provided sufficient room around the kiln for men to stand under cover while stoking the fires.

The economic need to produce better and finer high-fired pots, led to the coal-fired kiln being developed as a downdraft version. The heat in the downdraft was prevented from being drawn straight out of the top of the kiln and was led down and through the wares, thus building up the temperature inside the kiln. It was then expelled through a flue located at the bottom of the kiln rather than at the top.

In most kinds of burning-fuel kilns, the pots were protected from damage by the flames by placing them in a fireclay holder with a lid; this was called a saggar.

Wood-burning kilns

Wood was the first fuel used for firing kilns.

To fire a kiln successfully with wood there has to be a wide fire mouth and a large grate to hold the fuel. There should always be a space below the grate to collect ashes so that these do not block the flow of air to the fire and reduce the heat input. A large, wide, high chimney will draw the flame and heat through the kiln while additional flues can provide an extra supply of air or release gases when required.

Oriental kilns built on a hillside, such as the Japanese climbing kilns, acted as natural chimneys, the additional flues in the sides and top adding more draft when needed.

If there is a good supply of local wood and a suitable site to build the kiln, then this method of firing can be a cheaper alternative to other fuels. The disadvantages are the amount of wood needed to achieve the temperature required and the type of wood available. Soft woods burn quickly and fiercely. The wood has to be cut into workable lengths and split into a size which, when placed on the fire, will generate instant heat — rather than smoldering before igniting. To achieve this ignition successfully, wood has to be cut, split and, most important of all, dried out before commencing firing. You will also need storage space for keeping the considerable quantity of fuel that is required.

Modern ceramic fiber kilns

The basic design and construction of kilns has not altered greatly since the first kilns replaced the bonfire. The most radical change has been the relatively recent use of gas and electricity as fuels instead of wood and coal. These fuels are comparatively simple to use.

The use of ceramic fiber materials in kilns has increased dramatically over the last ten years. Their popularity is due to the much improved insulation properties and their light weight. They can be used alone or combined with insulation brickwork, and may be fired with electricity or with gas.

These are far more efficient pieces of equipment. This is largely due to the accuracy of the temperature controllers which govern and indicate temperature (and the rate of its climb) switching on and off at given signals. They include digital displays and safety cut-off devices. These kilns can be programmed for the whole firing cycle — providing much greater sophistication and control.

Gas-fired updraft kiln with single burner
Heat tends to be concentrated near the floor area.

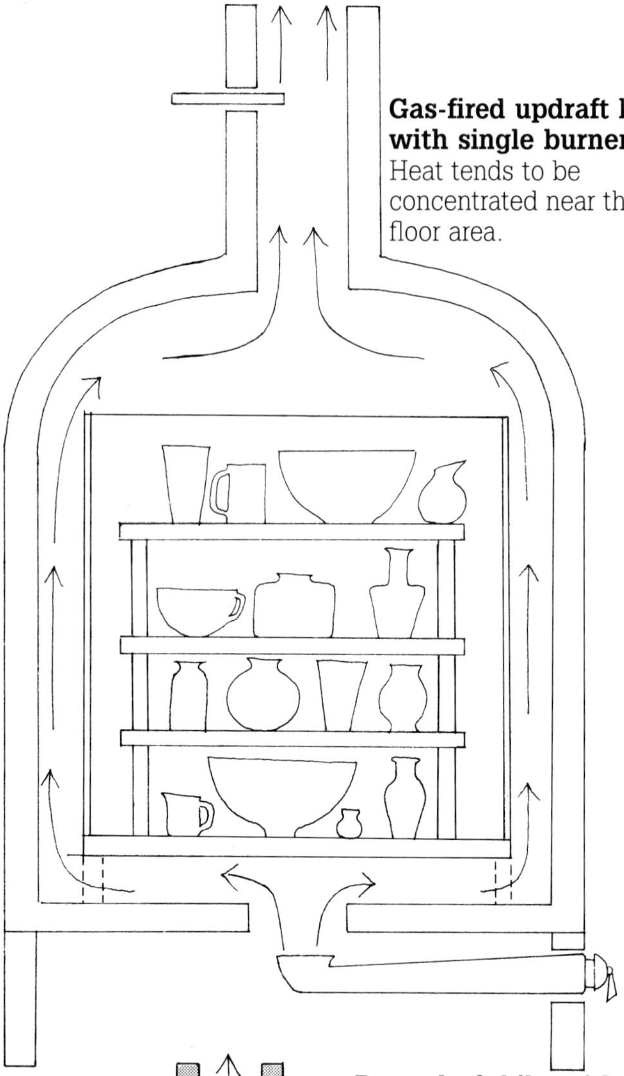

Downdraft kiln with single burner

damper

Gas-fired downdraft kiln with multiple burners
Heat is distributed fairly evenly throughout the kiln.

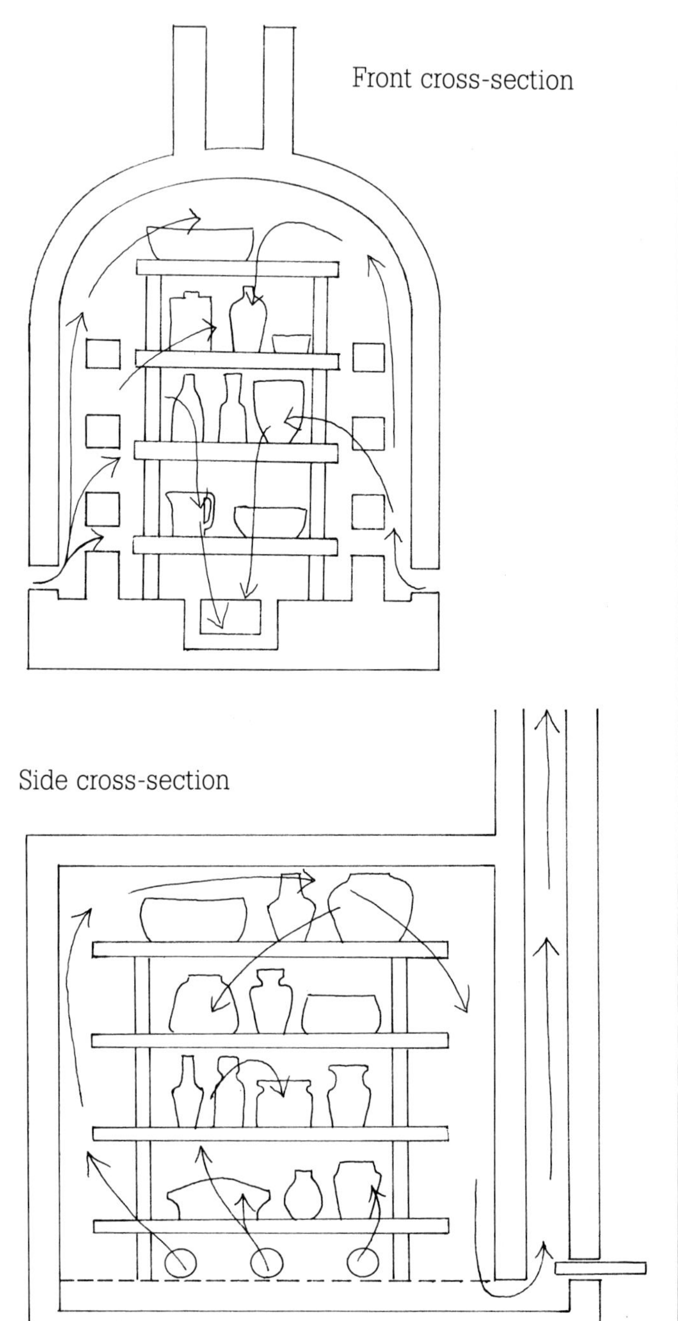

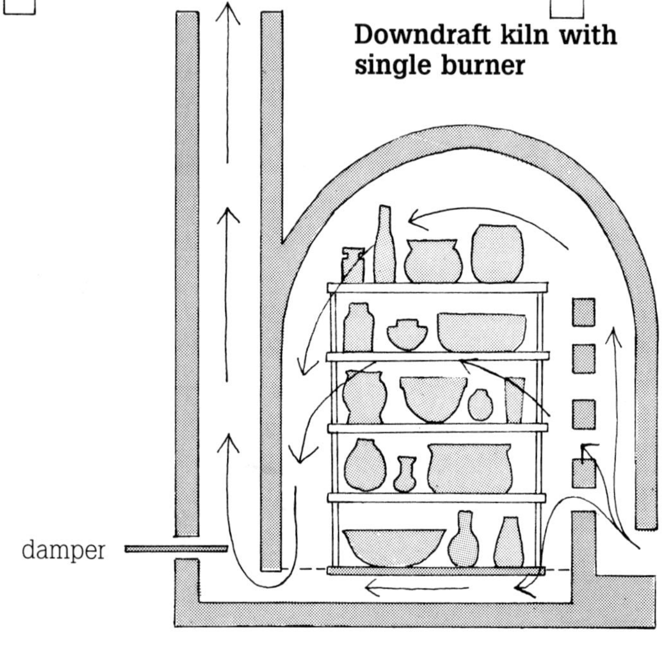

Wood-firing kiln

Stoking a wood kiln can be a long tedious business but interesting effects can result from the ash, and the fuel is relatively cheap.

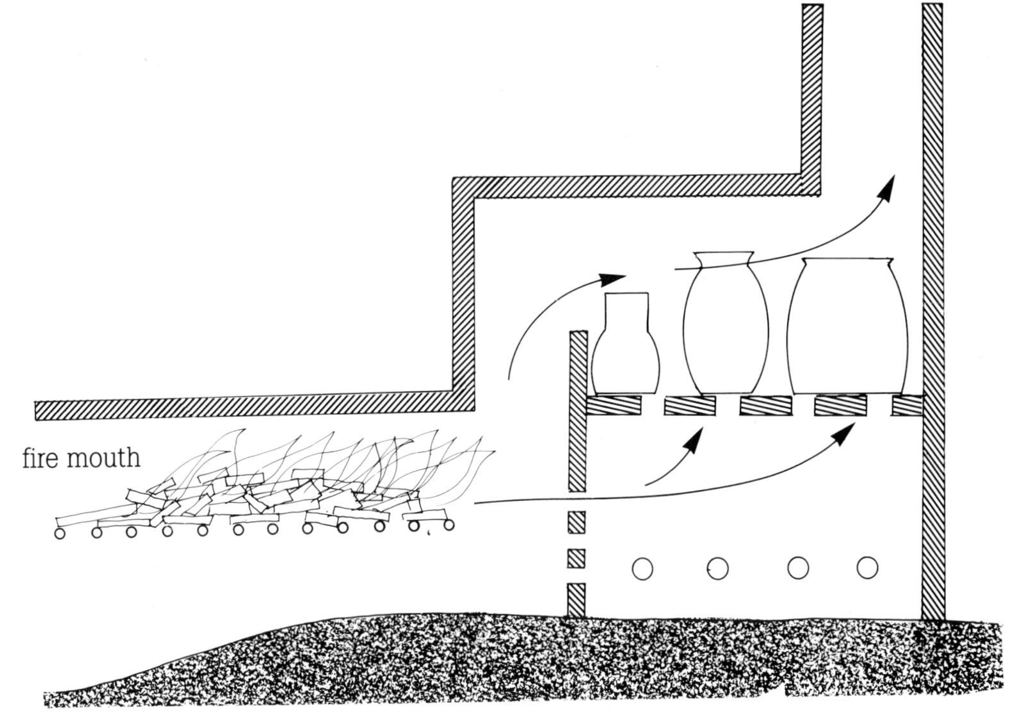

A modern electric kiln
Elements heat the kiln

Firebrick provides good insulation

Safety locks prevent door opening during firing

Light indicates when kiln is on

Temperature control

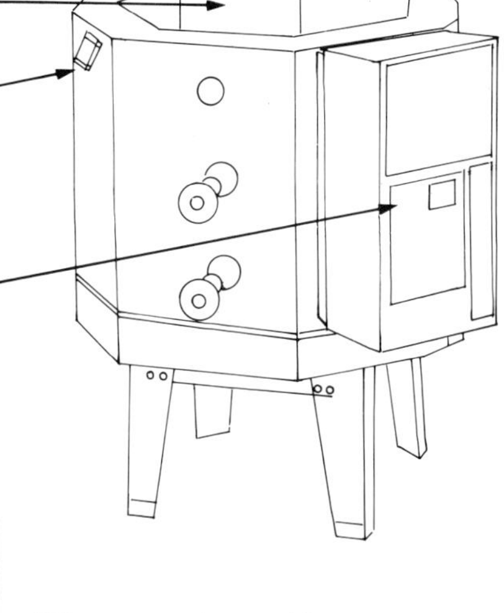

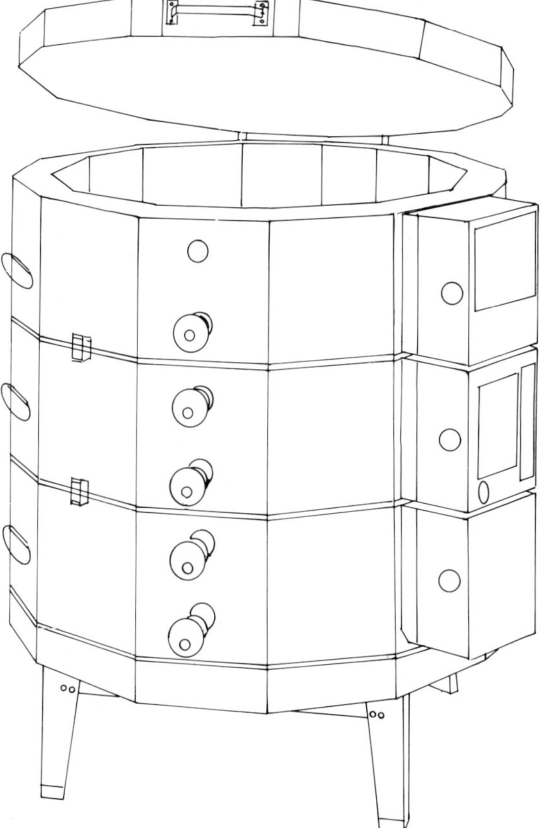

An electric kiln, heated by coiled elements within the side walls, is simple to operate and there are no problems with control of heat imput and drafts. However, they are generally more expensive to purchase and run.

Kilns come in a wide range of sizes and levels of sophistication, to suit different potters' needs. They can be of one-piece construction or sectional.

Firing a ceramic fiber portable kiln

This firing was to be at temperatures to about 2370°F (1300°C) to achieve fast firing of porcelain glazed ware.

The interior dimensions of the kiln used for the trial were small, 7 x 9 inches (18 x 23 cms) only, although kilns of many dimensions are available to suit all potters' requirements. The kiln is a circular construction made from metal casing with an interior lining of thick ceramic fiber. It has a fiber lid with a vent hole in the middle, which sits on top of the kiln without fixing. The height of the kiln can be increased by adding a collar before placing the lid on top.

The five small cylindrical pots used for this experiment were made of translucent porcelain clay, thrown and then decorated with underglaze colors. They were biscuit fired in an electric kiln and glazed with clear shiny porcelain glaze. These pots were fitted carefully into the kiln, resting on the grooved fire brick base.

The temperature was monitored by a digital pyrometer inserted through the wall. The gas jet was lit at 9.30a.m. It was then gently adjusted to warm the pots and the kiln for 15 minutes, taking the temperature to 300°F (150°C).

At 10.45a.m. the gas pressure was increased for a further 15 minutes, raising the temperature slowly to 570°F (300°C). The pots and interior viewed through the vent on top of the kiln were seen to be reddening.

Firing of ceramic fiber kiln to 2390°F (1310°C).

Fahrenheit (approx)	Centigrade
2390	1310
2370	1300
2190	1200
2010	1100
1830	1000
1650	900
1470	800
1290	700
1110	600
930	500
750	400
570	300
390	200
210	100

Time: 9.30 9.45 10.00 10.15 10.30 10.45

off 1310°C 2390°F

Reduction

The gas pressure was increased and the temperature rise quickened to 1470°F (800°C) at 10.15a.m., and to 2010°F (1100°C) at 10.30a.m. At this temperature the top vent was partially covered with a fire brick until a flame could be seen leaving the vent; there was a change in note of the gas-jet noise and the temperature dropped slightly as reduction was started. The gas pressure was reduced slightly and the temperature started to rise again to 2280°F (1250°C) at 10.40a.m. and to 2390°F (1310°C) at 10.45a.m. with reduction taking place from 2010° to 2390°F (1100° to 1310°C). The normally clear shining glaze, fired to 2335°F (1280°C) in an oxidizing atmosphere, had a pleasant slightly matte finish and the underglaze colors were pale and slightly fired away where they had been applied thinly.

The pots were removed when they were cool enough to handle, within an hour of firing completion.

The easy removal and replacement of the lid during firing (provided the potter is wearing protective gloves and goggles) ◊ makes this an ideal Raku kiln which provides easy access for long-handled tongs to remove the red-hot pots. It is lightweight, portable and (because of its good insulation) cheap to fire.

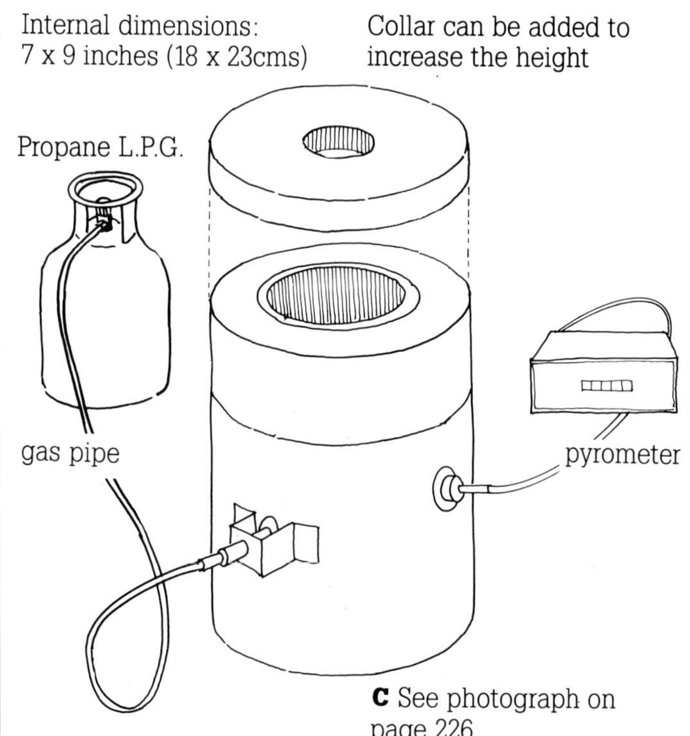

C See photograph on page 226

Basic firing techniques

Raw firing

It is quite possible to decorate and glaze raw clay objects so that their conversion into ceramic can be completed in a single firing. When using this method it is important to increase the heat input very slowly until about 1112°F (600°C) as in a bisque firing, to allow the water content in the clay to evaporate gradually. Some potters use this method as a way of cutting down the firing time and reducing the cost incurred when two firings are used. In the hand of an inexperienced potter, raw glazing pots can have disastrous consequences; pots may quickly soak up the glaze and crack if allowed to become too wet. Moreover, clumsy or careless handling of the raw pots when glazing and decorating can often result in damage to the ware. The safest form of firing for the inexperienced potter to adopt is to fire once to a biscuit temperature and then glaze the pots and fire a second time for the glazing. This is because, apart from the ease of handling, pots regularly fired at 1832°F (1000°C) will have a known absorbency rate and with this in mind glazes can be mixed for an even application and a reliable result each time.

Biscuit firing

Biscuit or bisque firing is the first firing of the raw clay. The pots are completely dried before being placed in the kiln. During the firing process their consistency changes from a dry, crumbly material into a hard, durable rock-like condition which is porous, ready to accept a glaze and can be safely handled when decorating and glazing. Dried pots (sometimes referred to as 'green ware') which are without glaze may be stacked in the kiln with the pots touching; if the pots are of an identical shape and size they may be stacked foot to rim to the full height of the kiln. Be careful, however, when stacking pots inside or on to shallow, wide-bottomed pieces like bowls or dishes because the weight can cause them to crack in the firing.

Biscuit firing should be started gently with low heat and plenty of ventilation in the kiln and in the kiln room. Leave bungs and spy holes open to let steam escape; a mirror or piece of metal held at the spy hole will collect condensing moisture to inform you if it is still present. The chemically combined water will be driven off between 932° and 1112°F (500°-600°C) and the particles of clay will be drawn closer together as the water is lost, so shrinking the pots considerably. The rate of firing may be increased fairly quickly after this stage to a temperature of about 1830°F (1000°C), although bisque firing can be anywhere between 1652° and 2012°F (800°-1100°C) — depending on the material and its intended use. Some potters soak their pots at the top bisque temperature, especially with porcelain, in order to burn off any carbon deposits which can cause bloating during the subsequent glaze firing. The kiln must then be left to cool down slowly to 212°F (100°C) before the door is opened slightly. The kiln should then be left open until the pots can be safely unpacked and handled.

Glaze firing

The temperatures of the different glaze firings indicate the point at which the glaze should mature. The main types of firing are:

Earthenware — mostly red clays firing to 1868°-2048°F (1020°-1120°C).

Stoneware — firing between 2192°-2372°F (1200°-1300°C)

Porcelain — 2228°-2480°F (1220°-1360°C)

Modern porcelain bodies are now available and these can mature in temperatures of only 2228°F (1220°C).

Earthenware

Glazed pots should be stacked so that they do not touch each other or they will fuse together. Pots fired to earthenware temperature will be glazed but the body beneath the glaze will remain porous: for this reason the whole surface of the earthenware pot is often glazed to stop the leakage of liquids from the pot after firing. To prevent the glazed feet of the pots sticking to the shelf of the kiln, the pots should be fired on stilts or bars of high-fired clay. Stilts are three-pointed stands which lift the base of the pot away from the kiln shelf. The points of the stilt are so sharp and fine that the marks they leave behind can easily be polished away so that they hardly show and do not break the surface of the completely glazed pot. Beware of the sharp spines left behind when the stilt is removed; before they have been ground down they can be very sharp and dangerous. Be extra careful also when removing slivers of kiln stilts from the base of pots; they can inflict severe cuts on unprotected fingers. It may be better to use a carborundum stone to remove splinters. ◇

Shelves should be coated with a thin bat wash of alumina and kaolin mix to protect them from pieces of stilts or running glaze (care should also be taken to see that no pot touches another or overbalances on to another).

The glaze firing should be started off slowly with some ventilation in the kiln to allow steam to escape from the drying glaze, up to 842°F (450°C) over three hours, say, followed by acceleration to the final temperature.

Staffordshire Bear Jugs c. 1740
The heads are detachable to form cups on these English salt-glaze pieces produced in the 18th century.

Raku pot by David Roberts (UK)
Multiple striations, both fine and slightly heavier, have been created by a white crackle glaze used in a Raku firing.

Blue vase by Tessa Fuchs (UK)
This piece has been given a biscuit firing at 1796°F (960°C), a glaze firing at 1976°F (1080°C) and a one-hour 'soak'.

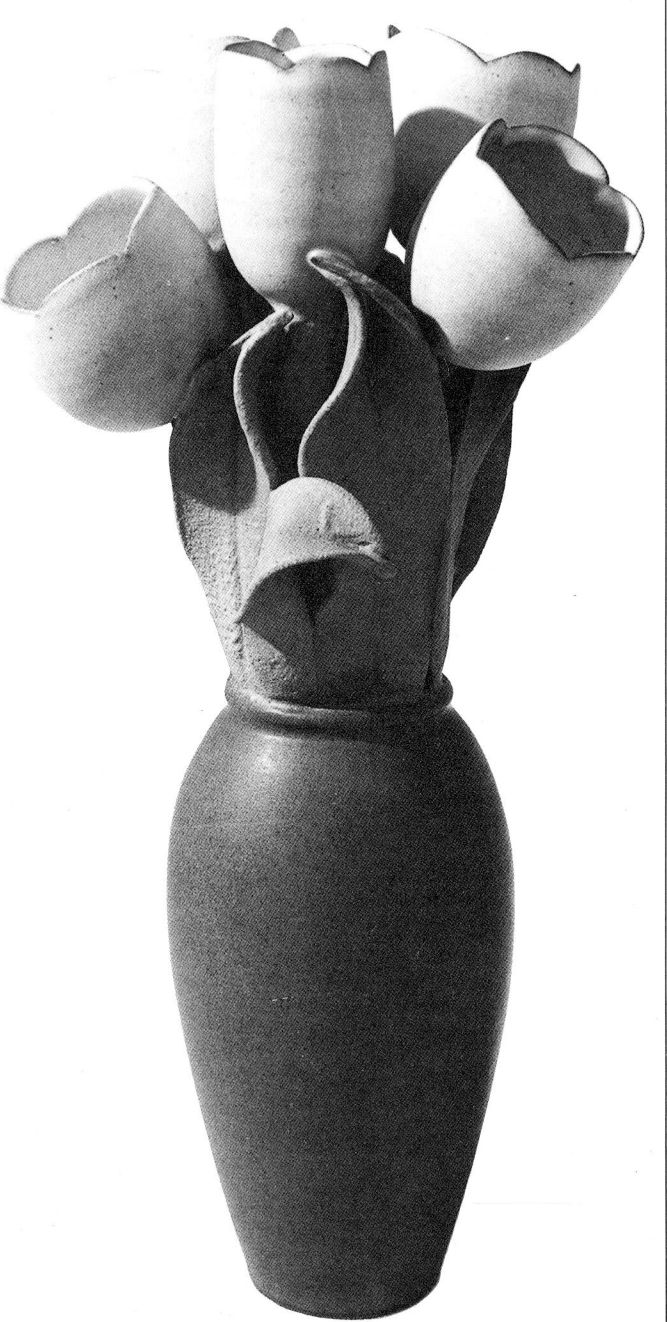

Smoked vessel
by Gail Russell (USA)
A simple line of metallic thread is the only decoration required on a porcelain surface where smoking has produced fascinating clouds of color and light.

Detail of smoked vessel
by Gail Russell (USA)

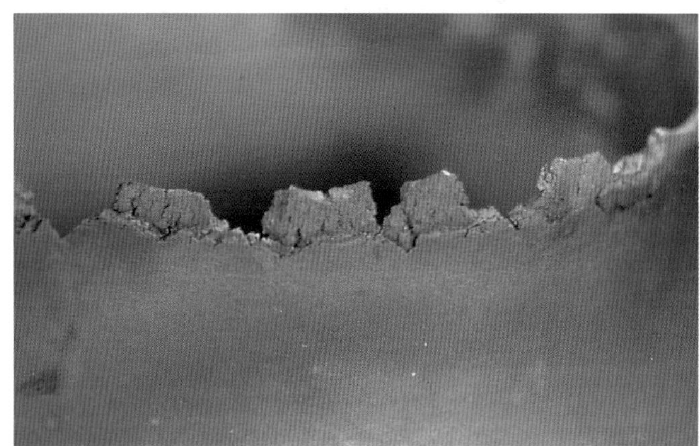

Bowl
by Susan Nemeth (UK)
Thin layers of clay and colored slip are press-molded and then biscuit fired to 1742-1830°F (950-1000°C).
Sanding down between this and the glost firing (Cone 8-9) reveals the underneath layers and can produce dramatic effects.

Luster glaze pot
by Greg Daly (Australia)
Keep careful records of your firing techniques and temperatures so success can be repeated. Here silver and copper have been used to create a beautiful speckled shiny surface.

Salt-glazed ware by Jane Hamlyn (UK)
Raw glazed ware has been once-fired in an oil-fired kiln.

Vase by Lana Wilson (USA)
This piece has been reduction fired on soft brick pedestals. A copper sulfate wash over a wax design has produced a gentle gray-blue design in the gas-fired reduction. If charcoal briquettes had been used then oranges, pinks and purples would have been introduced.

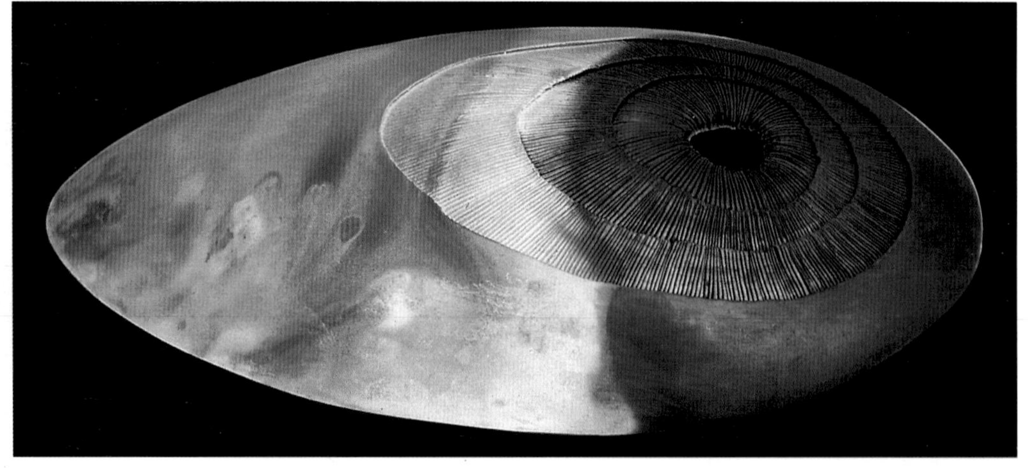

Eliptical vessel by Ray Rogers (New Zealand)
The subtle blend of gray, cream and ocher tones, pit-fired and burnished, combine with the 'fungoid' aperture on this vessel and together create a 'galactic' atmosphere.

These methods of firing are only a guide to the many variations with which one can experiment. Many astounding results are achieved by experiments with clay, glaze, atmosphere and kiln, but make sure you record the method of work and the type of firing, so results can be repeated.

Stoneware

Stoneware clays and bodies are fired at higher temperatures than earthenware ones and to survive, the pots must be placed in a stable position in the kiln on clean shelves which have been coated with a bat wash alumina and kaolin. This will protect them from any drips of glaze and prevent the feet of the pots from sticking.

Unlike earthenware each pot should be fired with a cleaned foot and placed directly on to the kiln shelf for firm support. Pots placed on stilts will be tend to warp and sink on to and around the stilt, welding it to the foot of the pot at high temperatures.

If there is a tendency for the glaze to run then it should be wiped off for a reasonable height up from the base of the pot and this should be placed on a bed of silica or sand or on a spare piece of broken kiln shelf or firebrick.

Properly fired stoneware pots are almost non-porous whether glazed or not.

Stoneware reduction firing

The atmosphere inside the kiln may be controlled and adjusted to produce extra qualities or to give a different effect to the finished glaze.

The normal atmosphere in a kiln is called an oxidizing atmosphere; this means that the fuel burns in the air which is drawn into the kiln where it is controlled by air ducts and dampers. In a flame-burning kiln the air supply to the flame inside can be stopped or reduced at source or by adjusting the size of the chimney with dampers and this action produces a smoky reducing atmosphere. Reduction firing or reducing the amount of air inside a kiln for a short time has the effect of changing the colors of bodies and glazes. As the flame is starved of oxygen it will seek oxygen in the oxides and clay of the pots and this change in the chemical composition alters the final color.

If body reduction is required then this should normally begin at about 1832°F (1000°C) before the glazes begin to anneal (set); if the reduction is started too early carbon may be trapped in the clay body and cause it to blister.

Too much reduction will cause a backing up of the flame; the noise of the burner will change and the temperature will drop. After reduction, as the kiln reaches maturing temperature, maintaining for a short period a clear oxidizing atmosphere will clean up the kiln chamber. Holding the temperature steady at this point 'soaks' the glazes and allows them to melt evenly. Each potter develops his or her own preferred firing programmes. There is no one correct way because the variables are so great. However, these guidelines will make a useful starting point for beginners.

Reduction and firing with flame-burning kilns will create toxic fumes and smoke so the kiln should be positioned away from the main studio with adequate ventilation of the chimney flues and kiln room itself. After firing, close all spyholes and entry ports to prevent cold air being drawn into the kiln.

In an electric kiln, reduction can be carried out by inserting small pieces of wood or moth balls through the spyhole but this will damage the elements if it happens too frequently and several oxidized firings must be done before another reduction firing is attempted — to prevent deterioration of the elements. Some electric kilns are available with the elements sheathed in ceramic tubes to prevent damage during reduction.

Reduction firings have been successfully completed in electric kilns but the methods involved are dangerous and can inflict expensive damage on the kiln elements. It is not a practice to be recommended.◇

Porcelain

The initial manufacture of porcelain objects must be carefully controlled as during firing any defects (such as dents or warping) will tend to recurr at the high temperatures involved — even though these faults may have been repaired by the potter.

In the kiln, glazes and bodies vitrify and fuse together at porcelain temperatures, becoming as one. Because the body vitrifies at about 2372°F (1300°C) it can in fact be fired without a glaze and the white bone-like surface perhaps polished afterwards with fine carborundum powder.

The degree of whiteness achieved varies according to the type of firing and is clear white or cream-white in an oxidized firing, becoming a blue-gray white in reduction. Coloring oxides will react differently too (e.g. copper and iron) and this can be exploited once the potter thoroughly understands how to control the kiln and fuels.

Pots completed at low temperatures, as in pit-firing or Raku, respond well to the action of the flames which can create flashes of color; and pieces are sometimes surrounded by combustible materials such as grass, seaweed or string to create patterns of burning. Porcelain is most suitable for this treatment because of its simple white ground, although it will remain rather fragile.

Salt-glaze firing

Salt glazing in the past has been used extensively on decorative stoneware, especially in the 19th century, but in industry now it is chiefly employed for building bricks, drain pipes and chimney pots. It was developed to a high degree in the Rhineland region of Germany during the 13th century and thousands of pitchers and flasks were produced there. Particularly famous are the 'bearded' Bellarmine jugs. Salt glazing has been revived by numerous imaginative studio potters during the last twenty years or so because of its unique decorating possibilities. It differs from other firings in that the glaze on the pots is introduced during the firing rather than beforehand.

Salting takes place when the kiln is nearing the 2282°F (1250°C) mark. Dampened rock salt (sodium chloride) is put into the firebox at intervals; it reduces to a vapor which is drawn into the kiln and settles on the pots. This combines with the silica in the clay to produce a mottled 'orange-peel' effect. In a reducing atmosphere the color created will be within a range of golden browns due to the iron present in varying amounts in most stoneware bodies — but you can also produce blues, yellows, oranges and greens, depending on the slips and oxides applied before the salt firing.

The act of adding salt results in clouds of hydrochloric acid being given off and therefore care must be taken to install the kiln away from buildings and people and to take the direction of the wind into account. ◊

Salt-glazing kilns are usually built for that purpose alone; the inside walls of the kiln become glazed with the salt deposits of successive firings and will leave deposits on pots if a normal firing is undertaken. These kilns may have a limited life due to the corrosive action on the kiln lining and furniture.

Local salting of an individual pot can be done if fired in a sealed sagger and surrounded by salt; the burning of the salt is contained in the sagger so that it does not affect either the kiln lining or any other pots present during the firing process.

H see page 287

Sawdust firing

This is a simple but effective way of firing.

The kiln can be constructed with loose-laid ordinary house bricks or from an old metal refuse bin with air holes punched in the sides.

A square construction with walls two and a half bricks wide on each side can be built by looselaying overlapping bricks before the pots are put in; or the pots can be added in layers as the walls are built up. First make the 'kiln' two bricks high, then tip in a thick layer of sawdust and place a layer of pots on the sawdust. Cover the pots with more sawdust to the top of the bricks, place a piece of wire mesh over the pots and the bricks and carry on building up the bricks, filling in each layer with sawdust and pots and then covering each layer with mesh. When the kiln is complete it can be covered with a metal dustbin lid or a kiln shelf after the top of the sawdust has been lit. The sawdust will burn slowly away and the mesh will stop the pots at the top falling on to the pots underneath. The pots will be soft fired, porous and colored black or gray from the smoke of the burning sawdust.

Try not to build a sawdust kiln in an exposed situation where winds will cause it to burn too quickly and so damage the ware. These wares are not fired hard enough to hold water and their value is merely decorative.

Pots with a burnished finish are particularly striking when fired this way. Variable surface treatments can be achieved by surrounding the pots with other combustible materials while packing the kiln. Salt, peat or shavings will all act on the surface in different ways (although salt is more usually applied when firing at higher temperatures).

Handbuilt pinched pots, small modeled figures, animal and bird shapes: all these lend themselves to a sawdust firing especially if the surface has been painted with oxides or burnished. The low temperature of the firing and the effect of the smoke and flames serve to retain the deep shine that would possibly be diminished if the pieces were fired at higher temperatures.

Raku firing

Raku is closely associated with the Japanese tea ceremony. The word 'raku' comes from the Chinese symbol representing concepts such as pleasure, enjoyment, comfort, peace and happiness. The technique has become popular with contemporary potters able to exploit the special qualities provided by the glazing and firing processes.

Raku body is usually white and should remain porous after both biscuit and glaze firing. Fireclay with about 10-30 per cent grog will provide a good basic clay, to which more plastic or shock-resistant clays can be added as necessary when larger pots are attempted.

Raku kilns can be bought ready made and fully portable for regular use. They are usually quite small and therefore quick to heat up; even so, this initial heating could take two to three hours with refractory brick construction. Ceramic fiber has made the construction and firing of raku a much shorter and more reliable process.

The insulation properties of ceramic fiber means that the kiln can be heated in a short time and quite easily kept up to temperature. Raku glazes are usually fired at around 1652°F (900°C). They can be applied directly to the raw pot but a bisque firing beforehand reduces the risk of pots exploding or cracking. The body from which most Raku pots are made is heavily grogged to reduce the risks of thermal shock. The kilns have removable doors or lids which allow the pots to be taken out (when the glaze is seen to melt) using long-handled tongs. The pots may even touch each other lightly in the kiln because the molten glaze will pull apart quite easily and heal over fairly well. The alternate cooling and smoking of the pot creates colored patterns in the open texture of the clay, lusters with some metallic oxides and interesting crackle effects in the glaze.

Raku is a fascinating means of introducing the student potter to all the elements of the art in one session: mixing the open-textured clay, making the pots, applying the glaze by dipping and pouring, then painting with glaze and oxides (copper being a particularly effective one). The dried pots are placed into the red-hot interior and, because the kiln has a removable lid or door, the student should be able to see when the glaze melts. The pot is then withdrawn from the kiln, glowing red, to be plunged into peat, damp leaves or sawdust. The resulting clouds of smoke, and even flames, create reducing conditions which can be controlled to some degree and so produce unusual and unique effects. (Some interesting color variations can be achieved by these post-firing techniques. For example, lustrous colors will often be obtained from glazes containing different amounts of copper oxides.)

Raku firing is one of the most dramatic of spectacles with the added bonus that the finished work will be on display at the end of the day.

Raku: post-firing reduction

The pot should be placed in a metal container and covered with sawdust, damp peat and leaves. The restriction of air and the burning of the sawdust (or whatever fuel is chosen) combine to change the surface appearance of the glaze and the oxides (in a reduction firing). A white glaze will usually crackle and the pattern of cracks will be blackened by the smoke, as will any unglazed clay surface, while copper oxides produce the lustrous effects described above. The process can be repeated several times to achieve the result required. These results are never wholly predictable and vary according to the temperature of the firing kiln, glaze composition, oxides and the amount of reduction achieved.

Raku firing is best done outside or in an open-sided shed so that there is plenty of ventilation. Great care must be taken when removing the lid from the kiln and when handling the pots. Articles of clothing have been known to catch fire as a result of incautious behaviour and carelessness could well mean that the potter suffers burns. ◊

Firing tools

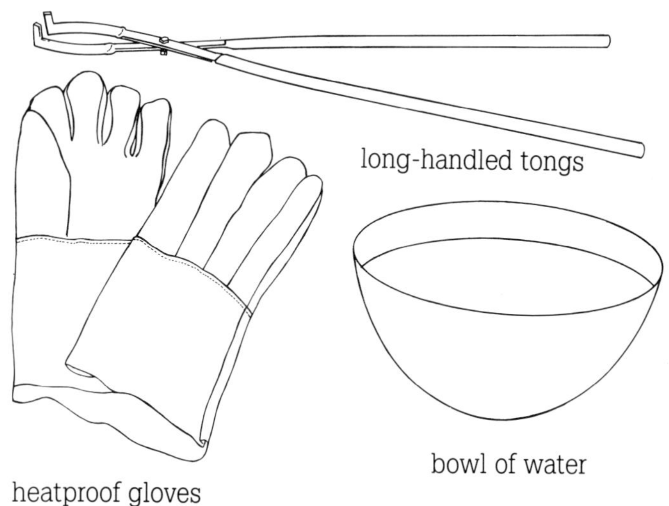

Suitable kilns for Raku

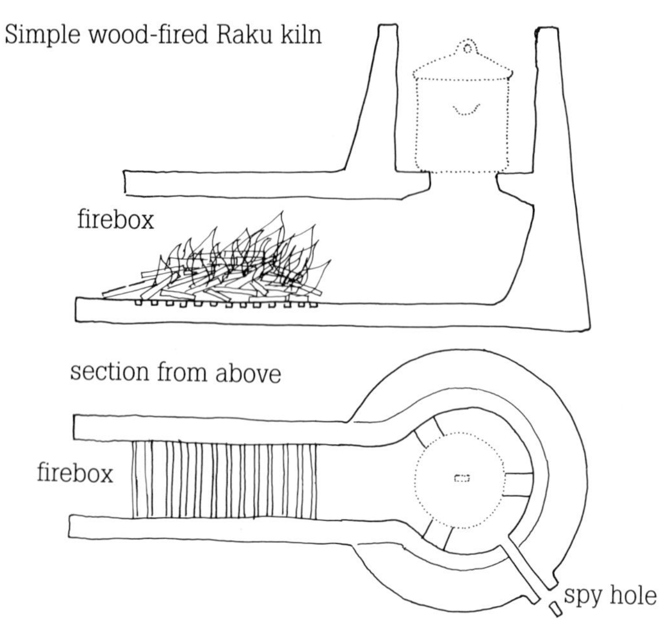

A brick-built Raku kiln has a door at the side for the placing and removal of pots.

A top-loading ceramic fiber kiln which is lightweight and ideal for Raku firing (see page 236).

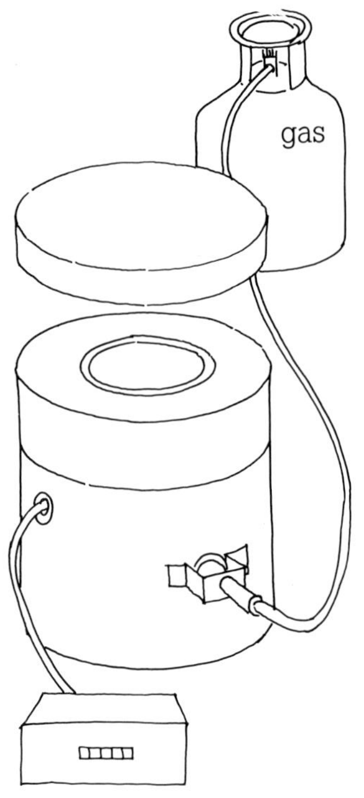

Sawdust firing

Sawdust kilns can be a square or round brick structure topped with a dustbin lid or kiln shelf to control the draft. Make sure each item is surrounded by sawdust.

Walls should be 2 bricks deep with a few gaps between them for ventilation (unless you decide to create a raised chimney on top — in which case bricks should have no gaps between them).

Make sure dustbin lid sits neatly across the top. (Kiln shelves or metal sheeting can be used instead.)

Lay a piece of wire netting (chicken wire) between each layer of bricks to support the pots.

A steel drum can serve as a kiln if holes are made to introduce sufficient air.

A simple dustbin can also be used for reduction after firing if filled with sawdust, strips of paper, shavings, leaves or peat.

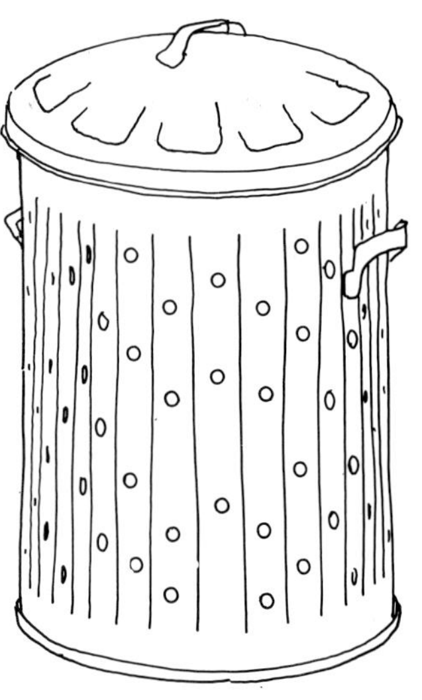

Monitoring the temperature

Pyrometric cones.

The temperature inside a kiln is most accurately measured by pyrometric cones (or bars) by a pyrometer or both. Pyrometric cones are small triangular shapes made of different proportions of ceramic materials designed to melt at given temperatures. They are each numbered, the number impressed on each cone indicating the temperature at which the cone will melt. These temperatures vary slightly according to the make of cone. If the same make is used all the time then the kiln can be fired with confidence, once the cones have been checked against a pyrometer.

Cones are often used in sets of three, one at the temperature below the one needed, one at the temperature required and one slightly above. This combination will give a warning when the critical middle temperature is approaching, and the third will indicate if the temperature has gone too high.

The cones are placed in a wad of clay (which must be dried thoroughly first) or in a ready-made holder, so they are leaning at a slight angle in the kiln in line with the spyhole. Place a small electric lamp in the kiln and shut the door to check the position of the cones before the firing. When cones are correctly aligned, remember to remove the electric lamp before you begin firing.

Pyrometers

Pyrometers are used to indicate the temperature rising or falling in the kiln. A pyrometer consists of a thermocouple sheathed in a ceramic tube or sleeve which is inserted through a hole drilled in the side or top of the kiln and is attached to a meter outside the kiln which displays the temperature. Modern digital read-out pyrometers indicate the temperature rise degree by degree and are particularly useful when glaze firing or when the atmosphere in the kiln needs to be adjusted.

Some potters use either cones or pyrometers but combinations of both may be used. Before these devices were available potters would include test pieces placed in an accessible position so that they could be withdrawn towards the climax of the firing to check that the right temperature had been reached.

pyrometer indicates temperature within kiln

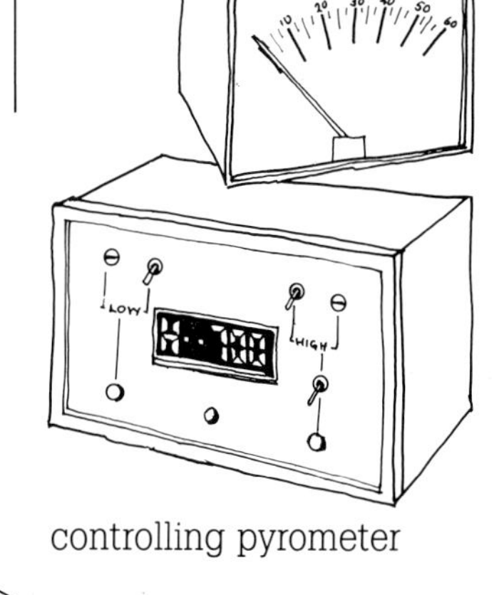

controlling pyrometer

Minute electrical currents are produced by the thermocouple as the heat rises in the kiln and these can be detected by the external pyrometer and a temperature reading given. More sophisticated systems will cut off the power at a set temperature.

Pyrometric cones melt at a given temperature and so indicate the temperature within the kiln.

thermocouple

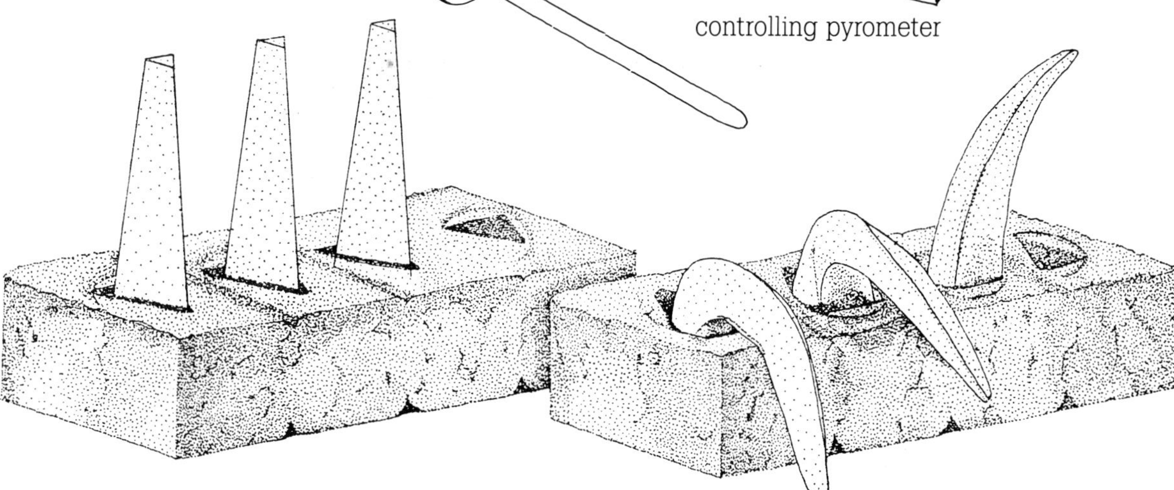

Packing the kiln

Unfired thrown pots are quite strong when bone dry and are capable of supporting fairly heavy weights as long as that weight is evenly distributed. If the pots are of various sizes, shapes and weights, then care must be taken to stack the heavier pots at the bottom of the kiln so lighter pots can be placed on top and small objects put inside larger bowls.

To make use of all the space in a kiln, fireclay shelves and supports are necessary to spread the weight and support the wares. The kiln shelves should be clean and dry with a coating of an alumina-based batt wash. Tiles can be fired vertically on edge; large flat shapes should be supported on layers of silica sand. Shelves are propped up at three points with round or square pillars; this helps to prevent shelves bending as can happen when four supports are used. If additional support is required for a very heavy piece, it can be cut from soft refractory brick with a coarse saw. Take care and wear a face mask if you do this: inhalation of the fine fireclay dust can be dangerous.◇

It is more economical and will make a more efficient firing if you include a wide range of shapes and sizes in the kiln, filling every small space to distribute the heat evenly; empty space is wasteful and requires more fuel to heat.

Stoneware and porcelain glaze firing

Place porcelain pots directly on a kiln shelf with alumina coating or protected with silica sand. Placing the pots on a thin disc of porcelain clay will prevent their sticking to the shelf.

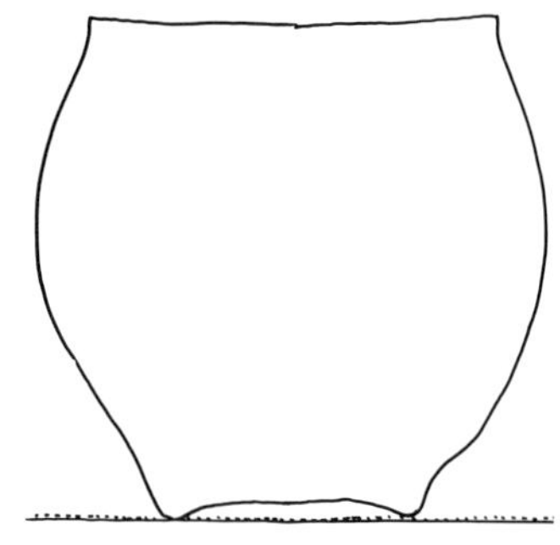

Do not use stilts at temperatures above 2100F (1150C) as pots may sag over these or stilts become embedded in the base. Keep the feet and bases free of glaze.

Make sure pots do not touch each other or they can fuse together.

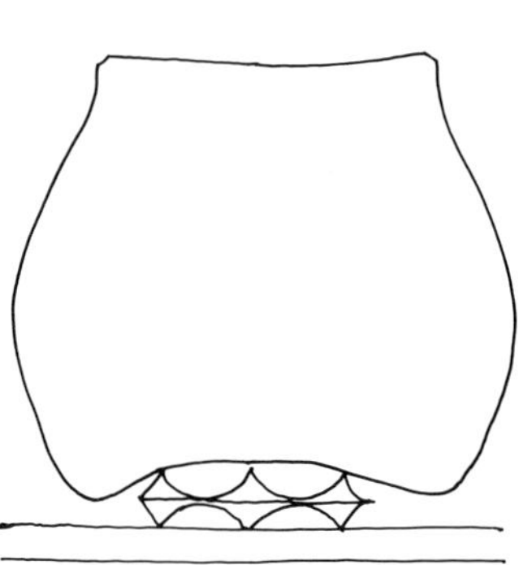

Kiln furniture

Three-pointed stilts or small spurs may be used to lift the base of the pot away from the kiln shelf.

Shelves (or bats) can be supported on cylindrical props.

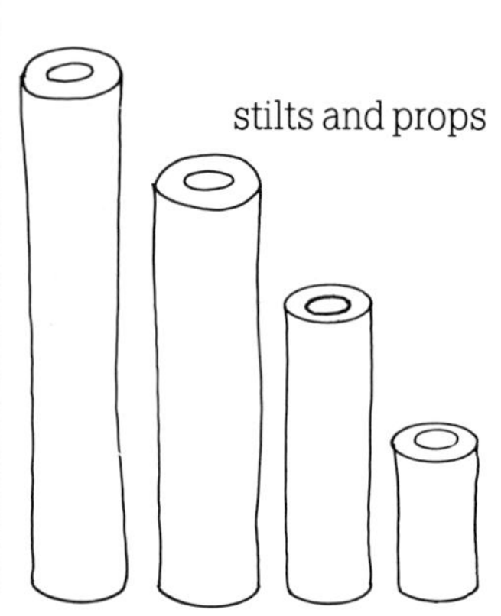

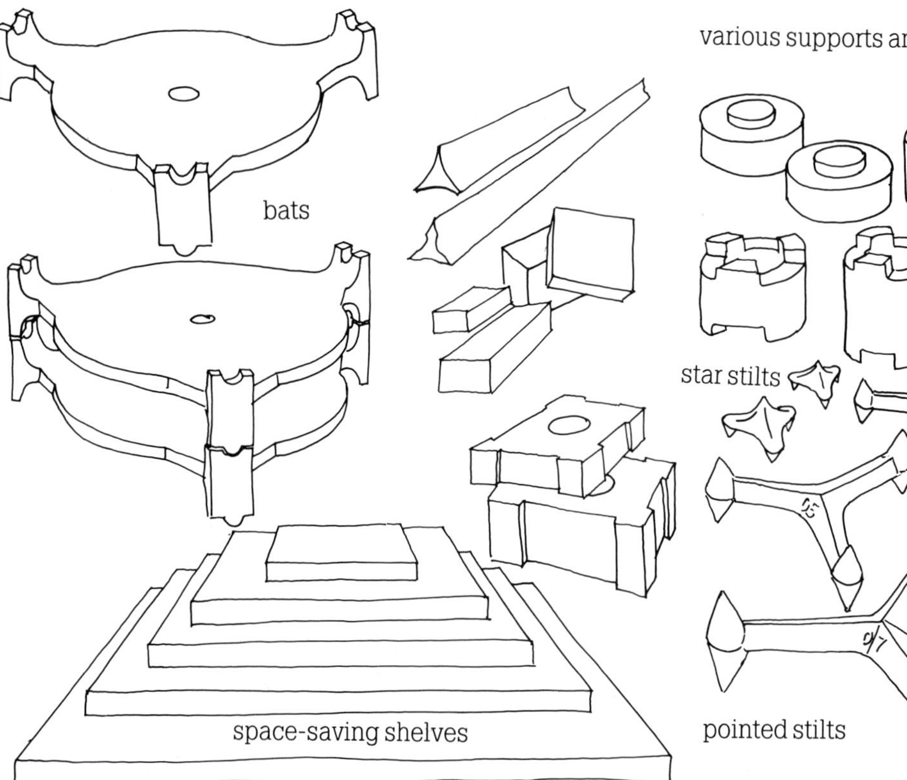

Earthenware glaze firing

A thin coating of bat wash will protect the kiln shelves from damage due to dripping glaze.

Pots with earthenware glazes are porous and need to be glazed all over. This means the pot should be placed on a stilt to prevent its sticking to the kiln.

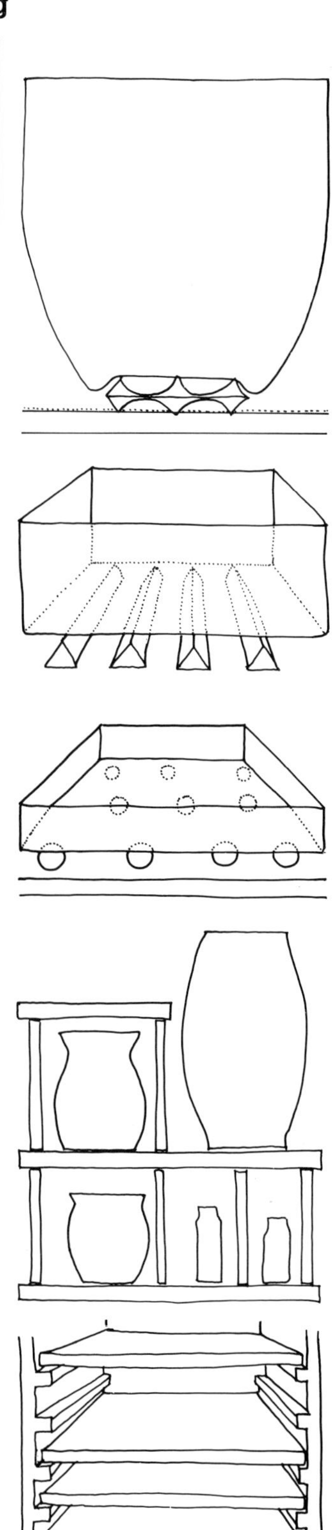

A large dish may need fire bars underneath to prevent its sagging.

Sometimes large pieces can be supported on clay dots that will support the whole area of the base.

Three supports under a shelf are more stable than four.

Put supports over each other as you build upwards so the shelves are all stable.

Large pieces may need an extra prop underneath to prevent the kiln shelf cracking.

Shelf supports are sometimes built into the walls of the kiln. (Fordham Thermal Systems, Newmarket)

Packing a biscuit kiln

Pots can touch each other. They may be stacked in 'bungs' if the sizes are similar and the weight is not too great.

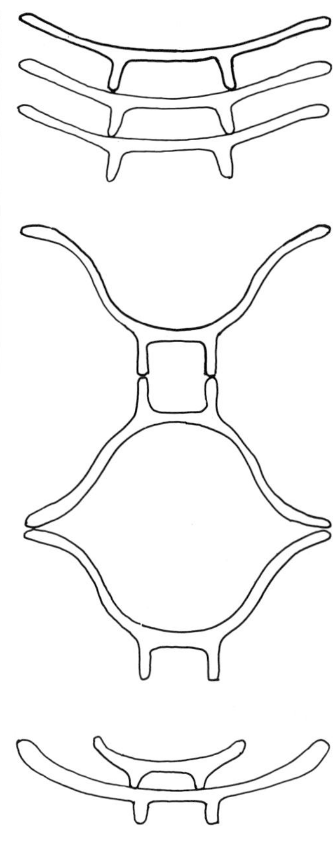

Bowls may be stacked rim to rim with smaller pots inside them.

Large heavy pieces should be placed at the bottom of the kiln.

A lightweight bowl may rest on thick platters or inside larger bowls.

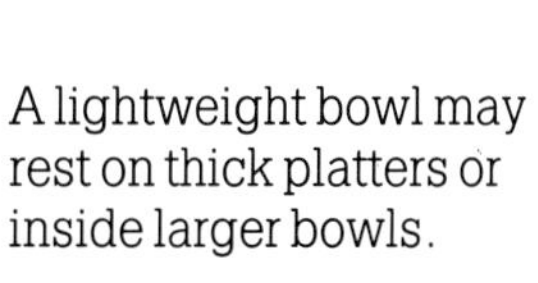

DO NOT wedge pots into the mouths of larger pots — the pot underneath will crack if you do.

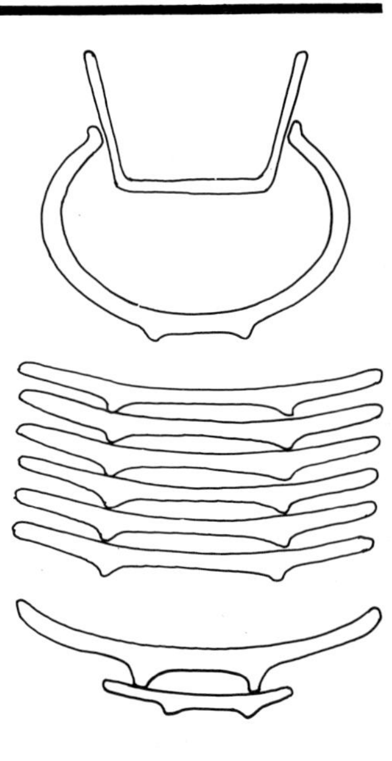

DO NOT stack too many dishes on top of each other; the ones at the bottom will crack if too much weight is placed on to them.

DO NOT attempt to do this.

Potter's roulettes from Uganda

8 History

Introduction

Fired clay objects have always provided a major source of information about early civilizations. Heat-hardened clay is resistant to decay and to disintegration; thus any excavation into sites of ancient origin reveals a good deal of information in the clay objects and pots fired long ago.

How progressive were the creators of these pots, how did they live and dress, what did they eat? Throughout the progression from handbuilt ware to pots produced on a flat revolving stone and finally on the wheel, the customs and lifestyles of past civilizations and cultures are represented in the various forms of pottery as well as on the surface decoration.

This decoration was painted, scratched, beaten in or added on to the pot's surface — a surface which might have been coated with fine slips (called terra sigillata) or given a polished exterior that shone like a glaze (called burnishing). These styles of decoration have not changed and are still used today.

Elaborate ritual, in particular religious and funereal offerings, have often made use of the potter's craft. So, from everyday functional purposes — the essential cup, dish or storage jar — to the formal rites of the grave, potters have contributed much to our knowledge of the history of the world's peoples.

Clay vessels have been made for at least eight thousand years, and probably longer — clay effigies and models are known to have been made many years before that.

The development of pottery and ceramics making has been gradual process. Mostly the potters were craftsmen producing vessels for everyday use. The spread of the use of the wheel, the advances in kiln-building techniques and the use of glaze can be traced from one country into another—as well as the different styles of decoration. Although these developments are often spoken of as though they happened quickly and frequently, in fact hundreds or even thousands of years may have passed by before they were effected.

As the use of different clays and materials (and the building of the kilns in which the pieces were fired) were discovered and developed, so these techniques spread from one place to another along the normal trade routes or perhaps being carried by war and invasion. Occupying forces often introduced their own style of ceramics to the lands they conquered as well as themselves absorbing local innovations and styles.

Neolithic jar c. *2500 BC*
Made in North-west China, this jar has been painted with pigments in an abstract design.

Egyptian urn
This predynastic urn has interesting crocodile relief decoration.

Initially the spread of culture and artistic ideas centered on the Mediterranean countries and the Far East, but wherever these developments took place the one constant factor was the skill of the craftsmen involved. Throughout these early years the methods used were skilful and well thought out, the decoration bold and appropriate to the form of the pot and the pieces practical for their intended purpose.

Firing was one of the main factors in this development. Greater sophistication in kiln building, the discovery of refractory clays and increased knowledge of firing techniques all produced changes, as did the use of the potter's wheel. Wide differences exist between the low-fired, simple and often rough pots made in Europe and the West, and the high-fired stoneware and porcelain from China and Japan in the Far East.

There has been no predictable progress through the centuries. Some cultures never developed their ceramics further than handbuilt, low-fired earthenware and yet they produced objects of great beauty.

Some mid-African countries and those in South America still make pots in the traditional way, by handbuilding, decorating with slips and polishing the surfaces to make the pots more waterproof. The pots are fired in a bonfire in an open pit, just as they have always been done. But the skilful manufacture, the treatment of the surfaces, and the low firing have developed a style with a unique beauty of its own.

Pots were used to store grain and water when tribes became less nomadic and started to cultivate the land and build permanent shelters. Bark and gourds were used to hold water and food, and the first pots reflected these natural shapes. The uses for which the pots were intended obviously influenced the choice of shape and the quality required. Holes and lugs might be needed for hanging pots or to retain a skin cover and keep the contents cool; the early potters made spouts for pouring and pointed bases which enabled the pots to be buried in sand to improve coolness and stability.

In the first place, hardening the clay was probably achieved by a combination of natural drying in the sun and being heated by fire.

It is thought that woven baskets were lined with clay to make them watertight. This coating would shrink and dry while the woven rushes or cane may well have been accidentally burned off to leave the clay lining low fired.

The Vestonice Venus
Clay effigies and models were some of the earliest forms of pottery produced. This female statuette is of baked clay and comes from the Upper Aurignacian period.

Introduction

Certainly a pattern of weaving, such as might be created this way, is a common form of decoration on Neolithic pots.

Initially, pots were made and turned on a mat or flat stone. In time this was replaced by a flat slab on a spindle set into the ground. As many as ten turntables may have been made; on each of these a partially finished pot would be coiled, smoothed by hand and then left to harden in the sun. Meanwhile the potter would work upon the pot next in line. This turntable was called a slow wheel and served as an aid to handbuilding rather than as a real throwing device.

The skills involved in making everyday items produced simple, uncluttered shapes, pleasing to hand and eye.

The first wheel-thrown pots are thought to have been made about 3000 or 4000 BC in Mesopotamia. The speed of the turning wheel, propelled by the foot or with a stick, meant that the pots produced could be thinner walled and more even in form. Moreover the amount the potters were able to produce was greatly increased. Flat stone wheel heads which have a hole in the middle for a spindle and date from about 2000 BC have been excavated in the Middle East.

Early examples of pottery were coarse textured and painted with pigments and colored clays—rather like cave paintings.

Designs were often bold and geometric; some pots were polished. As the firing techniques became more controlled brighter colors (which tended to blacken or disappear altogether in a direct flame kiln) were retained in the simple updraught kilns where the pots were protected from the fire and excess smoke.

Greater control of the atmosphere was developed when the pots were set in a chamber above the fire inside a kiln with a perforated floor. The heat rose through the pots and was controlled by a chimney flue at the top. In this way the atmosphere could be clear and oxidized (leaving the pots red) or smoky (in which case the pots would be blackened).

The skills with which the craftsman controlled the kiln and the constant search for different or better clays has been reflected in the fineness and variety of wares produced throughout the world.

Whatever the era, whatever the civilization, the methods used or the results, ceramics provide a fascinating study—for archaeologist and artist alike and for potters throughout the world who can learn much from their forebears.

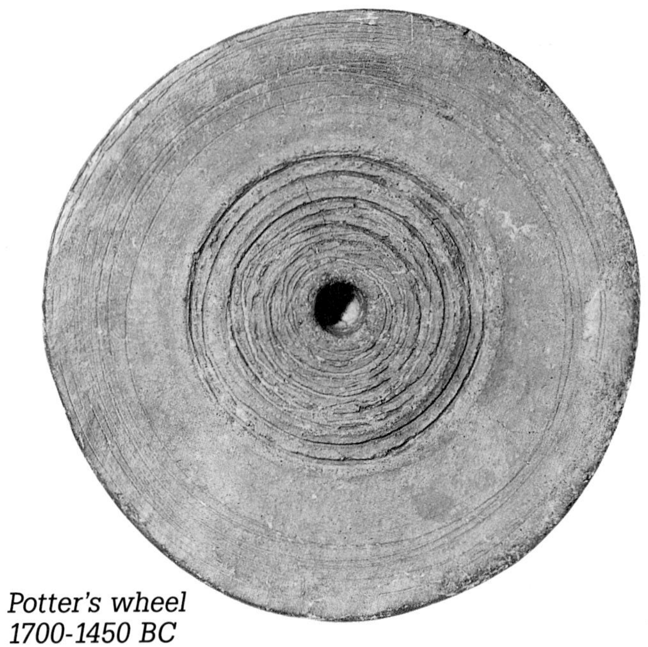

Potter's wheel
1700-1450 BC
Minoan potter's wheel made of terracotta; a pivot was inserted in the central hole.

Bronze-age beaker
2200-1400 BC
The beaker people introduced simple chevron and incized designs, as on this encrusted urn.

Large Minoan vases c. *1450-1400 BC.*
The excavation of the Palace of Minos at Knossos, Crete revealed huge vessels, used to store oil, wine and possibly cereals. These were set into stone cists in the floor.

Neolithic pot 3500 BC
Panels of decoration and scrolls can be seen on this sturdy pot found in the Southern Ukraine, USSR.

Prehistoric/neolithic vessel
A pottery vessel from Mildenhall, Suffolk with lugged handles; the simple scored decoration is still visible.

Chinese funerary jar c. *2000 BC*
This elegant jar illustrates how advanced ceramics were in China during the Neolithic period.

Mesopotamia

Sumerian seal
c. 2900 BC
Two ibexes are depicted on this cylinder seal impression from the Jamdet Nasr period. These impressions would have been rolled on to clay.

Some of the earliest known civilizations were found in the land between the Tigrus and the Euphrates rivers. The oldest ceramics discovered in the Middle East were in fact found in a Neolithic village in Anatolia and date from about 6800 to 5700 BC. By 6000 BC Mesopotamia was producing simple incised black ware followed by painted designs which were usually geometric and on red bodies covered with a matte cream slip.

By 4000 BC the design of kilns had improved and firing no longer eradicated the clear colors painted on to the pottery—which was harder and stronger in the higher temperatures reached. Also the development of the axle and the simple slow wheel required smoother clays. Coarse stone particles now had to be removed by 'levigating'—turning the clay into a liquid state and then tapping the finer clay when the coarser material had sunk to the bottom.

Perhaps the most dramatic change, however, was the development of a simple glaze. This exploited the properties of lead and developed from glass-making processes during 2000 to 1000 BC. A simple form of glass made from quartz and silica sand was at first powdered on to the surface of pots. Then lead was added to the glass frit which, when ground up, made the first real glaze, one that did not shrink so much on cooling. (Pots from this period have been found in Syria and have a blue-green glaze, produced mainly with copper oxides in an alkaline base.)

The glaze was also found to be opaque and white when ashes of tin were added to the mix. This, together with oxides of metals, was used extensively on tiles for the decorating of temples and on the walls of buildings. The doorways and gates of Babylon were particularly good examples of this. Alkaline glazes had been used before, as well as glazed frit, but until the discovery of lead glaze all the glazes then available would not adhere properly to the body of a clay pot because they contracted so much when they cooled.

These discoveries gradually spread across the other countries of the Mediterranean to influence the development of pottery throughout the world.

Egypt

Egyptian jewelry 1550-110) BC
Made of glazed composition during the 18th dynasty, these delightful floral garlands would have been presented to guests at a banquet and are collars to be worn every day rather than the grave finery which is more usually discovered.

Ptolemaic jar
This sturdy blue ceramic jar exhibits a better polished surface and handles — features not seen on very early Egyptian pots.

Egyptian coiled bowl c. 400 BC
Handbuilt from coils and fired upside down in a wood fire, this pre-dynastic bowl has a full globular form and painted spiral decoration.

Egyptian faience
Mandrake fruits made of faience which was developed originally to replace the soapstone products.

Pre-dynastic statuette
Egyptian statuette of a woman made in clay.

Egyptian shabtis
Two different treatments of similar forms: the smaller blue faience from the XXI Dynasty (faience could be used for small objects only) and the painted larger figure from the XVIII-XIX Dynasty.

The oldest Egyptian pots were dark and crude and made in the Lower Nile valley in Neolithic times. Rather more pleasing pots were being made by settlers in the Central Nile Valley around 5000 to 3000 BC. Here the Badari people produced well-made pots in fine red river clay. Mostly cooking pots, they were thin rimmed and were burnished with fine combed decoration. They had brown and red bodies and rims which were blackened and hardened from being fired rim-down in the ashes of the fire. Some beakers with flared rims have also been found from this period. They were decorated with a pattern of incised lines that were filled with white pigment.

Gradually the shapes became more refined and taller, with a better polished surface and more even coloring. As well as dishes and simple bowls, narrow-necked vases and vessels began to appear along with rotund flasks and the very occasional spout.

Strict governments controlled art as a reflection of religious belief, and allowed little or no change in the subject matter for decoration for almost 3000 years. Pots were not as restricted as tomb painting or sculpture, but were nonetheless restrained as a result.

In time, lighter colored clays were found in the desert valleys. These fired buff colored and so encouraged the painting of surfaces with pigments of iron and manganese; the surfaces imitated stone carved pieces which were then a popular art form as well as mimicking the natural patterns of alabaster. Freer decoration evolved and handles appeared for the first time in Egypt.

Two major changes in pottery making happened in the period 2500 to 2000 BC. First came the introduction of the slow wheel from Mesopotamia. Spinning the pots produced finer shapes with thinner walls. The designs often echoed the shapes of items previously only able to be made in metal, for the first time regularly exhibiting spouts, handles and lips in clay. There were bowls with inward-curving rims, tall libation vessels and high stands for elegant ritual ware. The Meydum bowls of the 4th and 5th Dynasties are especially renowned.

Predynastic 'reed-ship' jar c. 3300 BC
This pot is typical of the Naqada II or Gerzean culture with dark painted designs on buff clay.

The second major change was the discovery of a simple method of making a glazed surface. A glaze is usually regarded as a substance which is applied to a formed pot. However, Egyptian faience (or Egyptian paste) was a body made up of quartz, sand and soda mixed into a crumbly paste. As it dried the soda formed a crust of crystals on the surface which melted into a simple glaze in the kiln. Mixed with copper, cobalt or manganese, the alkali reacted with oxides to form a turquoise, a rich blue or a purple. It was initially produced as an improvement on the soap-stone and its development incorporated ideas from Mesopotamia.

The nature of the mixture made it unsuitable for anything bigger than beads, small pieces of jewelry or objects made in press molds. These could be quite complex in shape and design but would be no bigger than say 3 inches (8cms) high. Later on the same material would be used as an inlay to decorate furniture, coffins and the walls of temples. Turquoise tiles in the tomb of Pharoah Djoser at Saggara (about 2600 BC) are the first indication of the use of faience. It was not applied to pottery until around 1500 BC.

During the period 2100 to 1600 BC the increased trade imports from other Mediterranean countries (including the islands of Crete and Cyprus) influenced the form of Egyptian pottery.

Meanwhile the wheel became a far more efficient tool. Ancient hieroglyphics show potters using a raised wheel which was possibly pushed around by the feet or by hand. Shapes were thrown and then joined together to create complicated pieces with spouts, handles and pedestal feet. Pots were large—some over 3 feet (1 metre) high. They were well thrown with colors that were painted on, after firing, in rich designs of flowers and geometric shapes.

During the New Kingdom the customs of burial ritual underwent change and for some reason the pottery placed in tombs became rather more utilitarian. At the same time there were technical improvements. Lead glazing spread from other countries and the wheel developed into the more efficient upright shaft with a heavy flywheel.

This essentially Egyptian style continued to be produced until around 50 BC when the Romans invaded Egypt and introduced their own designs.

The Minoans

The Minoan civilization of Crete was unique. Named after the legendary king Minos (whose throne can still be seen at Knossos) it developed from an independent community which was established in about 3000 BC and then flourished for about 1800 years.

Being an island race, the Minoans knew that the sea was their main resource and means of survival; most things were influenced by the surrounding Mediterranean and by fishing. Protected by their navy and without the stranglehold of a dominating priesthood, the Minoans knew that trade with other communities was necessary to maintain their economy and culture. Two other products which were of prime importance to the economy were wine and, most important of all, oil from the many olive trees on the island.

Technically the Minoans were not as advanced as the Egyptians but the greater freedom of their culture and their prosperity meant that their pots were far superior with more glowing variation and freshness of approach. The designs and decoration that the Minoans used on their pottery reflected the sea's influence on their race. They made much use of bold, freely interpreted designs based on sea creatures such as dolphins and octopus, all of which were meticulously observed and then flowingly applied in rich colors. The pots were finely made and the potters were honored members of society, highly regarded by the rulers of the island who valued the fine ware they produced.

The shapes and styles of the pottery reflect the occupation of the islanders; huge storage jars for wine and oil which were nearly 6 feet (1.80 metres) high, elegant vessels from which to pour and drink wine, small bottles for exporting oil and perfume. Often the whole surface of the pot would be decorated and the rounded vases and oval bottles were an aesthetically pleasing form on which to paint these fluid designs. Applied raised patterns were also used.

The rich civilization of the Minoans was destroyed in about 1400 BC. The reasons for this are not fully understood but it may have been due to the tidal waves which swept the north coast of Crete after the massive volcanic eruption on the nearby island of Santorin: equally it may have been brought about by invasions from the Greek mainland or by an earthquake which destroyed the capital Knossos. The high standard of the pottery and the rich culture of the Minoans never recovered their former glory but the vitality of the ceramic work, revealed only comparatively recently by the excavations on Crete, remains a lasting tribute to this unique society.

The influence of the sea on the island race of Crete is evident in the subject matter chosen for the Minoan's flowing bold designs.

Greece

Attic vase 740-720 BC
Intricate geometric design was at first set between bands of slip around the neck and foot but gradually grew to cover the entire pot in stylized patterns.

Greek amphora c. 540 BC
Found at Rhodes, this style of Greek decoration shows the emergence of the human figure in design — in this case 'a running man'.

By 1500 BC pottery techniques were refined and well practised. The use of the fast wheel was widespread and employed skilfully to produce shapes which were joined and turned. The styles and techniques of Crete, Mesopotamia and Egypt all had their influence on the ceramics of the Hellenistic Empire and many of these earlier ideas were absorbed into the Greek ware. To begin with, pots were decorated fairly crudely and the shapes reflected objects made in metal—which was regarded as a more worthy material, being the major source of wealth.

By about 1000 BC, however, with the emergence of the Greek city states, the more familiar Greek styles of pottery emerged. Pottery was now highly regarded being made not only for funerals and monuments but to serve everyday functions such as water carrying, wine dilution and cooking. The State was the patron of the arts and as such actively encouraged the production of decorated and functional ware for grave monuments and as prizes for successful athletes. The bulk of the pottery produced for home use—the purely functional, undecorated—but pleasing pottery which was made locally—has largely disappeared and the painted pottery, conserved in the grave, is the familiar Greek ware we see in museums today.

Athens and Corinth both had good deposits of fine red and yellow clays so naturally they became the main centers for the production of the red and black ware. The smooth

Greece

Athenian clay was excavated from the borders of the city and fired rich red while the Corinthian potters used their buff-yellow clay which was lighter in color. Over the years there was inevitable competition between these two centers — which between them dominated the evolving history of Greek pottery.

Although many of the fine pots we see today in museums and galleries were found in funeral sites, in fact most pots were made for a specific use and had names which reflected this.

Hydria were water-carrying pots. They were large enough to hold a good supply of water drawn from a well but not so large that they could not be carried by a woman for some distance to and from the fountain or communal well. They had three handles—two for lifting and carrying, and one to be held when pouring or when the vessel was empty.

A krater was the name given to a large bowl used mainly for mixing wine and water—a favorite drink.

Amphora
Vessels used for storage often had pointed bases that could be sunk into the ground.

An amphora (from the Greek word meaning to bear or carry) was a large pot used for storing wine and oil. Sometimes it incorporated a wide foot for stability or a pointed foot for ease of storage or burying in sand.

The best clays were those used by the two main centers of Athens and Corinth. A fine clay slip with a high iron oxide content was painted on and retained its slight shine—which increased during firing.

The iron slip would fire a dark red in a clear (oxidized) atmosphere but by controlling the amount of air in the kiln (through the use of the damper in the chimney) and burning green wood to create a smoky atmosphere, the pots turned black. A short period of firing in a clear oxidizing atmosphere at the end of the cycle meant that those parts of the pot not covered by the iron slip would fire red again. This critical stage of the firing had to be carefully controlled; if it lasted too long the black parts of the pot would also turn back to red (known as terra-sigillata).

The shape and decoration of the pots was carefully worked and skilfully executed. They were thrown in sections, joined at the leather-hard stage and then turned to a smooth finish. Fired in an updraught kiln, the pots were supported in a chamber that was set above a perforated floor and had a chimney flue to control the atmosphere.
The first style of decoration, called geometric, consisted of carefully worked patterns set in between bands of black

slip around the neck and foot of the pot. The designs were almost mathematical and very stylized arrangements of circles, spirals, triangles, zigzags, and swastikas. By 900 BC the whole pot was covered by these regular strips of pattern.

Geometric patterns gradually gave way to designs which incorporated figures, animals and horse-drawn chariots. At first these were very 'formal' stick-like figures that were almost abstract but eventually the silhouette shapes became the main part of the decoration while the geometric patterns were still added to the neck and foot. The figures were painted in outline with the slip and the details were scratched or cut through to the red clay surface.

At first the shapes were largely of plants and animals such as deer and goat. In time a wide range of creatures such as bulls, griffins, lions and eagles appeared and the potters from Corinth developed wonderful friezes with an oriental flavor. There were floral borders and complex designs incorporating many creatures—with those from myth and legend appearing increasingly frequently.

The Athenians concentrated more on the human figure, depicting heroes and gods as well as contemporary scenes of races, battles, and religious and temporal celebration. The black profiles were always of men and white was used for women. The black-figure painting was at its best in about 500 BC but was then gradually overtaken by a style called 'red figure'.

This was a technique which allowed the Greek artists to depict the human figure in a more realistic way than was possible with the black-figure method. Instead of painting the figures in black, the background was painted black while the figures were left red, with details painted in black line. A further three-dimensional quality was added with other colored slips and stains.

The painting was freer and the subject matter less formal too, to include for instance, scenes of daily life. Clothes and hair could be seen in greater detail, the figures no longer kept to the base line and more colors were introduced.

A further refinement, reserved for small bottles that held precious perfumes and oil, was to use a rather fragile white painted ground with colors painted on top.

These techniques gradually died out as the Roman occupation of Greece took hold and the Romans adapted the Greek potters' methods in order to promote their own styles of work.

Greek figurine 200 BC
Terracotta woman playing a cithara.

Black-figure decoration 530-510 BC
Odysseus blinding Polyphemus is depicted on this Caeratan hydria from Eturia. Scenes from myth and legend became increasingly popular.

The Romans

The Roman Empire was the dominating force and culture for many centuries. Gradually its great influence spread throughout Italy and into neighboring countries. Romans conquered the Etruscans and the Greeks, large parts of the Islamic Empire in the Eastern Mediterranean, North Africa, Spain, Southern Europe and England.

Roman pottery was produced wherever the Romans conquered and settled, imposing their own styles and technical skills on the local population and then proceeding to somewhat overuse those indigenous styles that they thought worth copying. The Romans had no traditional ceramic knowledge, but they imparted their own 'flavor' to the pottery of many widely different areas. The main style used for everyday use was adapted from Greek terra-sigillata, high-gloss, fine slip covering and was known as 'red gloss' ware (sometimes referred to as samian ware). These pots, often decorated with a raised pattern of figures and friezes, became glossy in the clear oxidized firing.

Raised patterned bowls were often made in molds (for speedy repetition and production) with rims and feet added later. These molds were crisply carved with detailed designs which appeared in relief on the finished bowl but are somewhat metallic in style, failing to exploit the clay's plasticity. Some of the largest centers of production were in France, Germany and England.

A particularly fine, raised heavy slip was produced at Castor, near Peterborough in England. One style from France had patterns that were cut out with a v-shaped tool and looked rather like the decoration on cut glass.

Arezzo, near Florence, gave rise to Arretine ware around 30 BC. This pottery was especially artistic with uncluttered lines and classical figures. The simple elegant decoration was often inspired by the then highly valued metalwork designs.

Lead glazed ware such as had been used in Egypt and Mesopotamia spread through Europe—to Italy, France, Germany and then to western England.

In Egypt the glazed frit ware was used and adapted by the Romans. Overlapping patterns of scales, delicate underglaze and the flowing of the brilliant glaze into the carefully incised design show just how great a control the Roman potter had evolved in exploiting a very difficult medium. In a modest way this was the forerunner of the later Islamic ware in blue and turquoise.

Roman kilns were very similar to Greek designs but had a temporary cover, rebuilt each time it was fired. Built into a hillside, they were round in shape with a firebox which led the heat through a perforated floor and around the pots to escape via a flue in the temporary dome of clay. This was dismantled after each firing; it was some 4-6 feet (about 2 meters) in diameter.

Under the Roman yoke, pottery for the vast empire—for its armies and its cities—was mass-produced in a way that was not to be equalled until the Industrial Revolution. As the legions moved even further from Rome so its centralizing influence diminished and the original styles became freer and, in many cases, debased.

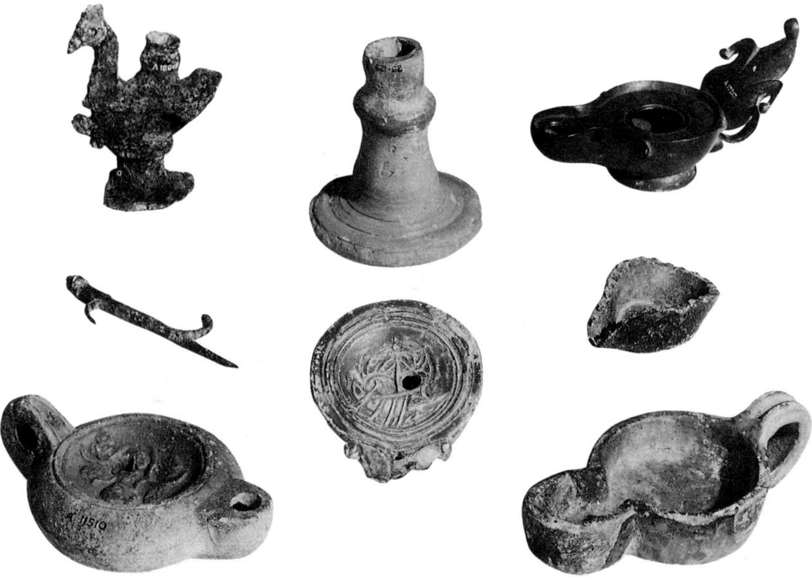

Household items
Ceramic lamps, lamp fillers, pins and candlesticks of Roman manufacture.

Fountain niche
1st century AD
Roman garden scene from Baiae, Italy, intricately constructed from mosaic tesserae in delicate tones of brown, blue and turquoise.

Roman vessel
This simple pot has been enlivened by the face caricature.

A group of simple Roman jugs

China

Early Neolithic Chinese pottery, which was made as early as 4000 BC, is unlike all the ceramics produced at that time by other civilizations. It is very sophisticated for the period and decorated with carefully worked swirling patterns in red and black that were made from earth pigments. The pots have thin walls and were fired at 1830°F (1000°C).

The first pots to show a degree of wheel work were made in about 2000 BC in the Shantung province, north-east China.

Bronze and jade were the valued materials. In fact, the high firing of clay and the building of the necessary high-firing kilns is related to the heating and melting of bronze. During the Chou dynasty (1155-255 BC) high firing of fine white clays was made with glazes that produced a finish which resembled jade. The emergence of kilns that were capable of firing pots to these high temperatures was one of the most significant developments during this period.

Pots were fired in linked cave-like excavations sloping upwards in hillsides of heavy clay soil. A hole higher up the hill cave created a draught and controlled the atmosphere inside the kiln. Insulation for the kiln was provided by the thick banks of soil surrounding the kiln, these banks hardening further the more the kiln was fired. The higher temperatures achieved and the ability to create a smoky reducing atmosphere led to the first high-fired stoneware and to the earliest simple ash and feldspar glazes — resembling jade.

Excavated graves have revealed many examples of pottery from this period. Pots were buried with important rulers—along with their wives and servants. Confucius (550-480 BC) denounced this barbaric sacrifice and so clay or wooden substitutes were used instead. The famous army of life-size ceramic warriors discovered in 1974 would have been created as a result of this change in policy.

Neolithic Yang Shao jar
A symbolic human figure and geometric circles have been simply painted with brown pigment on to this narrow-necked urn.

Ming pilgrim vase c. 1580-1595 BC
Blue and white porcelain vase made for Philip II of Spain. The two circular faces are decorated quite differently.

Sung vase
Ju kiln ceramic ware: controlled firing allowed the glaze to crackle into an overall pattern of fissured lines.

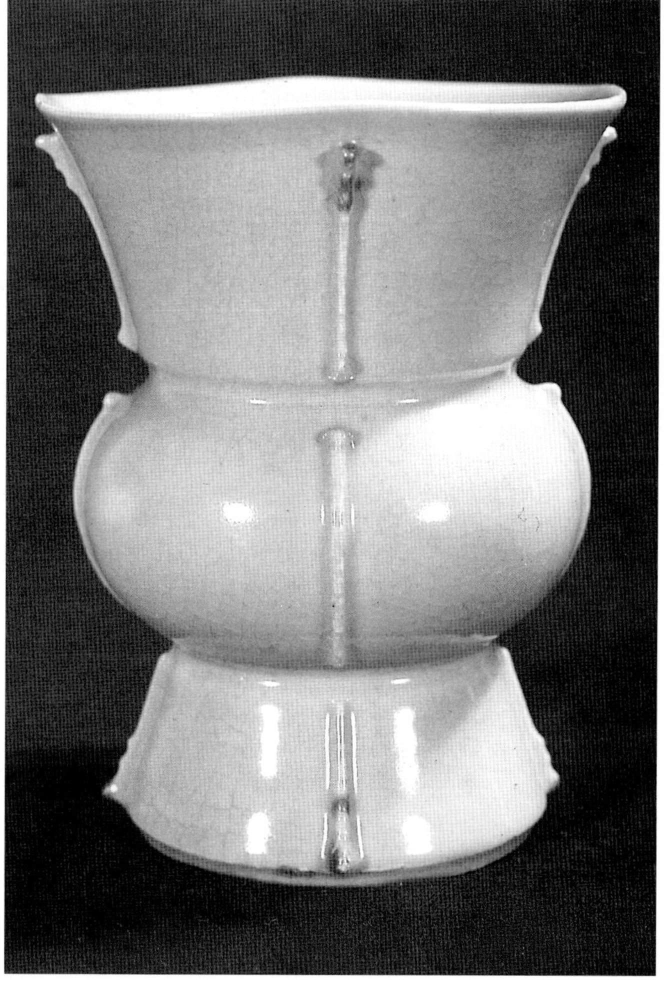

The Han dynasty (206 BC-220 AD)

This was a time of increased prosperity and unification. Bronze still influenced techniques and style but pottery began to be the chosen medium for offerings within tombs. Stoneware pots with feldspathic glazes which used light-colored clays were widespread by the beginning of the Han dynasty (206 BC-220 AD).

Stoneware and lead-glazed earthenware were both made during this period, lead glazes having been brought back from the Middle East by traders. The green lead glazes now available meant that the color of bronze could be imitated well and this added to the metallic appearance already suggested by the shapes.

Particularly interesting were the 'hill-censers' (or 'hill-jars'); their lids were developed into the shapes of islands and mountains to represent the mystical Taoist Island or Mountain of the Blest. Many forms and animals associated with the spirit world were seen in the decoration at this time.

The use of fired light-colored clays and further improvements in kiln construction made this one of the most prolific periods in the early production of glazed stoneware. Kilns were constructed on a sloping surface in a large domed shape which curved down to a narrow aperture near the bottom of the chimney. The heat was drawn across through the pots, reaching temperatures of 1200°C (2192°F) near the firemouth. The cooler part of the kiln near the top was used for firing lead-glazed earthenware.

Tripod vessel
206 BC-220 AD
Many pots from the Han dynasty imitated the sought-after bronze ware and this earthenware vessel even has metal-like rings modeled on the outside.

T'ang dynasty (618-906 AD)

A settled period with a central government led to confidence within the country and a belief in its strength and position in the world. This allowed for greater tolerance of other cultures and the opening up of trade routes which brought outside influences to the arts. Ch'ang-an was the northern capital and a center for both the commerce and the arts of Asia, acting as a crossroads between the East and the West.

The T'ang period is typified by its fine white earthenware with colored lead glazes of green, amber and yellow. These are rich in pattern on full rounded shapes that seem to swell almost to bursting. The T'ang three-color style simply glowed with a kaleidoscope of color and pattern. Sometimes the free-flowing glaze was applied to the top half of the pot only, probably to prevent the glaze running down on to the kiln.

High-fired stoneware was also produced. The soft gray-green celadon glazes were made with small quantities of iron oxide in the glaze; they were fired in a reducing atmosphere to create, not brown or yellow as in oxidization, but blues, grays and greens—glazes that mimicked jade and were used over incised patterns on light-colored clays.

There was a great variety in the types of pot produced in this dynasty—ewers, cups, bottles, vases and flat dishes with simple but effective designs incised into the surface.

The production of tomb figures reached its zenith in the early 8th century with detailed models of dancers, musicians, warriors, stern officials, horses and camels—all executed with great delicacy and finesse.

It was during this expanding prosperity that the first type of porcelain was made—using the white clay kaolin (China clay) and petunze or China stone (a feldspathic mineral). Firing to temperatures between 2280°F (1250°C) and 2370°F (1300°C) produces a finish that is strong, hard and, when sufficiently thin, marvellously translucent. White, the Chinese color for mourning, was much sought after in ceramics and may have promoted the experimentation which led to the eventual emergence of porcelain.

T'ang woman 618-906 AD
Delicate figures were produced with their simplicity of form heightened by fluid colored glazes.

Earthenware horse 618-906 AD
This horse from the T'ang dynasty has been modeled with great sensitivity.

China

Sung dynasty 960-1279 AD

This dynasty supported a long period of excellence in ceramics, unprecedented in the world before. A fine white porcelain was made in the Hopei province in north China from carefully prepared minerals and clay. It was fired in technically advanced kilns and showed off to great advantage fluted designs and incised patterns. At last real porcelain the first to exploit its finest qualities and potential was being produced; the potters of the Kiangsi province achieved pots of delicate shape and great translucence.

White slip-covered stoneware decorated with dark painted patterns was also made. This was a quite restful style of stoneware in celadon and tenmoku brown. Its glaze had a heavy iron content that broke into orange around the edges and on the rims.

The patronage of the Court (including one emperor, Hui Tsung, who was himself a talented artist) supported a period of great skill and experimentation. No longer did pottery need to mimic metal; it was explored for its own sake and all aspects underwent great changes and developments. Carefully controlled firing introduced many subtle effects; glazes became far more sophisticated and innovative: running into pools of color, crackling into fissured lines; streaked and patterned; or with silver 'oil-spot' flecks. The most celebrated celadon glaze was Ju ware in an opaque blue-green (made only for the Court). The new technique of sgraffito (scratched) designs made the most of the simpler glazes, highlighting the light and dark areas. Other pots had rich elaborate decoration; they were painted in floral and foliate designs, peonies, lotus sprays and birds and dragons.

By this time the porcelain factories at Ching-te Chen in the Kiangsi province became increasingly important and grew in size and domination of the ceramic industry. Many white wares were produced, including those with 'hua' (secret) decorations which were carved or painted and then covered with a white glaze, their secret to become visible again only under special lighting conditions.

Ming dynasty (1368-1644 AD)

Underglaze painting and enameling became the most popular decoration of the Ming dynasty. At various times previously a cobalt of an inferior quality, found at that time in China, had been used. Now it was replaced by a purer cobalt imported from Persia. This gave the rich color known as Mohammedan blue. The color was easy to apply once thinned with water; it could be used pure and dark or as a thin wash for paler shades.

The initial designs were largely geometric arrangements of bands or panels filled with plants and flowers which covered the entire surface of the pot. By the 16th century human figures and landscapes were illustrated with greater freedom. Some of the imperially commissioned vessels were an immense size.

A thin wash of another oxide copper fired in a reducing atmosphere turned a bright red. The designs were of simple fish or fruit as control of the color was difficult. Polychrome wares (called the 'five-color' style) employed many rich colors and by 1500 the imperial yellow was particularly effective. The clay deposits at Ching-te Chen were practically exhausted by the end of this dynasty.

By the 17th century large quantities of blue-and-white porcelain were being shipped to Europe by the Dutch East India Company. These exports were of mixed quality.

Ch'ing dynasty (1644-1912)

Weakened by rebellion, the Ming dynasty collapsed and control was assumed by the northern Manchu tribes. This long dynasty brings us virtually into the changing modern world—however, there was rather less change than in previous dynasties. Imperial patronage continued and maintained an antiquarian approach to styles with decorative effects largely dominated by the traditional floral designs and motifs, and the ubiquitous dragon! A thriving export trade via Canton was fed by the imperial factories and European influences began to infiltrate. The versatility of clay when used to imitate other materials (such as shells, ivory and lace) was explored with great finesse. Celadon glazes were combined with carved designs. Cobalt blue and lustrous black backgrounds were overpainted with gilt. By the eighteenth century teapots, cups and saucers and statuettes in Blanc de Chine (especially of the Buddhist goddess of mercy, Kuan-yin) were popular exports.

One of the most influential changes during this dynasty was the emergence of the delicate *famille rose* porcelain, a rose-pink enamel of European origin which was suffused and softened with opaque white. Colloidal gold was used in the glaze to produce fine and elegant designs.

Chun ware
A pale lavender-blue glaze enhances this stoneware jar from the Sung dynasty 960-1279.

Tzu-Chou type stoneware bottle
Buff stoneware has been given a white glaze and painted with a plant and butterfly design. This is probably from the Sung dynasty.

Ming fishbowl 1567-1573
Five-clawed dragons in pursuit of flaming pearls are featured in cobalt blue on this fine Ming piece which is a full 29 inches (74 cms) in diameter.

Persian hawk
12-13th century
Luster printed earthenware hawk from Rayy where ceramics of an exceptional standard were produced.

Persian bowl
9th century
Made of earthenware, this bowl has been glazed red with decoration that has been incized through white slip and printed in contrasting colors in a stylized eagle design.

Turkish panel of tiles late 16th century
Painted earthenware tiles in a rich interwoven pattern of red, green and blue — from the baths at Istanbul.

Early Islamic

The style of the early Islamic pottery did not change from the soft lead and alkaline glazes used for centuries until traders introduced into the area T'ang ware with its lead glazed pots and porcelain.

There was no awareness of the possibilities of pottery techniques as there had been in China, no kaolin and no high-firing kilns. Unglazed water pots had been made for centuries. Because the buff clay was porous, seepage of water through to the outside and its evaporation there helped keep the water cool.

By 700 AD the Arabs, following Muhammad's teaching spread their empire south and west through Egypt and North Africa to Spain, to India and the boarders of China.

The teachings of the Muslim religion said that there were to be no graven images of people and that no precious metals were to be used. Adherence to these two laws brought about two types of decoration: a painted pattern of shapes known as arabesques and the use of the more humble metals to form lusters.

The use of lusters originated in Egyptian glass production; metallic salts with resin are added to glazed and fired light-colored surfaces. The pottery is then refired to a low temperature in a reduction kiln. Burnishing removes the dull metallic film and the glittering surface is then revealed.

Many colors were used in the early years but by the tenth century the rich golds and rubies had been replaced by plain brown or yellow ware which was less complicated to produce. Fine lusterware was exported from Baghdad and reached as far west as southern Spain.

The great tradition of Islamic painting on plates and tiles was brought about by their attempts to copy the Chinese porcelain. At first, rough clay was covered with white slip and lead glazed. Then additions of tin oxide to a lead glaze made the glaze white; this provided a smooth white background for oxide painting and lusters.

Although the white ware was a fair imitation of the Chinese porcelain, the styles of the Arab painters soon gave it a unique character of its own, with the dominant colors of blue and turquoise. This painted lusterware was to influence the styles of Italy, Spain and eventually all the countries of Europe. Samarquand ware, made in eastern Persia, was of an excellent quality with slip-painting that was ahead of its time in that country.

There was no one style which covered all that great area. Styles varied from country to country. Kashan, south of Teheran was an important center where potters passed on their traditions from father to son for some four hundred years. Iranian pottery was often decorated with Kufic script or by their traditional bird or animal motifs. To begin with, large expanses of white were left but in time the surfaces were covered with decoration. An area which produced ceramics of a particularly high standard developed in Northern Persia around the center of Rayy; especially famous was the Seljuq ware. Designs were carved through a white slip with a monochrome glaze over the top. The use of a white semi-vitrified body, decorated with rich painted designs and covered with a clear glaze was a feature of this region — as were the beautiful alkaline glazes in rich deep blues and turquoise.

Crushed quartz crystals or pebbles with a frit of borax or potash were added to clay to produce a strong but fine translucent white body—often pierced with tiny holes so it looked especially delicate. Laqabi ware was carved with raised lines and low-relief scroll work to prevent color pigments running together. Minai decoration, which looked rather like illuminated manuscripts, used enamels to produce complicated and sophisticated designs and miniaturized paintings of scenes from the court and legends.

It seems the Mongols may have spared some of the Persian potters when they invaded for craftsmen continued to produce lusterware there into the 14th century. However, Cairo and Damascus became the most important centers for pottery production under the Mamelukes when many artists fled there. Vast numbers of oriental spice jars and medicinal containers were exported—usually decorated with blue and black designs.

Later Islamic

Timur the Lame (1336-1405), a ruthless but strong Turkish leader, made Samarquand his capital and from there dominated the Islamic empire. As he was a lover of the arts, his rule encouraged a brief revival of crafts and architecture but the true renaissance took place after 1453 during the Ottoman empire.

Large white slip plates and dishes painted in black and heavily lined cross-hatched ware were replaced by the highly decorative Isnik ware produced in Asia Minor from the 15th to the 19th centuries. Once again deep colors including cobalt blue and turquoise were painted on to white slip under a transparent glaze. Realistic floral designs of vines, carnations, hyacinths and roses were used—along with arabesques and scrolls.

With the decline of the Ming dynasty in the 17th century attention was turned to the production of blue-and-white wares. These had been made since the late 14th century but were now created in far greater numbers to exploit this new gap in the European market. Many Chinese designs were adapted, often being outlined in black, and export continued for two centuries. Chinese green-glazed ware was also emulated.

Gombroon ware became very fashionable in England. This pottery (named after the port) was made with a fine, translucent and surprisingly sturdy white body. It was often incised, and pierced with many delicate-looking holes and then covered with a clear shiny glaze.

In the late 17th century lusterware also had a revival. Ruby or yellow-brown designs were painted on to a white or deep-blue glaze which gave a coppery reflection.

Eventually the cheaper mass-produced European porcelain and faience undermined Islamic production of pottery. Despite the mingling of Byzantine, Persian and Chinese cultures it had always retained its own individual beauty. The many changes and variations in style and technique were reflected in the highly decorative ceramic tiles created throughout the whole Islamic period.

Japanese ewer late 18th-19th century
Ko-kutari ware from the Tokugawa period with an animated dragon design.

Japan

Japanese pots inevitably inherited their style and appearance from China and Korea. Neolithic pots were made by hand, coiled and beaten. Known as Jomon ware, they were low fired in primitive kilns, and often decorated with impressed patterns of plaited rope or mat work.

In time, Chinese influence introduced the wheel and the high-firing kiln; their influence was also apparent in the shapes, colors and glazes implemented. However, the Japanese ware had its own more spontaneous and sensual style achieved with a far less severe approach and enriched perhaps by the Pacific influence. The feel of the natural clay and its individual qualities were no longer restrained but allowed to influence form and decoration.

From the 13th century onwards Japanese stoneware was produced in large quantities. The discovery of kaolin deposits and of fireclay for kiln-building made for greater progress, and the shapes of the pots were inspired by the culture of the Korean immigrants, especially in some of the early blue-and-white ware.

By the 14th century six main centers for ceramics were established and the increasing popularity of the tea ceremony spread from Zen Buddhist monks to wealthy emperors, the nobility and then the rich merchant class.

The tea ceremony was carefully structured, with pots and dishes used at all stages of the proceedings. This ware was specially created for the ceremony and was highly prized. There were, for instance, tea jars and bowls, cake trays, water jars and even vases and incense boxes. The tea-masters dictated the shape and style of all the ceremonial pieces that were used and some were presented by them as gifts for appreciated services.

Seto was one of the most important ceramic centers and produced gray, opaque and black shiny ware covered with a thick lustrous glaze. Bizen ware is still produced today by descendants of the original potters (such as Kei Fujiwara). The pieces have the same rough-textured heavy style and are made of a smooth reddish clay which is often left unglazed.

The Raku style of pottery developed as an integral part of the tea ceremony. These Raku pots were made with a coarse clay which could withstand the thermal shock of being placed in a red-hot kiln, briefly fired and then removed to cool quickly. Sometimes this took place during the tea ceremony itself and the distinctive type of ware created was used immediately.

Terracotta figure 7th century
This striking figure of a kneeling man is an example of Haniwa pottery from Yamato-mura, Ibaragi-ken.

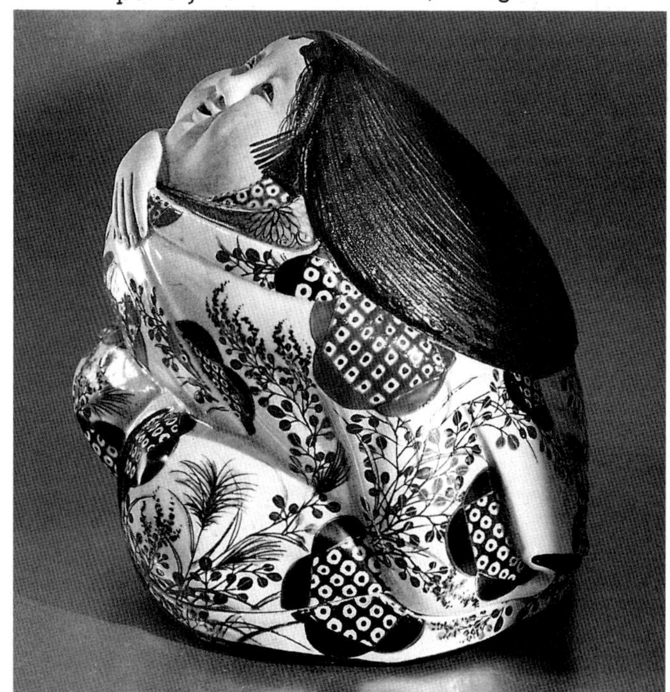

Incense box c. 1830
A delightful piece of Awaji ware, this imaginative box has been decorated in soft reds and blues on a milky white glaze.

Japan

The Japanese discovered several new clays during the 17th century. Their styles changed accordingly to incorporate the porcelain products they could now produce. The colors of nature changed too: in their designs blossom was green, fruit was indigo and water red! One of the most famous Japanese styles of porcelain exported to Europe was Imari ware which was decorated in reds, golds and indigo. It was ornate and vivid.

The Kakiemon family developed the use of overglaze enamels in soft restrained reds, yellow and blues; the designs were often asymmetrical, having clear expanses of milky-white porcelain in a style that reflected the delicacy of Japanese paintings. The growing cities of Edo (modern Tokyo) and Kyoto (which produced excellent earthenware and elaborate enameled ware) developed in both size and importance from the late 17th century onwards.

From 1716-1735 Nabeshima pottery (named after the local lord for whom they were created) was most delicately painted and amongst the most refined ware produced at that time. Delicate blue drawings rested usually in shallow dishes, enriched with glowing enamels.

The blue-and-white porcelain of the 18th century was some of the finest ever made and was copied by the factories of Derby and Worcester in England.

Many individual artist-potters made their own styles famous. This was then a new phenomenon—only to be found in Japan.

Some of the old-established centers of ceramics in Japan employ master potters whose ancestry can be traced back to the great periods of porcelain production four hundred years ago. These potters do not consider this to be a very long passage of time; they have merely maintained the same working traditions with the deep spiritual ritual that is so much a part of the creation of Japanese pottery.

Tea-ceremony vessels
Coarse-grained Raku pots which could withstand rapid firing were formed, fired and then used during the tea ceremony: this tradition is still taking place today.

Byzantine (4th–14th century)

When Emperor Constantine claimed the ancient city of Byzantium as his own Constantinople, he can scarcely have realized that it would be dominant for so long — 'the crossroads of the world' between East and West. Inevitably the styles and techniques of the ceramics produced during this period in Europe reflect this mixture of cultures and exhibit a rich oriental influence as well as their Greco-Roman heritage.

The use of lead oxide as a very simple means of obtaining a glaze was one such Roman inheritance — one that continued to be implemented widely especially during the 8th and 9th centuries.

Here there was no Islamic bar on precious metals so gold and silver was much in demand at court and pottery not nearly so highly regarded as in the Muslim countries. Designs were largely practical for use in the home or on the farm. It was only when war and strife reduced the general wealth during the 13th and 14th centuries that clay emerged from its role as the material used for earthenware pots for the masses, to become a valued medium.

Decoration on pottery tended to be a formal stylized mix of the different cultures' designs. Doves, the Christian symbol, appear alongside Egyptian-inspired ducks and fish, Western heraldic leopards and Persian lion and hares. One much-favored creature was the eagle who represented strength with goodness. Human figures were only occasionally represented and were then rarely naturalistic.

Under the Macedonian dynasty the dominant white ware emerged. Made of a whitish clay, it sported molded designs of rosettes, crosses and creatures of the bestiary. Similar motifs appeared on polychrome ware; they were painted on directly in vitreous colors and then covered with a transparent glaze. Sometimes these motifs would be outlined in black.

Decorated ceramic tiles were also made from the white clay and used as architectural refinement, being designed for specific structures in palaces or churches.

Meanwhile undecorated ware, covered in a yellowish glaze, was being made for domestic use. Effective external molding was created out of clay pellets and pressed into the sides of a pot to look like petals or fish scales.

Red-bodied slip ware became very popular in the 10th century. The use of a white slip on a red clay body echoed the effect of Islamic lusterware and was especially pleasing when used on the inside of bowls.

Early Byzantine flask 6th century AD
The relief decoration in the center depicts the nativity.

Sgraffito was developing too and was used on the bowls under transparent, sometimes colored, glazes. Dots, scrolls and geometric designs were incized through the white slip. In time, somewhat freer illustrations of animals, trees and birds appeared but all were still executed with a fine point and with detailed borders surrounding the central design.

Crusaders carried home such wares and introduced them to the West. The excellent Syrian ware was especially widely spread. With mixed styles of decoration that rendered it equally suitable for Saracen or Crusader, it was distributed to the Aegean and even Italy as well as into Western Europe.

By the 12th century, flat wedges or gouges were being used to remove areas of slip (and sometimes the body as well) to accentuate the designs and render them slightly raised and in relief, reminiscent of medallions.

During the 9th century lead-glazing techniques had been greatly improved and had spread into northern Europe. The method was simple and inexpensive and was to form the basis for the most popular type of decoration throughout Europe. Incized decoration was highlighted with rather runny oxides — which tended to spread under the glaze. Potters in different areas took up the method and then used it on their own local clay to produce many varying styles and designs.

The potteries of Constantinople, Salonika and Cyprus were renowned and exported their distinctive wares to many neighboring countries and beyond. Throughout the Byzantine period, its incorporation of East and West and its transition from the Ancient to the Modern world produced a fascinating blend of cultures.

Spain

Celtic invaders settled in the east of the Spanish peninsular and by the 6th century BC the Iberian culture was established and producing distinctive painted ceramics. They used iron and manganese oxides to create a wine-red finish to their sturdy ware. The Romans held sway for some time but after their withdrawal the techniques they had introduced were lost and it was really not until the Moorish invasion in the 8th century AD that Spanish pottery began to develop towards the richness and quality that could be attributed to the Islamic influence but which was individual and new, a fine derivative from this most westerly point of the Muslim world.

Trade with the other surrounding cultures introduced the Byzantine glazes and lusterwork from Mesopotamia. During the 13th century Spanish potters began to use a white tin glaze on biscuit-fired pots. Once the surface had been fired it was decorated in rich luster and fired yet again. Lusterware was being produced in great quantities in Andalusia by the middle of the 13th century, largely because by then Christian armies had conquered much of Spain and the Muslim contingent had converged in this small area of the south — only to be joined by Iranian potters who had escaped there from the Mongol invasion of Arabia. Thus there was a valuable pooling of knowledge: silver, sulfur, copper-red ocher and vinegar lusters were used and cobalt-blue decorated ware appeared.

This was to be a thriving industry for some three hundred years with exports even in the early days reaching such widely differing markets as England, Egypt and Sicily.

Generally the designs were geometric, foliate or in Kufic script in two different shades of blue on a white background — with a pale-golden copper luster as the final finish. The decoration was often inspired by metal engraving work and by the influence of such Islamic centers as Rayy.

'Cuerda-seca' or the 'dry-cord' technique was sometimes used to separate various colored glazes, especially in Valencia. In time this was replicated by using outlines drawn in manganese and grease.

Some of the vessels were huge with vases up to four feet high and large plates that had to be placed on edge in the kiln. The largest and most complex vessels were formed as separate pieces and then fired in their entirety. The Alhambra palace housed several such vast pieces.

Multicolored ceramic tesserae were used to produce complicated designs on important buildings in the South.

Lusterware bowl early 15th century
Here the Mudejar potters have depicted an ocean-going ship used by Portuguese explorers in a finely detailed design.

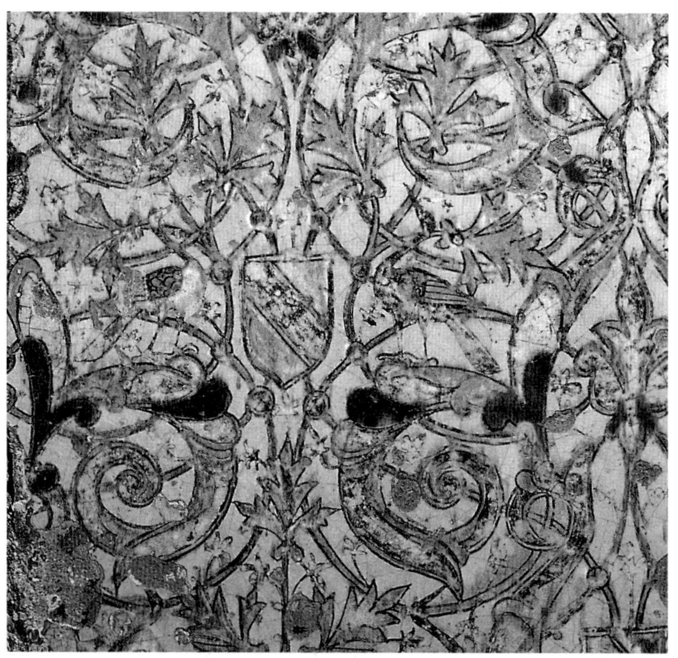

Ceramic tile 14th century
The arms of the Nasrid sultans are enclosed within the interwoven foliate design of this rare ceramic tile.

By the late 15th century the Italian Renaissance began to effect the Spanish ware. If anything the vessels became over-elaborate with their more simply decorated undersides often more attractive than their garlanded scalloped repoussèd tops! Lusterware was still highly regarded and the large central serving dishes used at table became especially ornate.

With the unification of Spain under the Catholic rule of Ferdinand and Isabella in 1491, precious metal was once more acceptable and lusterware could not compete with the gold and silver seized during the New World conquests. Moreover, such associations with the Moorish aspects of Spain were now discouraged.

Hispano-Moresque style ware was to be resurrected in Valencia during the rococo period of the 18th century and maiolica ware to be developed under the Italian influence during the 16th century, especially in the form of vivid picture tiles and panels.

However, the Talavera ware made in central Spain (at Talavera de la Reina) remains as an individual style of Spanish pottery which emerged in the 15th century and continued to be produced through to today — although its quality diminished during the 18th century. Bright colors were painted on to or under a gleaming milky-white glaze.

In 1575 royal patronage promoted the distribution of Talavera ceramics and the sturdy large dishes with vivid animal motifs and sweeping foliate borders were produced even in the new Mexican colonies.

In 1487 the Catholics reconquered Andalusia and the Muslim potters fled, taking their lusterware techniques with them to Valencia. Here there was a strong Gothic influence (due to the area having been earlier dominated by the Catholic church) and decoration styles changed accordingly. More naturalistic plants and foliage as well as a wide variety of birds and animals formed rich interwoven patterns and heraldic devices were also very popular.

Manises became an important center for the ceramic arts in Spain, especially famed for its lusterware, and potters produced beautiful work to commission as well as taking their personal expertise abroad to the neighboring courts and nobility of Europe. This Gothic-Muslim (Hispano-Moresque) mix produced the 'mudejar' style with iridescent yellow, red, blues and gold mingling with deep cobalt underglaze paintings — often of bryony and vine leaves (a *Mudejare* means an unconverted Muslim). Gothic inscriptions such as *Ave Maria gratia plena* mixed with arabic forms of decoration. The popular Manises ware was exported to Italy in Majorcan ships and was consequently known there as 'maiolica'.

Valencian bowl late-15th century
This beautiful Hispano-Moresque bowl has been painted in gold luster and blue in richly decorative designs.

Italy

Italy was inevitably dominated by Roman influences for a large part of its early ceramic production. Few changes emerged until the impact of lead glazes, brought about by the Byzantine predominance, meant that plain amber and green glazes began to be replaced by a greater awareness of color. In fact up until the 12th century the range of ceramics was limited to simple earthenware practical pieces and some basic decoration; bowls were imported from Muslim potteries — and because of their pagan origins, were relegated to the exterior of churches.

The use of white tin glaze on red-bodied pots was first seen in Sicily in about 1200 AD and eventually spread into northern Italy. Copper green oxides, firmly outlined in manganese brown or purple, were painted on to the raw glaze in geometric and animal-inspired designs with foliage and sometimes coats of arms — all rather static but restful to look at.

The maiolica ware imported from Spain on Majorcan ships stimulated the Italian potters' development of tin-glazed ware and by the 15th century their 'severe' or formal style had blended the Gothic and oriental designs into a wider, more sophisticated repertory which varied from one region to another, but which all reflected a new aesthetic approach and sensitivity to the subject matter concerned. Designs were controlled but none-the-less vibrant and implemented with rich primary colors. Tuscany, and in particular, Florence, was producing fine distinctive decoration. The import of Valencian ceramics inspired the 'Florentine green family' with formal designs reminiscent of peacock feathers. Two-handled, small round jars, decorated with oak leaves, were made in Tuscany and disported a rich ultra-marine blue applied so thickly that the design was virtually in relief.

Unable as yet to produce lusterware, the Italian potters developed imaginative color effects to mimic this. As well as fine pottery, the artists applied themselves to rich ornamental pavements — some of which can still be seen today, glowing with color and intricate design.

By the end of the 15th century contact with the Eastern cultures had revealed the existence of Chinese porcelain; the floral motifs of the Ming dynasty were aligned with Chinese subject matter (such as junks or pagodas) which reflect its impact on decorative treatment. At the same time the Della Robbia workshop was applying maiolica techniques to three-dimensional work and producing sculptural pieces of outstanding quality. The white-glazed madonnas set against pale-blue backgrounds are reminiscent of marble.

Earthenware istoriato plate 1580-1600
Latona changing peasants into frogs: vivid narrative majolica decoration painted by Alfonso Patanazzi of Urbino.

Italian maiolica
Both the 'Virgin and child' figurine and the tin-glazed plate with cross-hatching and outlined design are from Orvieto.

Decorative Italian maiolica
The clay pot with its portrait bust and peacock-feather design, and the dish with highly stylized foliate pattern in Gothic-floral style are lovely examples of Italian decoration.

Adoration of the Magi late-16th century
Detail from a maiolica group by Faenza potters.

Italy

With the enormous impact of the Renaissance on the world of art in general between 1500 and 1600, the Italian schools of painting exerted great influence over ceramic decoration, with the form and shape of pieces being suddenly regarded as less important than the painting techniques. This is a reversal of the norm, possibly to the detriment of the pottery produced. However, the new pictorial effects were still perforce maiolica orientated and the high value placed on artistic merit, backed by wealthy patronage, encouraged the emergence of individual ceramic artists.

Pharmacies were popular meeting places and as drugs were then stored in ornamental maiolica vases the potential of the medium was recognized, especially in the production of vast decorative display dishes.

Mythological as well as religious scenes were illustrated in the '*istoriato*' (story) style which replaced the more abstract subject matter of previous decoration. From its origins in Faenza to its further development in Sienna the revolutionary ideas spread. By the time the Medici family were lending their patronage to the Cafaggiolo potteries (at their best from 1500 to 1525) deep-blue backgrounds and vigorous brush strokes made the style quite distinct from its ceramic predecessors and the influence of Botticelli and Donatello is unmistakable. Borders and backgrounds were sometimes filled with grotesques.

Zoan Maria Vasara took the Faenza-inspired styles into the Duchy of Urbino — where they were to mature and develop yet further with individual ceramic artists introducing a fresh approach and a new vigor to the *istoriato* concept.

Meanwhile lusterware was at last being produced in Italy. As the Catholic conquest of Spain set Islamic potters to flight, their skills were absorbed elsewhere and by 1501, Deruta (near Perugia) was producing fine polychrome wares with a ruby luster and a golden mother-of-pearl effect. At Gubbio in Umbria luster decoration reached new heights in the middle of the century. Iridescent gold, silver, and rich reds were used, and luster was applied over finished painted work as an additional embellishment.

Venice and Ravenna were also producing lead-glazed sgraffito ware and by the 16th century they were making quite complicated pieces with modeled borders and raised ornament. Venice was renowned too for its delicate painting and lavender-gray glaze.

Baroque styles of decoration began their development by the turn of the century and by 1600 a smooth, white Faenza tin glaze appeared which better enhanced the shapes of pieces produced and decoration became supplementary — rather than dominating the overall design. Painting was simpler and more delicate, with the range of colors limited to blues, yellow and orange. Exported ware was called faience (after its origins in Faenza) and the techniques were taken by Italian potters to Rouen and Nîmes in France, and to Antwerp in Belgium

Complicated Baroque shapes which imitated prized metal objects and were laden with heavier decoration had appeared by the 17th century — heralding a general decline in the regard for ceramics.

Capodimonte vase c. 1745-50
This soft-paste porcelain vase has been painted with enamels in delicate chinoiserie style.

Capodimonte figure — mid-19th century
Capodimonte soft-paste porcelain restored realism to porcelain sculpture.

Liguria was an area where merchants and navigators, as well as travelling artists, converged — most especially at Genoa. Chinese imports encouraged the development there of blue painting in several successful styles during the 17th century.

Nothing, however, could compete with porcelain and although attempts were made to produce this in Pisa and Florence in the early 17th century it was not until the 18th century, inspired by German and Austrian styles and success, that Italy really resumed an important role in the world of ceramics by creating fine porcelain pieces.

Francesco Vezzi and Geminiano Cozzi founded factories in Venice which produced hard-paste porcelain but it is the soft-paste porcelain of Capodimonte which is justifiably the most renowned.

The Capodimonte factory was founded in 1743 by Charles IV, a King of Sicily whose marriage to a Saxon princess no doubt led to the introduction of Meissen ware; its influence is evident. When Charles became King of Spain too, the factory was removed to Madrid and the Spanish influence (in both clays and style) was inevitable. The throne of Naples was left to Charles' son Ferdinand who opened up another porcelain factory. There the influence of the excavations at Pompeii and Herculaneum in the 1740s meant that the factory initially made vessels reflecting classical antiquity but by 1779, under a new director, ceramics depicting lively groups of ordinary townsfolk, street-traders and middle-class citizens were being produced, bringing realism and humour back into ceramic sculpture.

The Capodimonte paste was white and, considering its softness, could be worked remarkably thinly to create delicate translucent ware such as tiny painted cups and saucers. By contrast, perhaps the most spectacular work resulting from the Bourbon factories was the construction of three complete rooms of porcelain for the palaces of Portico (from the Capodimonte factory), Aranjuez and Madrid (Buen Retiro factory). Rich in rococo fantasy and created by thousands of interlocking porcelain pieces, they show an unrivaled extravagance in the use of ceramics.

France

Circular tazza 16th century
An example of the highly decorative effects achieved by Bernard Palissy.

Ceramics in France seemed always to have come under the influence of other cultures, from the Roman domination of Gaul to the Muslim potters who fled there from Spain — while the earliest French maiolica (known there as faience) was painted with geometric shapes in manganese purple and copper green which obviously reflected early Italian influence.

By the 14th century lead-glazed earthenware was being produced — small objects from this period, such as money boxes and toys, have been found in the moats dug along the old city wall of Paris — as well as being discovered in the sewers! Early use of slip can be detected in many of the fine tile pavements. The influence of French heraldry motifs in the 13th century was replaced by scenes of contemporary life as human figures were increasingly depicted. Glazing techniques were greatly improved by the 16th century. Beauvais and Saint-Porchaire earthenware were especially fine. They were prized exports to the Low Countries and England, Beauvais stoneware making especially good drug-jars.

Bernard Palissy

The master potter, Bernard Palissy, was born about 1510 and first acquired his skills working in stained glass. Commissioned to make an Italian-style rustic grotto, his work was seen in progress at Saintes by Catherine de Medici who was captivated by it and promptly established the potter in Paris. He was able to achieve rich effects with polychrome glazes, producing brightly marbled backgrounds. He used many natural-looking designs of flora and fauna (such as fish, frogs, snakes, shells and leaves) for decoration, taking casts from life and reproducing these in relief. He died about 1590, after imprisonment for heresy as a Protestant. His work was widely imitated for some three-hundred years and his molds reused time and again.

Nîmes, Nevers and Rouen

At Nîmes in Provence and in Rouen immigrant Italian potters had settled and were teaching their skills to French artists. They were influenced by the work of the Royal goldsmiths and silversmiths and produced some fine examples of faience with French styles and subject matter gradually deposing total Italian dominance.

In Nevers, especially, French inspiration was drawn from engravings while informal pastoral scenes reflected the themes of fashionable novels of that time. By the end of the 16th century the industry there was well established and renowned, with the 'bleus de Nevers' especially popular. These were decorated with yellow and white sprays of flowers and birds on a blue ground (or the reverse, occasionally). Imported Chinese blue-and-white ware was so popular that this style too was adopted, and so in time were the Dutch Delft designs.

Towards the very end of the 17th century the costly French wars led to the melting down of gold and silver ware so

Earthenware dish 16th century
Bernard Palissy took casts from flora and fauna to create very realistic relief decoration.

Vincennes teapot c. 1753
The royal works at Vincennes produced richly decorative porcelain.

French faience received an unexpected boost and the original styles produced at Rouen became justifiably popular. Typically Louis XIV, the 'style rayonnant' (based on metalwork patterns) consisted of a rich embroidery of lines enriched with touches of polychrome color and resembling festooned drapery.

Further south, at Moustier and Marseilles, the Clerissy brothers were producing excellent faience decorated in blue in the classical style.

As well as smaller ware such as spice boxes, sugar dredgers and tobacco-graters, huge dishes were produced in Rouen, capable of bearing some twenty partridges or fifteen chickens! Around 1740 the rococo style was adapted by the Rouen potters who depicted rural scenes and amorous subjects that imitated the styles of the contemporary painters Watteau and Boucher. Delicate Chinese scenes and motifs supplanted the 'style rayonnant'. Garlands, grotesques and fantasy figures swirl on plates, trays and dishes in harmonious animation — usually framed by rich border designs.

Strasbourg

By the middle of the century a new technique of decoration was introduced from Germany into Strasbourg. Instead of 'in-glaze' painting directly on to the raw glaze, colored enamels were used on a surface already glazed at a high temperature (grand feu). Now a lower temperature glaze only was needed (petit feu). This meant a much wider range of bright colors could be employed.

At the same time the German influence brought to Strasbourg (and thence to the rest of France) huge and realistic ornamental objects — such as tureens shaped like vegetables or animals. The naturalistic details made these very strange tableware indeed. For instance a boar's head complete with tusks, realistic teeth and fur and gleaming eyes might well have been set before you during a banquet.

The faience produced in the Strasbourg factory under the brothers Paul and Joseph Hannong was especially fine. However, the latter's attempts to make porcelain proved ruinous. After the death of his patron, an old Cardinal, who died in 1779, Joseph died in poverty and debt but, under his influence, faience in France had reached its apogee.

Meanwhile manufacturers in Marseilles had also been producing excellent high-temperature and enameled ware. Its status as a sea-port is reflected in the images of fish, shells, fruits de mer and fishing tackle on these pieces.

Egyptian service
1810-12
Napoleon was a great patron of the Sèvres porcelain works and here the decorative gilding and enamels depict landscapes and hieroglyphs to commemorate his Egyptian campaign.

By the end of the 18th century faience was no longer so fashionable as the porcelains, hardwares and English creamware which then came into vogue.

The secret of porcelain

The formula for making porcelain was a highly prized secret. Louis Poterat of Rouen had worked alone and his knowledge died with him. The Chicareau family of St-Cloud exploited their knowledge of the secret under the patronage of the King's brother, the Duke of Orleans, and produced early gilded and polychrome ware.

In due course the manufacturing processes were no longer able to be kept so cloaked in mystery. In 1726 the Prince de Condé established a factory in Chantilly. He had assembled a wonderful collection of Chinese and Japanese porcelain — an instant source of inspiration to the painters there. In time their individual French tastes and style emerged with delicate roses, convolvulus and cornflowers painted or modeled in relief on the porcelain. Porcelain produced at Mennecy reflected these transitions in style but specialized in production of small trinkets — cosmetic jars, patch boxes, ink wells and so on.

Soon however, all this was to be eclipsed by the emergence of the royal works at Vincennes and Sèvres. Sèvres was placed. It was set on the road from Paris to Versailles and near to the château built for Madame de Pompadour (Louis XV's favorite mistress and a patron of the arts who appreciated porcelain).

This was to be a wonderfully creative period in ceramic development. The best artists of the day were summoned to work at Vincennes. They produced delightful models of children while forty-five women and girls were working on the production of artificial flowers. (Madame de Pompadour was to receive the King before a whole bank of these — sprayed with perfume to enrich a winter's day!) Lightness, elegance, grace — these were the order of the day. The use of colored grounds became more and more dominant — first various rich blues, then daffodil-yellow, apple-green and Madame de Pompadour's own bright pompadour pink.

Porcelain production was still expensive and precarious. The King had to purchase the factory in 1759. He was sufficiently interested in the business to become involved in the sale of the porcelain, through exhibitions held at Versailles.

In time the styles become more grandiose, with porcelain plaques used in furniture. In 1793, after the Revolution, the factory became state property and regained its former prosperity.

Meanwhile from 1772 Sèvres was also producing hard-paste porcelain which can withstand higher temperatures and so can be used for large-scale work. This eventually displaced the soft-paste porcelain. Napoleon placed large orders for both his palaces and for use as diplomatic 'handouts' and the glories of his regime were depicted on many pieces. The countries through which he marched are all shown too — the Egyptian campaign receiving special attention, with hieroglyphs and Egyptian landscapes depicted on tea and coffee services.

The Sèvres factory was long to continue to reign but at Limoges, the discovery of kaolin, in about 1771, set into motion the work of a factory to produce fine porcelain bodies. This led to the manufacture of hard-paste porcelain throughout France.

The importance of Delft *Helmet jug and two jars*

From Italy to Antwerp came the much-prized maiolica. Its arrival was the initial spark of inspiration that led to the establishment of potteries there in the early 16th century. The ware became known as Delft and was to remain a 'tour de force' in the world of ceramics for over two centuries and was still being produced at the beginning of the 19th century. The art of tin-glazed earthenware was to be spread throughout Europe via the Antwerp connection.

Delft is actually a town near Rotterdam with easy access to the major waterways. Here the industry was centered, almost exclusively, towards the middle of the 17th century. To begin with the designs were Italianate but then the rival Floris factory produced an intereresting form of decoration based on scrolls and coils of strapwork. Many lovely tiles were made, as well as boldly painted dishes and plates decorated with borders of raised knobs and often the much favored milkmaids laden with pails on a yoke.

The blue-and-white influence

The blue-and-white Chinese porcelain imported into Europe after the founding of the Dutch East India Company in 1609 produced a wealth of exciting new ideas soon to be adopted by the Delft potters and imitated by them rather better than most, especially as they were capable of throwing thinner ware than some of their competitors. It was therefore closer to the delicate texture of porcelain.

Outlines were drawn in manganese purple on to the fired unglazed clay and then the blue was applied. A final coat of lead-glaze, known as 'kwaart', was added to emulate the high polish of Chinese porcelain.

When the Ming dynasty declined towards the middle of the century the pottery production at Delft expanded dramatically — both to fill this gap in the market and the one left by the decline in the Dutch brewing industry — many disused breweries being adapted to become pottery works.

Under the directorship of Adrianus Kocks, wares of the deepest and purest K'ang Hsi blue were created: some that were over three-foot high were made for William and Mary at Hampton Court. Beautiful tulip vases and other many-spouted flower holders were produced in fascinating shapes and forms. Fine artists from other spheres of work were commissioned to produce appropriate landscapes, designs and portraits. Skills in painting and decoration were considered of paramount importance. 'Pricking' through a drawing or engraving and rubbing pumice

Delft ware 1689
The decoration on this tin-glazed earthenware dish reflects the Chinese styles, so well imitated by Delft potters.

powder through the master copy allowed works of art to be reproduced as decoration.

The De Roos factory produced blue and white plates which illustrated New Testament stories in a distinct Dutch style of painting, as well as elegant cupboard sets of vases — three baluster shape and two beaker shape — that were a 'must' on the top ledges of every wealthy household's baroque cupboard.

Tiles

Tin-glazed tiles were produced throughout Holland to be used as wall decoration. Initially the tiles were decorated with overall patterns, then bright central designs, followed by a period when the decoration of the corners and edges had to be of prime importance. Later they were used to create both small pictures and vast decorative designs, often with central figures of ships, monsters, horses and men-at-arms. Biblical scenes were ever the most popular although simple rural landscapes were also very much in fashion in the 18th century.

By this time red, green and yellow were added to the original blues and purples. Japanese Imari ware was imitated with gold brocaded oriental designs, while attempts were also made to mimic the effects of lacquer.

Baroque patterns and rococo figures of shepherds and shepherdesses were introduced, mingling rather strangely with the typically Dutch landscape backgrounds.

In time the Delft industry could no longer compete successfully with the German porcelain and English creamware. Their work had lost its originality and became somewhat repetitious and rather more clumsy. Eventually hard-paste procelain was being made in factories at the Hague but the glory of the Delft pottery produced around 1630-1700 — from the everyday household container to the great ornamental pieces and elaborate wall tiles — remains all over Europe as a reminder of the richness and versatility of ceramics.

Germany

In the wake of the Romans and bereft of their technological expertise, development in this area was fairly slow, but by the 8th century a new type of unglazed pottery emerged with squat, almost spherical vessels being produced in the Rhineland. With the improvement in kiln and firing techniques in the 9th century stronger, more resilient ware was produced and pots made at Pingsdorf and at Badorf in the Rhine Valley were being exported into Britain by the 10th century.

The use of lead glaze eventually reached central Europe from Byzantium but, on the whole, unglazed ware was preferred, perhaps because of the emergence of the harder more vitrified stoneware ('steinzug'), which was already impervious to liquids. Moreover the clay vitrified at higher temperatures than the lead glazes were able to withstand. This early stoneware was usually blue-gray, decorated with roulettes, incised grooves and foliage designs (in some cases reminiscent of Roman pottery).

Salt-glazing and stoneware

Salt glazing became immensely popular and successful. Its subdued colors and texture were interestingly variable and as it always provided a fine thin coat it heightened the delicate designs inspired by metalwork.

Many beakers, tankards and jugs were produced, with the rich clay deposits on the banks of the Rhine and the convenient transportation and trading facilities there making this the most important center of stoneware manufacture. At Siegburg, pots with stamped relief designs and a yellow-to-brown salt glaze were sometimes incised with scrolls of holly-leaves.

Eventually the pots became taller and more slender or tapered, heralding a change in style and the coming Renaissance period. During the first half of the 16th century the most exciting stoneware was produced. In Cologne, pots decorated with fascinating human faces were created as well as intricate spouted jugs and elegant ware. Ceramic artists worked from models provided by the finest engravers of the time as well as hollow molds, such as were used for producing decorative sweetmeats like marzipan at a time when the court confectioners had to produce real masterpieces for the table.

Meanwhile a quite distinctive and different style of ceramics was developing in the use of lead-glazed earthenware by stove-makers. These stove tiles grew to be veritable frescos with wonderful molding and sculptured figures — usually finished with a deep green glaze. Nurem-

Siegburg jugs c. 1550
Stamped relief designs of twining leaves and a bearded caricature are typical of the salt-glazed Bellarmine jug.

berg became the most important center for this work, with the introduction in the South Tyrol of tin-glazed earthenware allowing the stove maker to create white tiles for stoves, walls and floors.

After the decline of the industry at Cologne the Westerwald area became the focal point in the production of salt-glazed ware especially as many potters fled there to escape The Thirty Years War. But it was at Kreussen in Bavaria, where low-temperature glazes or enamels were used by glass painters to decorate the ware in bright colors, that the most expensive ceramics were produced.

German salt-glaze ware was justifiably regarded as some of the finest ceramics then available and was created with great technical skill and with the assistance of fine artists. The pottery was held in sufficient regard for Queen Elizabeth I to be presented with it.

The impact of maiolica

Henceforth maiolica (or faience) ware was to make a rich contribution to German Renaissance pottery. Italian and then Dutch and Chinese designs influenced its initial production but there were many Germanic specialities. Lidded jugs shaped like owls were popular as well as cups shaped like birds and designs based on tulips or lilies.

Dutch religious refugees had taken up residence in Hanover and in 1661 a factory was founded there which produced faience till the end of the 18th century. Typically the emergent European design developed there with patterns of scattered baroque flowers and exotic birds on tall-necked jugs with handles of a plaited-rope design.

In Frankfurt too, maiolica ware flourished. Highly decorative and covered in a clear 'kwaart' glaze, it was more 'showy' than the Hanover ware and aimed at the wealthy purchaser. Most of the pots were painted in blue tones with stylized Chinese landscapes and motifs such as butterflies and birds. The Kassel factory founded in 1680 produced work with distinctive neat painting in pale blue. At Nuremberg styles of painting developed for the decoration of windows and glass had some influence over the styles at the turn of the century.

With the fashion for drinking tea, coffee and chocolate becoming more widespread in the early 18th century, it was inevitable that the type of ware produced would adapt to accommodate this — as well as the arrival on the market of the competitively low priced but high-quality English creamware. The ordinary cylindrical mug became the working man's vessel and has ever since been the shape used for holding beer.

Meanwhile Johann Friedrich Böttger of Meissen discovered how to produce porcelain and despite the attempts of faience workers to reproduce this fine ware, their days were numbered. As yet, however, the factories continued — introducing the 'Strasbourg flowers' decoration as well as ornamental engraving and 'leaf and strapwork'. By 1750 rococo hunting scenes became popular and Indian styles of decoration were introduced. The work now imitated porcelain — especially the tea and coffee services, the designs of Chinese figures and the use of enamel colors and gold lacework. But it was with Meissen porcelain that the future lay.

Meissen plate c. 1818
This porcelain dessert plate is part of the Saxon service and depicts Apsley house which was designed by Robert Adam.

Meissen porcelain

Böttger's career was initially inspired by his attempts to turn metal into gold. This dabbling in alchemy led to his arrest and subsequent transfer to Meissen where Augustus the Strong (ruler of Saxony) hoped the young man would indeed produce gold or, at the very least, semi-precious stones. Böttger's experiments resulted in his stumbling on the secret of producing hard-paste porcelain by mixing refractory clay with fusible earths. He modeled some beautiful pieces from the new material, cutting, polishing and engraving it — employing all the know-how of the glass industry to mimic the qualities of precious gems. Glazed and then decorated with gold and lacquer, the work was outstanding and received with acclaim.

When Böttger then went on to use a white clay (or kaolin) found locally in 1708, the Chinese porcelain had its first true rival — much to the envy of the rest of Europe.

The earliest pieces reflected their Eastern predecessors, including 'blanc de Chine' figures, but by 1710 Böttger was employing metalwork styles of decoration that were undoubtedly European. Curiosity about the bizarre and the exotic inspired by travellers' tales meant that there was a taste too for the grotesque, with dwarfs and cripples the subject matter for small models in porcelain. Augsburg goldsmiths decorated the Meissen ware and distributed it along with their own products.

Böttger died in 1719 when his invention was only just beginning to be fully exploited. Artists such as Hörldt, Heintze, Lowenfinck and Klinger were soon to be producing outstanding work at Meissen. Some forty painters were employed there in 1731 producing chinoiserie, landscapes, harbour scenes, designs from Indian and German flowers — as well as insects — and at last the much sought-after underglaze blue of the K'ang Hsi ware.

At Dresden, in Augustus the Strong's new Japanese palace, the porcelain was to be displayed in princely surroundings. As well as table finery, Augustus commissioned large-scale figures and vases to be created — in the first instance by Gottlieb Kirchner and then by Joachim Kändler, a young pupil of the sculptor. His work is truly outstanding, especially considering the then primitive state of porcelain technology. He produced lifesize figures, the most animated animal and bird sculpture, table services and utensils, small groups of figures for table decoration and wonderful pairs of lifelike lovers, beggars, hawkers and characters from the Commedia dell'arte. The nobility of Europe and ambassadors from all over the world came to buy his renowned pieces of work. Throughout his life, amid changing patterns of art and fashion, Kändler retained his own style and expertise and set standards that his successors all aimed to maintain. He died in 1775 but his models are still used today.

For a period of some forty years the Meissen factory monopolized the porcelain secret. Eventually Vienna and the other German factories discovered the formula and were manufacturing competitive pieces. In many respects they mimicked the Meissen porcelain but were still very fine. Franz Anton Bustelli produced some wonderful work at Nymphenburg — as did Peter Melchior at Höchst and Wilhelm Beyer at Ludwigsburg.

European porcelain in the 18th century was produced with great skill and artistic finesse. The pieces were expensive; they were rich and opulent in style. In short they reflect their time in history. Led by Meissen in Germany, porcelain production was no longer an intrinsically Eastern product but from now on became a part of Europe's culture.

Meissen-inspired group
1770
The porcelain shepherd and shepherdess were produced at Höchst.

England

Early pottery

As an island race the British were influenced rather less than most by migration or invasion and new techniques often reached its shores somewhat later than the rest of Europe. None the less it was in time to boast one of the most prolific ceramic industries in the world.

The earliest pots (produced about 2000 BC or earlier) were handbuilt with rims and incised decoration.They were succeeded by vessels with shoulders and eventually cinery urns. The late neolithic beaker folk (who possibly came from southern Spain) introduced their distinctive beaker vessels about 1600 BC. The simple incised decoration provided a better grip as well as adding surface interest.

Larger urns were produced in the Bronze Age but it was the Iron Age that witnessed a much greater sophistication in ceramics — especially when the simple but effective potter's wheel was invented.

Life in Britain changed dramatically with the Roman conquest. Their fine red-glazed pottery was produced there as well as a whole range of pots which exploited local clays. When the Roman legions left, Saxon pottery reverted to much earlier forms. The highly organized ceramic industry disintegrated and the Dark Ages saw few changes or improvements. Only in East Anglia did new techniques penetrate; Ipswich, in particular, produced ware which reflected European influences — with spouted pitchers like Rhineland vessels. In Stamford, Linconshire, glazed ware appeared for the first time and was traded widely throughout the Midlands.

The late Saxon styles continued for some three-hundred years until the 12th century. Medieval cooking pots were larger and more squat while the pitcher was replaced by jugs with handles and pinched-out lips. Primitive kilns were still in use but the fast wheel had arrived and forms became more variable.

Apart from the highly decorated ware produced at Winchester the surface of pots which was treated very simply until the 12th century. Now strips of applied decoration, incising and relief work became fashionable. Faces appeared on the necks of jugs, and at Cheam and Rye, slips of contrasting color enlivened the surfaces. Glazes, sometimes tinted green with copper but often yellow-brown, provided the most common decorative effect.

Jars, jugs and cooking vessels formed the largest part of the potters' output but water bottles (sometimes shaped like

Harvest jug 1813
The slipware and sgraffito decoration includes the following inscription:

The potter fashioned me complete
As plainly doth appear
For to supply the harvest men
With good strong English beer.
Drink round my jolly reapers and
When the corn is cut
We'll have the other jug boys
And cry A Neck A Neck

Abel Symons 1813

Ram aquamanile late-13th to early-14th century
Made in buffware and decorated with green glaze and pellets, this was discovered on a pottery kiln site in Scarborough.

Minton tiles 19th century
Fierce heraldic beasts are part of the decorative theme on these tiles produced by the large Staffordshire pottery firm of Minton, Stoke on Trent.

animals) pilgrim bottles, urinals and flasks were made. There were no plates or cups as these were sill being constructed in wood.

1500-1700

By the 16th century temperature control of kilns was much improved and potters were able to produce a smoother clay finish and a greater variety of shapes. At last cups and plates became part of their repertoire. A rich green or speckled underglaze was used in Tudor times as well as roses and shields of arms in relief. German salt glaze was imported and imitated while the Benedictine monks produced fine hard Cistercian ware in a dark-red body that was glazed dark brown until the Dissolution in 1540.

Slip-trailed decoration became popular and this craft reached its peak during the reign of Charles II. Good local clays and readily available fuels for firing made Staffordshire particularly important in its production. Slipware was made in Wrotham and London too but at Staffordshire the use of liquid clays was exploited to the full.

The Toft family produced especially fine decoration, usually on large flat dishes. White slip covered the reddish clay body and the designs were outlined and filled in with rich patterns and often surrounded by a trellis border. Royal portraits, mermaids and pelicans filled the center while combing the slip into feathers or arches and marbling were all explored.

Earthenware remained the basis of English pottery until the late 18th century. Tin-glazed and salt-glazed ware from the Continent had remarkably little impact; the English lead-glazed ware produced mainly by local potteries reflected the traditions of each area and there was much variety in the styles and range.

Inevitably the Toft ware found its imitators. At Bideford and Barnstaple in North Devon local plastic red clay was dipped in white slip and decorated with sgraffito designs scratched through to expose the red body below. They were then glazed yellow. Large harvest jugs were very popular and much of this ware was exported to the American colonies.

In Europe the techniques of firing at high temperatures had produced stoneware and (by introducing salt at maximum temperature) salt-glazed ware with its glossy finish.

In 1671 John Dwight, previously a lawyer, obtained a patent to produce stoneware in opposition to the German imports

Tigerware jug 1573
This beautifully glazed Elizabethan jug has been capped with an elegant silver gilt lid.

English slipware 'tyg' 1649
This unusual lead-glazed earthenware cup with four handles has been decorated with slip-trailing and stamped white clay panels.

— especially the bearded Bellarmine jugs named after an unpopular Cardinal. He went on to experiment with figure work and created some excellent sculptural work, the most striking being the representation of his little daughter Lydia. John and David Elers became his rivals, producing unglazed red stoneware rather like the recent Chinese Yi-hsing imports with a fine finish and delicate touches inspired by their previous work as silversmiths.

Meanwhile Jacob Jansen, an immigrant Flemish potter founded the Aldgate pottery and began to produce white tin-glazed earthenware. Clay from Norfolk and Suffolk was used and the Cornish mines were an excellent source of tin. Despite its tendency to chip, the tin-glazed ware was used as a substitute for Chinese white porcelain and lent itself to printed decoration. There followed a parade of kings and queens, Adam and Eve, generals, bright tulips and many freely-drawn designs on a large variety of pots, jars, bowls and vases. English maiolica had developed its own style. Cups and mugs for tea, coffee and chocolate became fashionable while the ports of London, Bristol and Liverpool became centers for tin-glaze production.

1700-1850

Introducing white Devonshire clay and calcined flint to the Staffordshire clay enabled this to withstand higher temperatures and be used to create a whiter stoneware — which was lighter in weight too. By 1740 Staffordshire potteries were using this in molds to produce thin-walled and ornate designs as a challenge to the fine white porcelain from China. It could not attain the latter's deep rich glaze but the delicacy and decorative refinement were greatly improved and the fine bodies unrivalled. Relief decoration on these was especially effective. Touches of cobalt were used: sometimes the cobalt was implemented to highlight an incised design in a form of decoration known as 'scratch blue'.

Other Midland potteries were producing salt-glaze wares with an iron slip under the glaze to create a deep-brown lustrous finish. The Morley family of Nottingham produced excellent examples of this ware which soon became very popular.

Delftware was being made in Bristol, Dublin and Scotland but the British products failed to achieve the brilliance of the Dutch 'kwaart' glaze nor could they quite achieve the fine quality of the Chinese imports at this time. However, by the 1740s English landscape painting came into its own, to be represented in the ceramic field by decoration that reflected the achievements of Watteau and Gainsborough

Face jug 14th century
This buffware jug made in Bishopgate, London, has been decorated with polychrome colors and applied yellow pellets — no doubt in imitation of European imports.

in its style and quality. Some fine specimens of painting emerged.

The Bristol and Liverpool potters were also producing stylized floral decoration — as well as their delightful ship-bowls commissioned by sea captains to inspire, they hoped, success for their voyages!

In about 1740 enamel painting on salt-glazed ware was introduced by two Dutchmen and was used to great effect, especially on applied reliefs which were a very popular form of decoration then.

The potters in the early 18th century were eager to experiment with new ideas and by mixing clays in different ways produced interesting agate and marbling effects. They also discovered that adding manganese to both clay and glaze produced 'shining black' ware which was very effective with its high-gloss finish.

Creamware

The changing conditions in England brought about by the Industrial Revolution inevitably affected the world of ceramics; mass-production and the formation of larger units of industry, rather than small family concerns, became the order of the day.

The needs of the time were to be met by the evolution of creamware as the most successful product. Salt-glaze ware was abrasive and wore away silver cutlery; Chinese porcelain was expensive; tin-glaze ware chipped all too easily, and country earthenware was considered too 'down-to-earth' for the middle classes. The requirements of the mass market (for Europe, not just England) were to be satisfied by English creamware.

It had already been discovered that the addition of the whiter Devonshire clay with calcined flint produced excellent salt-glaze ware. By being low-fired instead and glazed with lead or (galena), before this single firing, the new cream-colored earthenware had a brilliant golden sheen. The method of production had to be changed to prevent the injurious intake of dust by potters. Grinding flints and ore underwater and using a fluid glaze on biscuit ware (which was then fired) solved these initial problems and the creamware industry was launched.

Thomas Whieldon (1719-95) of the Fenton Law pottery was a renowned potter — especially famous for his agate and 'tortoiseshell' ware (made by dusting coloring oxides on to the glaze). Many aspiring potters came to him to learn and

Long-eared owl
A delightful saltgalze stoneware figure of an owl — probably initially inspired by German saltglaze work.

Salt-glazed teapot c. 1770
Famille rose decoration and a 'crabstock' handle complete this Staffordshire teapot.

Staffordshire creamware squirrel c. 1790
English creamware swept the European mass market after the Industrial revolution.

Circular basket c. 1770
A delicate piece of Worcester porcelain, this basket has been finely decorated with a floral motif.

Derby shepherdess c. 1790-95
Modeled at Derby by Spangler, the details of this gentle figure are readily appreciated because of the simplicity of the fine biscuit porcelain.

England

Transfer printing late-18th to early-19th century

Littler's blue vase c. 1750

Toft slipware 1671-77
Mermaid design: slipware dish with latticed rim by Thomas Toft.

English Delftware posset pot c. 1695

improve their skills and in 1754 Josiah Wedgwood (at that time only twenty-four — he had become an apprentice potter when fourteen years old) formed a partnership with Whieldon. This was to prove very fruitful with many technical improvements resulting from their five years together.

Wedgwood developed several colored glazes (the green glaze was especially fine). With William Greatback he went on to produce a quite delightful series of original wares based on fruit and vegetables— such as cauliflowers and pineapples.

From 1760 onwards Josiah Wedgwood undertook the development of creamware in order to produce it on an industrial scale. He managed to obtain the patronage of Queen Charlotte and went on to improve the quality of creamware (now renamed Queen's ware) by adding china clay and china stone to create a blue-white, thin but resilient body called pearl ware.

As in Italy, classical forms were inspired by the excavation in the Pompeii area with applied reliefs as decoration.

By now Leeds, Liverpool and Bristol were also producing creamware and a major industry was underway. Many smaller towns in the north Staffordshire area manufactured this popular commodity and then shipped it down the Grand Union Canal to reach far-flung places around the world.

Other developments

Enoch Booth had apparently introduced biscuit-firing of earthenware in about 1750 and this made the pots more robust and their surfaces were better able to absorb the lead glazes.

Transfer and molds

In 1753 transfer printing (possibly invented by John Brooks) encouraged the mass-production of quite detailed designs. Begun in Battersea and then taken up by Sadler and Green of Liverpool, the technique was soon absorbed into the Wedgwood ceramics with Josiah himself often choosing the engravings. It was quite late in the century before Wedgwood produced his own printing. Elsewhere, transfers of exotic and rural scenes all appeared in turn — the familiar blue-and-white willow pattern making its mark as an English response to the still pervasive Chinese influence.

Peacock tiles by William de Morgan 1839-1917
Earthenware tiles have been decorated with polychrome decoration under glaze in the 'Persian' colors, blue and green predominating, and are typical of this artist-potter's style.

Meanwhile the possibilities of pouring liquid clay slip into plaster of paris molds were beginning to emerge and had been introduced by Ralph Daniel to the Staffordshire potteries in the 1740s.

Enamelling

Enamelling of salt glaze appeared about the same time and its effectiveness when used on creamware was being exploited by the 1760s. David Rhodes produced fine figures, landscapes, flowers and border designs on the Wedgwood pottery. The Wood family of Burslem produced wonderfully restrained color glazes and went on to explore enamel-finished figures.

Unglazed black stoneware (called 'basaltes') and red stoneware were soon in general production, reflecting the great variety of the Wedgwood wares. They were joined before long by jasper ware (a white stoneware which could be colored right through using metal oxides) but the bluish pearl ware was to be the most successful product and was particularly suitable for the very popular blue printing.

The Leeds pottery

Second only to Wedgwood, the Leeds pottery produced some fine work in the second half of the 18th century. David Rhodes developed his enamelling techniques there and the underglaze blue printing on pearl ware was of a very high standard. At the turn of the century there was a sudden interest in lusterware and Leeds produced excellent examples of this, especially in silver (from platinum) and pink (from gold). Its iridescent metallic effect, especially when combined with 'resist' patterns or transfer printing, was a feature implemented by many factories.

Imitation and inspiration

Ale mugs, chamber pots and pitchers were decorated in 'mocha' style, using slips to create tree, feather and moss patterns while molded jugs were created to represent a variety or of themes; ornate stags, apostles and tree-trunks were typical of this fashion.

Increasing interest in seaside holidays with the expansion of the railways meant that holiday souvenirs opened up a new market for ceramics, while the Great Exhibition of 1851 showed off all the new developments and techniques, which sometimes tended to be regarded as more important than style or good design. The French Sèvres porcelain was imitated and generally the work was over-decorative.

First-edition copy of the Portland Vase

Wedgwood 'cauliflower' teapot c. 1760-65

Soft-paste porcelain was produced in Chelsea in 1745 and, before long, hard-paste porcelain in Plymouth and Bristol, Staffordshire and Worcester (which in particualr produced excellent decorated ware and was the first porcelain factory to use transfer printing on a large scale). Chelsea went on to produce some of the finest and most luxurious English porcelain, Meissen-inspired undoubtedly, but wonderful figures and tableware emerged. Joseph Willems, a highly skilled modeller was responsible for much of this.

English bone china, creamware and porcelain

Josiah Spode is credited with the invention of English bone china in 1800, but Heylyn and Frye may well have been using it at Bow well before that. Bone ash, although not a glass, reacted with the clay ingredients to create a firm translucent body which was far more stable in the kiln. The Bow factory produced some excellent quality porcelain aimed at the middle-class purchaser.

Maiolica became known as majolica at about this time with Minton's factory producing vegetable and floral shapes as well as green-glazed plates with molded leaf designs.

Many country-style earthenware pots and salt-glaze wares continued to be produced in local styles but it was the creamware which dominated. The European market was forced to adapt accordingly and manufacture its own style of creamware while the wheels of the British industrial Revolution rolled on. Steam-powered machinery, jiggering (shaping clay on plaster molds using templates), plaster-of-paris molds and transfer designs all contributed to the products being made quickly and cheaply. Handpainted ware became a rarity. In time porcelain became less associated with gracing the table and became instead to be associated with the mantelpiece or cabinet. The Derby factory produced many Meissen-inspired figures and the standard of painting these was especially high.

William Billingsley developed a beautiful glossy white porcelain and, after breaking his contract at Worcester, established a factory first at Nantgarw and then at Swansea where a natural style of floral painting developed that was fresh and unusual.

In time the Coalport factory bought the Swansea molds and adapted the decorative styles there too.

Styles came and went — Imperial, rococo, Sèvres Renaissance — and then the new 'parian' porcelain emerged which resembled marble and was used widely for statuettes and sculptural embellishment.

With the approaching 'modern' age in ceramics there were to be many changes. Individual countries were far less isolated; industry and new discovery changed methods far more rapidly than could have been imagined just a few years before. The versatility and range of ceramics adapting to the fashions and improved techniques made a sharp contrast to the slow development of English pottery back in Saxon times.

Jack Tar 1854
With the Industrial Revolution the Staffordshire potteries came to the fore, producing pieces for the mass market. This jaunty figure decorated with clear bright colors is a typical example.

Aesthetic porcelain teapot 1882

America

For many centuries civilizations and different cultures had flourished in America, isolated from the rest of the world and totally unknown to the European communities. When at last explorers, conquerors and settlers arrived there in the 16th century they were to discover and then reveal to the outside world distinct styles of pottery that were excitingly new and alien.

In this seemingly primitive society the potentiality of the wheel had not been exploited at all — even for transport (mules, sleds and litters were use) and certainly not for ceramics. The pots were all handbuilt, using coils, hand modeling or molds, but achieved many complex and lovely forms and were decorated superbly.

Kilns were fairly basic but sufficiently developed in some cases to produce a reduction firing. Inspired perhaps by basket-making techniques, the pots were largely rounded and squat with stylized (rather than naturalistic) decorative designs. They were used for the storing, cooking and serving of food and drink — sometimes to serve such specialities as hot chocolate or tortillas. Pottery was also used in the manufacture of musical instruments such as drums, panpipes, rattles and flutes. Many vessels were buried with the dead and have provided a vital source of information.

Pottery vase from New Mexico
In the south west of North America (Arizona and New Mexico today) the Pueblo Indians produced distinctive decorative pots, usually black and red designs painted on a pale gray background.

Deer and sunflowers (for the deer to eat) were often depicted on Zuni pots.
Even the internal organs are indicated within the deer on this typical highly decorated vessel from New Mexico.

North America

There was little pottery to be found in the more northerly parts of the American continent, where tribes were largely nomadic. Better-quality ceramics were found within the Mississippi area but some of the finest work was created by the Pueblo Indians in the south-west of North America. Initially they were all coil-built in grey clay but later pink or red clay was used and clored slps (mainly black and white) employed in geometric designs.

Each tribe developed its own style of decoration. Zuni pottery (from Little Colorado) used black and dark-red paint on a pale gray ground with stylized birds, deer and geometric patterns swirling over the whole pot. The top section of Acoma pottery had bands of floral patterns and Hopi pottery used black designs on yellow.

The Hohokam Indians of Arizona built pots with characteristic sharp shoulders and by AD 600 were painting buff-colored pots with red outlines — sometimes of animals and humans. Mimbres ware from New Mexico is quite distinctive and lovely; abstract insects, animals, birds and human forms were painted in black on a white slip ground. Often the bowls buried with the dead had been punctured. These small holes may have been made to deter grave robbers but could have been created to allow the 'soul' of the object to escape along with the human 'soul' to whom the pottery had been presented.

Hohokam dish
Excavated at Snaketown, Arizona, this dish was made by the Hohokam tribe (who used the paddle-and-anvil method of pot construction).

Mimbres bowl c. 11th century AD
Oval bowl with black painted design based on childbirth. These bowls were buried with the dead and the hole may have been made to release either the owner's soul of that of the plate for inanimate objects were sometimes believed to possess souls too.

Mexico

In the early period of Mexican ceramics (which is considered to be from 1500 BC to AD 300) the pottery that was produced included many bowls, long-necked bottles, and jars with no necks, stirrup spouts and three feet. Hollow legs shaped like breasts were often used and the ware was usually slip-decorated and burnished. Incizing and wax resists were used to effect mainly geometric patterns.

The classical period which followed and lasted until AD 900 was centered around the independent cities with their great temples and pyramids. Polychrome color was used in black, dark-brown and cinnabar combinations as well as designs in stucco with carved plaster and colored clay.

Living in south-east Mexico, Guatemala and the British Honduras were the Mayas — the most advanced civilization in Ancient America. By the classical period wonderful narrative painting was appearing on their straight-sided wares, depicting processions and ceremonies, while everyday scenes of cooking and hunting were the inspiration for plaques and figurines, many of which embodied whistles. Animals and hieroglyphics were often used as decoration, while sacrificial cups were sometimes made in the shapes of animals. Crocodiles, snakes, deer and monkeys were incorporated into the designs.

By AD 950 decoration became more abstract and in time careful kiln control produced a black, shiny hard ware called plumbate.

Tenochtitlan (Mexico City) was founded by the Aztecs in 1325. Based on fearsome religious ritual involving human sacrifice, the society dominated earlier cultures and the variety of ceramic styles reflects this absorption of other traditions. By the time of the Spanish Conquest, decorative and sensitive designs based on wild life and painted in black lines were being used most effectively on the inside of bowls. The pottery was largely orange-colored and thin-walled. Bi-conical cups and chalices were made for the drinking of the fiercely alcoholic 'pulque' by the elders.

Terracotta lovers 500-100BC, found at Puebla

Jaina figurine made in terracotta by the Maya — the most advanced Ancient American people.

Figure of the rain god, Tlaloc, 14th century
From the mountainous area of Southern Mexico the influence of the Mixtec potters spread far north.

Mixtexa pottery was decorated with several colored slips and was of an excellent quality — the only kind, in fact, that King Montezuma would use. The designs were busy and interesting. Many pieces of pottery were sold at the great market of Tiateloco, and these impressed the followers of Cortès.

The valley of Mexico saw the development and decline of many cultures: Olmec, Tiatilko, Zapotek, Mixtec and Teotihuacán are just some that could be mentioned. Each had its own characteristic ceramics with Teotihucán in particular producing fine ware which sometimes reflected the architectural carving of the temple pyramids. The pottery was lively and expressive. In western Mexico where there was no dominant priesthood or temple hierarchy, pottery figurines were made to be buried in the tombs as companions to the deceased. These pieces seem to be more 'down-to-earth' and reflect a culture that was evidently enjoying life and comparatively carefree.

The Olmecs produced jade figurines, enormous stone heads and hollow figures (which usually had mouths turned down at the corners). The Mixtec specialized in seated figurines where detailed decoration and ornamentation probably reflected their equivalent skills in fine jade and gold jewelry. The pottery was highly polished and the polychrome painting excellent. The Zapotec built huge terracotta funeral urns depicting both human and animal gods. Gradually the impact of all these different and quite advanced cultures spread northwards to influence the Pueblo Indians and others in the south of North America.

Pottery pitcher c. 1100 AD
A Mogollon or Anasazi pot found in New Mexico with beautiful geometric black-on-white decoration.

South America

South America encompasses a vast area of coastal grassy zones, the high ridge of the Andes, river valleys, grassy plains and Amazonian jungle. Such a geographical mix led to the development of many quite separtate cultures — with thriving cities in the central Andes. The ceramic styles are quite different to those of central and North America.

The earliest pottery produced about 3000 BC was black or gray and round-bottomed. By 1000 BC the Chavin people from the north, whose massive temple was devoted to cat worship, had developed characteristic gourd-shaped pots in a light color that sometimes incorporated red inlay patterns. The stirrup handle (so-named because of its resemblance to a sadle stirrup) became a regular feature. It was not purely decorative as it kept the liquid contents clean, as well as providing an efficient handle and spout. These spouted bottles were made in human, animal, vegetable and house shapes. Rocker stamping and incized patterns were used to decorate pots.

Stirrup-spout bottles were also made by the Mohica people of the north coast. This was an advanced civilization who produced a good deal of distinctive pottery — often using fired-clay molds and then adding modeled features. Their styles are Egyptian-like with human figures drawn in a strange mix of both frontal and profile aspects and with an evident reverence for cats permeating their work.

Pottery was often shaped like fruit such as pineapples and some are 'portrait' jugs which are obviously close studies of real human characters. Always their form was strong and an important element to the potter.

Much of the people's daily life (as well as their architecture, ceremony, warfare, clothing and surrounding wild life) is documented very realistically in the decorative detail. Musical instruments (both wind and percussion) were also made from clay as well as whistling jars. These were shaped like birds and whistled when water was poured through the spout.

Meanwhile in the Nazia valley on the south coast brightly colored slips (as many as five to a pot) were used to decorate pots with intricate designs — often outlined in black.

The Tiahuanoco culture was highly organized with great walled cities and a rather formal, severe ceramic style. Egg-shaped jars with human features were produced in the Chancay valley, while the Chimu people (whose capital was the great city of Chan Chan) produced smoked black ware and the ever-prevalent stirrup-spout jar.

By 1450 the Incas' monopoly of power in the Andes became evident. Controlling over three thousand miles north to south, this remarkable administration was helped by the building of a vast network of roads; their power was to be unchallenged — absolute — until the arrival of the Spanish expedition under Pizarro in 1533.

The Incas' bureaucracy seemed to keep the ceramic styles rather static. Technically the pots were much improved, being strong yet thin walled. Narrow-necked water jars, beakers, shallow plates with bird-head handles and flat-based cooking bowls and jars formed the standard shapes. Decorative geometric shapes and stylized plants, butterflies and bees were carefully painted on to these.

With the European conquest of South America much was lost or destroyed but what remained or has since been excavated is fresh and exciting to us today; how much more so must it have seemed to the first Europeans who conquered the New World four centuries ago.

Peruvian stirrup vessel
The stirrup-shaped handle spout was produced for some 2,000 years. This intricately decorated Chimu vessel depicts the dragon god.

Portraiture vases 6-7th century AD
A flautist, potter and warrior are portrayed in these delightful burnished vases — finished with cream slip and iron oxide decoration. The potter figure is in fact depicted creating yet another portrait vase!

Painted scene reproduced from around a cylindrical Mayan vase. Classic period
Narrative painting was skilfully employed by the Mayan potters — this vase comes from Guatemala.

Later American developments

From the moment that European settlers landed on America's eastern shores it was inevitable that the early indigenous styles would be supplanted. In fact, only a few of the remote Pueblo Indian tribes continued to produce ceramics inspired by their own cultural heritage and did so until late in the 19th century. By the 16th and 17th centuries vast numbers of Europeans arrived as immigrants, bringing with them all the ceramic traditions that had evolved from the middle ages.

This new society had perforce to be adaptable to the demands of the environment and the newly-established communities as well as learning to use different sources of raw materials. The Puritan origins inevitably succumbed in time to fresh ideas, frontier inventiveness and regional development of style.

The New England colonies (especially Massachusetts and Connecticut) produced a fine red earthenware with simple incized decoration and natural lead-glaze colors.

The English kick-wheel and the simple Continental wheel were soon introduced along with the import of lead glazes and other coloring oxides. (Sometimes copper-oxide glazing was achieved by burning unwanted copper utensils in the kiln with the pots — and lead was oxidized from the linings of old tea chests!)

Gradually kiln technology improved and became more sophisticated while the demand for pots quickly accelerated with the growth of rapidly expanding communities. Pottery making became a worthwhile profession and young apprentices were recruited for an arduous seven-year indenture period. These new societies required a different range of pots, and pie dishes and bean pots joined the output of conventional jugs and vessels.

By the early 18th century the well established ceramic industry was skilfully producing red earthenware pots enhanced by trailed white slip decoration. However, the discovery of the dangers of lead poisoning decreased the demand for lead-glazed articles and the popularity of the more robust stoneware was assured instead.

Anthony Duché (an immigrant French Huguenot) was making stoneware pieces in the 1720s. Rather less common high-firing clays were needed for these products so New York City, which boasted suitable clay beds, became one center for stoneware production.

Pueblo-style pots
In the American south-east traditional Pueblo pots continued to be made. The black-on-black pot was produced by Maria Martinez who discovered this technique in the 1920s.

Redware bowl 18th century
A simple redware bowl with dark brown glaze and stylized bird decoration.

Gilt-edged jug, 1828
Painted with polychrome flowers and gilt, this is a product of the Tucker factory. The initials are presumably those of Elizabeth Slater, the original owner.

German potters, including William Crolius, introduced the Rhenish tradition and William Rogers of Virginia an excellent brown salt-glazed stoneware. In some areas the import of stoneware clay proved too expensive but elsewhere, (in Philadelphia, for instance) high-quality stone and salt-glazed ware developed.

A variety of new glazes and decorative ideas added an extra fillip to the interest in these products and American folk pottery became everywhere an outlet for artistic expression, especially once the Federal Government was established after the Revolution and import duty, which raised the price of foreign goods, rendered home-produced pots all the more desirable.

Calligraphic and incized decoration were superceded by floral and pictorial designs. These included many fish and bird motifs and an animated woodpecker design was created in the Shenandoah valley.

Although earthenware was still used for baking dishes (because it was better able to withstand the thermal shock of being 'cooked') stoneware became ever more popular and was produced in many local styles.

While the wealthier members of society continued to import the fine English and Dutch wares, new factory techniques using patterns and molds led to the production of pots *en masse* for the general market — a fast-growing industry that rapidly spread from coast to coast. Luster pitchers, red and black glazed coffee and tea pots, molded stoneware and rather poor imitations of the white ware from Europe were produced.

Meanwhile there was much interest in discovering the secrets of successful porcelain production. Back in 1739 Andrew Duché had experimented with its creation, using china clay from Cherokee Indian territory. By the 1770s Bonnin and Morris of Philadelphia were producing fine white American china, using clay from the banks of the Delaware mixed with calcined bones. Fruit baskets, bowls, jugs and plates were produced with painted and cut decoration, as well as underglaze-blue transfer designs.

However, the most famous 19th-century porcelain was to be created by William Ellis Tucker (also of Philadelphia) who began his career as a decorator of porcelain for his father's import shop. Tucker constructed a small kiln behind the store to pursue his experiments with feldspar and kaolin and then continued his work in a waterworks which he leased and converted into a pottery in 1826.

Captain John Norton was producing earthenware and then stoneware in Bennington, Vermont at the turn of the century. In 1837 he was joined by Christopher Webber Fenton, often designated the 'Josiah Wedgwood' of America. A wide variety of decorative pieces was produced including Rockingham ware which had a lovely speckled brown glaze. Different oxides were introduced to create many colored mottled glazes. By 1853 Fenton had established The United States Pottery Company and was producing white and yellow earthenware, slip cast or molded Parian porcelain (which resembled sculptural marble) and specialities like 'lava' ware. Fenton discovered that brushing blue slip on to the molds prior to introducing white slip created color just where it was required which then fused with the body when fired. This was a great improvement on cheap surface painting.

However, perhaps because the factory had overextended itself with such an ambitious range (making just about every conceivable domestic object from coffee urns to footbaths) it collapsed in 1858.

By the mid-19th century there were many rural potteries. Symbolic motifs such as the American eagle and ears of corn were very popular as well as political embellishments. Images of George Washington were common as were statements of support for particular candidates. Equally interesting were ideas perpetuated by Cornwall Kirkpatrick later in the century. Many of his exciting pots were elaborately entwined with surrealist snakes and Hieronymus-Bosch-type imagery.

A further development from the cast Parian ware first produced by Fenton was the delicate egg-shell porcelain called Belleek ware. Based on styles produced in the Belleek works founded in Ireland in 1863, the work was initiated by Ott and Brewer of Trenton, New Jersey, and was much influenced by Japanese design.

During the civil war there had inevitably been a reduction in pottery production while the country was preoccupied with warfare. Sadly, the coincidental improvement in road and railroad communications led to the collapse of many small workshops; they could no longer compete with the output from flourishing factories — products that could now be transported so readily, even to the remotest of regions.

Most of these factories were steered by potters trained in the Staffordshire potteries of England and, generally, the products were over-fussy late-Victorian monstrosities. One exception was the elegant Lotus ware by Knowles

Bennington Vermont poodle 1849
Lyman Fenton made this mottled-brown glazed poodle.

Rockwood earthenware jar 1900
The delicate relief slip decoration has been painted under a colored glaze by Hattie E. Wilcox.

Taylor and Knowles of East Liverpool, Ohio. This was made of thin bone porcelain and had a delicate texture and fine applied decoration. Porcelain was not to prove a major American forte and the rather more utilitarian simpler pieces were in fact the most successful to emerge from this period.

Eventually the Arts and Crafts Movement reacted against the mass-produced factory products. Initially inspired by Morris and Ruskin in England, this movement led to a revival of the smaller ceramic commercial enterprise, and during the 1880s and 1890s many studios and workshops embarked on the production of art pottery.

Meanwhile in the isolated rural south the folk tradition had survived, with local needs still being met by the production of simple but sturdy pottery. Here, the work of the black potter influenced the styles, especially in the creation of grotesque face jugs used for carrying water to work. These were reminiscent of African effigies and sculptural design.

In time the emergence of the studio potter and the acceptance of ceramics as a true art form led to its acceptance in the academic world. Potters such as Charles Fergus Binns (1857-1934) introduced more thorough training courses for the student potter which embraced the ideas initiated by the Arts and Crafts Movement but which covered the practical aspects as well. Pottery was incorporated into the university curriculum and was firmly established as an inspiring art medium.

From the simple tribal traditions of south-west America and the homely earthenware cooking pot made by the first colonists the world of modern American ceramics had evolved.

In the 20th century this was to explode into an enormous variety of form, color and imagery. Sculpture, 'funk art', Zen Buddhism, expressionism, abstract design and hyper-realism are just a few of the diverse artistic persuasions for which ceramics were to become a means of expression.

'Electra' pedestal 19th century
This 42-inch (108 cms) high pedestal has elaborate, Wedgwood-style sprigged decoration.

Africa

There is little information available about ancient African cultures: rock paintings and engravings in stone indicate a gradual evolution from a hunting to a more pastoral existence. Migration distributed the inhabitants of this great continent into different areas and cultures but generally they remained primitive societies. Apart from North Africa, where the Islamic influence was strong, the use of the wheel was never to be discovered and, on the whole, tribes were dependent on locally available raw materials.

There was, however, trade between the ancient African kingdoms of Kush (Sudan), Punt (Ethiopia and Somaliland) and Egypt, and iron-making spread westwards to reach the Nok people of northern Nigeria by 500 BC.

Bronze casting became established on the Niger River and in Nigeria (then Benin) it was developed into a fine art form, especially in the production of royal heads!

Wood and stone were popular for sculpture but terracotta too was used, some especially realistic figures being made by the Nok tribe in about 500 BC. On a smaller scale, tiny clay figures and stylistic effigy heads of great character were produced.

Pottery tended to be made by the women, generally hand-built from coils with the gourd shape predominating the production of containers. The hot climate and scarcity of water meant that much importance was placed on such vessels and huge collared vases to contain food or water were fired in open pits buried in straw and twigs. Only a few early examples survived till today as this firing did not produce very hard or robust ware.

A basket-weave design was sometimes added which lent interest to the surface and iron oxide designs in red or black were occasionally used but in the main these containers were functional rather than decorative.

In the late first century the Romans brought their culture to northern Africa and evidently it was there that the renowned red-gloss ware was made.

Apart from such outside influences in the north, the lack of machines and the simple tools used meant that the production of ceramics remained largely a functional pursuit with artistic expression being conserved for the development of form — such as the mouths of vessels being shaped like animal faces.

The bartering of goods led to the exchange of ceramic ware for other articles or food; this happened especially in areas

Kabyle dish from Algeria
An attractive painted vessel from North Africa, the only part of that continent where the wheel was used — as a result of the strong Islamic influence.

Zulu pottery vessel
Wide-mouthed and spherical with raised decoration on the shoulders, this simple burnished pot is typical of South African ware.

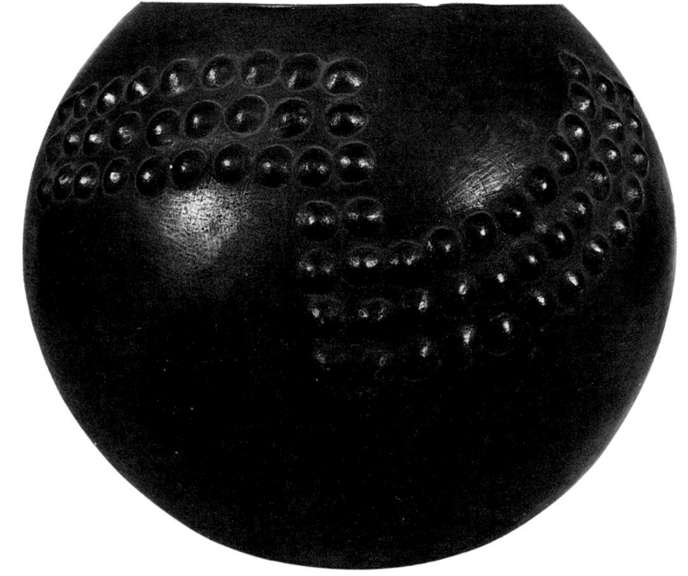

where no suitable clay for the production of pottery could be discovered so that clay objects had to be 'imported'.

Much has remained unaltered. The African cultures escaped most of the changing patterns of style and development and the impact of Eastern cultures which so influenced Europe. Nor did kiln development or the use of glazes penetrate to its primitive societies. Thus today subsistence economy and the simple techniques employed mean that ceramics are produced much as they were many centuries ago.

Although toys, musical instruments, ritual figurines and symbols, jewelry and pipes are made from clay, it is still the vessel or container that dominates ceramic production.

Terracotta head from Nigeria, 13th century

The shapes of these tend to be full and rounded and are built to be hung, to stand firmly on the ground, or to be most readily carried on the head. Water vessels are given the extra attention their importance merits and are often painted in reds and blacks. Coiling, modeling from 'quoits' and molding around another pot (or inside a basket) are the usual techniques while the final surface treatment and decoration employs the simplest of tools — such as fingernails, pebbles and shells. Clay slips are used and sometimes a simple vegetable, gum or resin varnish.

Inheriting ritual skills passed down from their ancestors, the potters whose craft this has become have had little incentive to change or develop their skills. None the less, Western objects such as areoplanes and cars are sometimes incorporated into the decoration — which is generally more localized and has shown greater inventiveness and individuality than other aspects of the work.

However utilitarian the purpose of the pot and however rudimentary the techniques used to create it, the end result is often quite lovely in its simplicity. Economy of form and decoration have produced articles whose stark beauty has an immediate impact.

Pottery bird from Suto, Lesotho
Possibly based on a guinea fowl, this work is sandy colored except for the red-painted legs and wings. The head has been roughly made and is a separate piece.

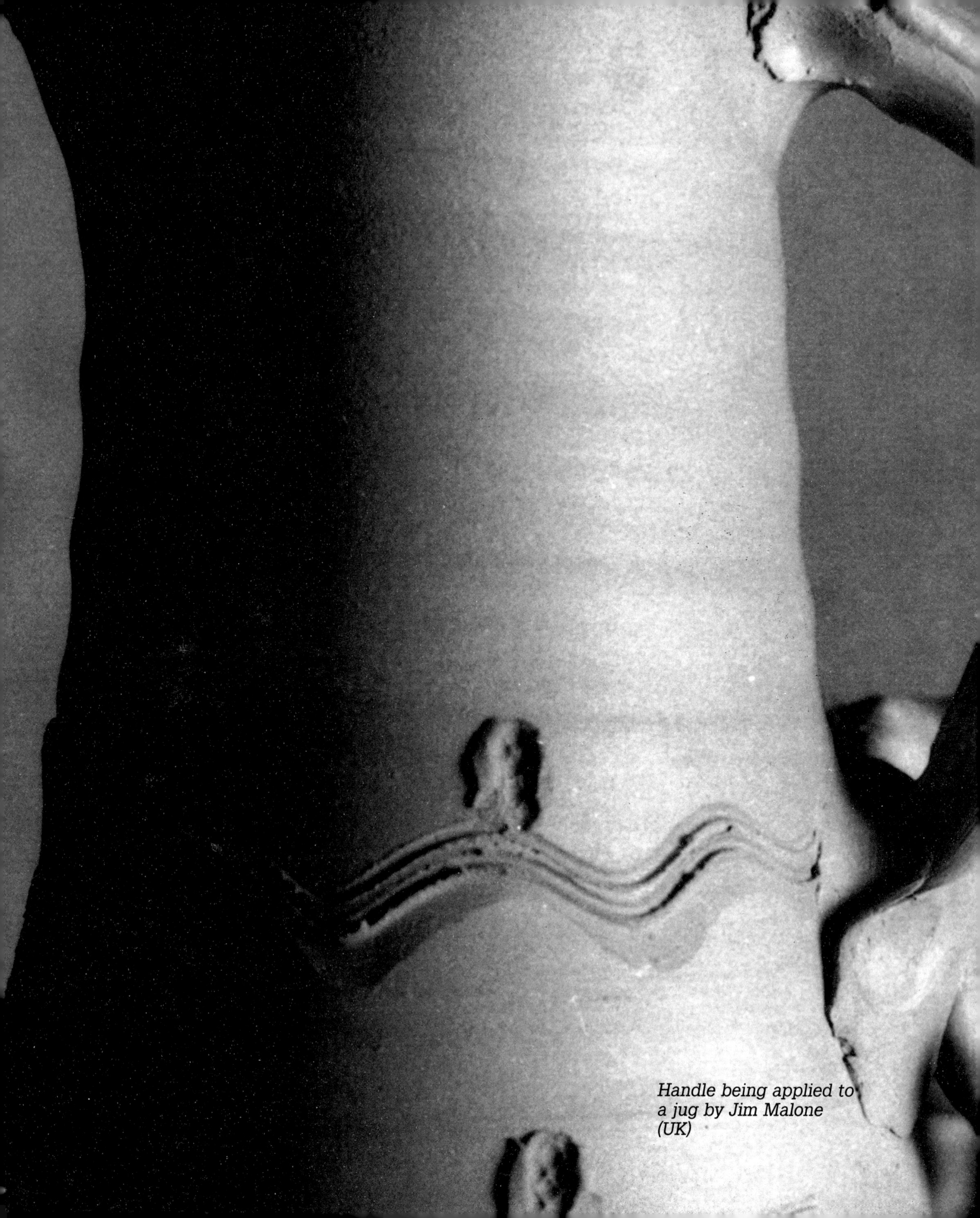

Handle being applied to a jug by Jim Malone (UK)

Useful facts
9

How to measure accurately

Measuring lids

(see also pages 96-9)

Measuring accurately is largely a matter of common sense and just taking a very little extra time to produce a better finish. If linking two or more sections of a piece together, such as a lid to a pot, it is vital to measure each part very carefully. You will need to use a pair of calipers, setting them to the required width of the first section of the pot. If fitting a lid, the calipers should be set to the width of the internal rim of the pot: the lid is then made to match this span. Always keep the calipers ready to hand in order to check everything as you proceed. Do not be tempted to rush this stage as an ill-fitting lid will spoil the completed item.

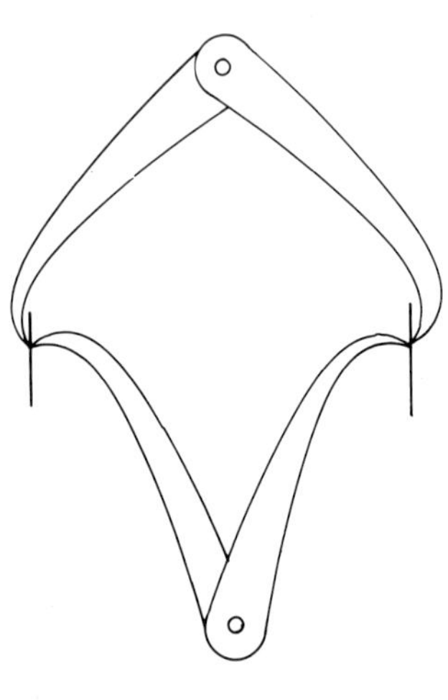

Fitting 'cap' lids

Make sure the full *external* diameter of the pot's rim measures exactly the same as the diameter of the lid's *internal* rim. The lid will then rest neatly and securely in place on the finished pot.

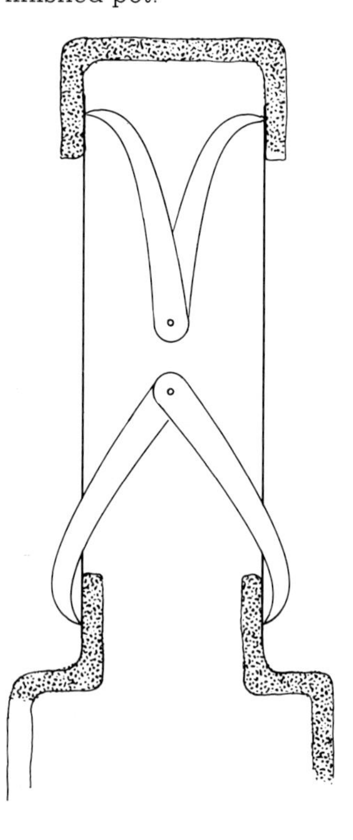

Gallery lids

The diameter of the pot's *internal* rim should be identical to the diameter of the lid's *external* rim. The diameter of the pot's *internal* gallery should be equal to the diameter of the lid's base.

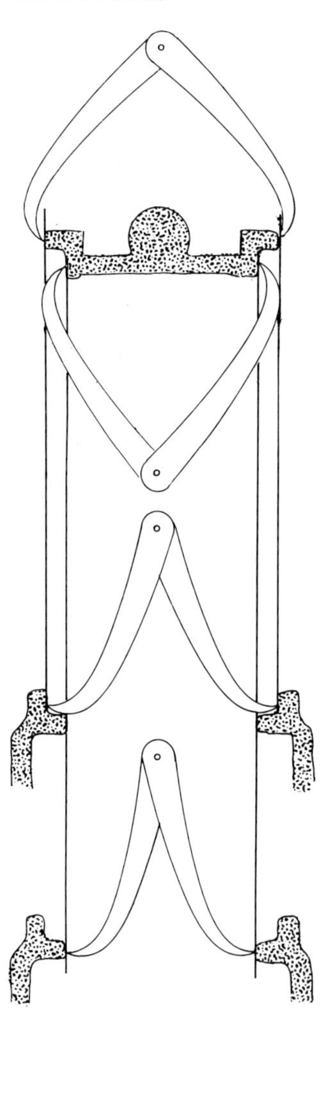

Flanged lid

Here the diameter of the lid's base should measure the same as the diameter of the pot's *internal* rim: the diameter of the lid's *external* rim must equal the diameter of the pot's *external* rim.

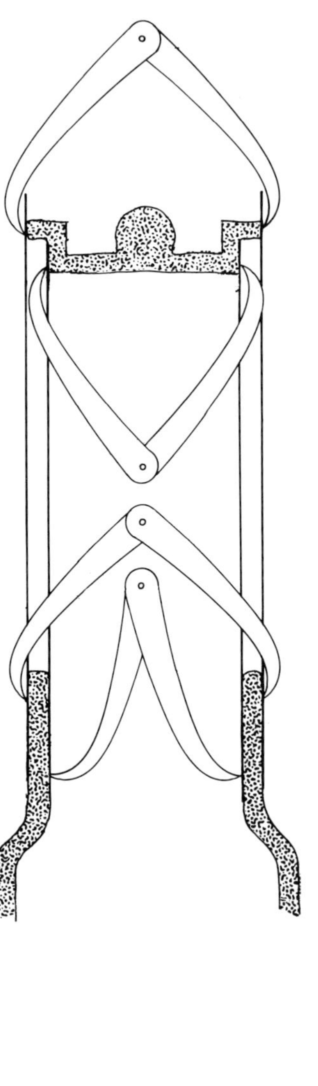

Flanged lid on a gallery

Make sure the diameter of the lid's *external* rim is the same as the diameter of the pot's *internal* rim. Also, the diameter of the lid's flange must equal the diameter of the pot's internal gallery.

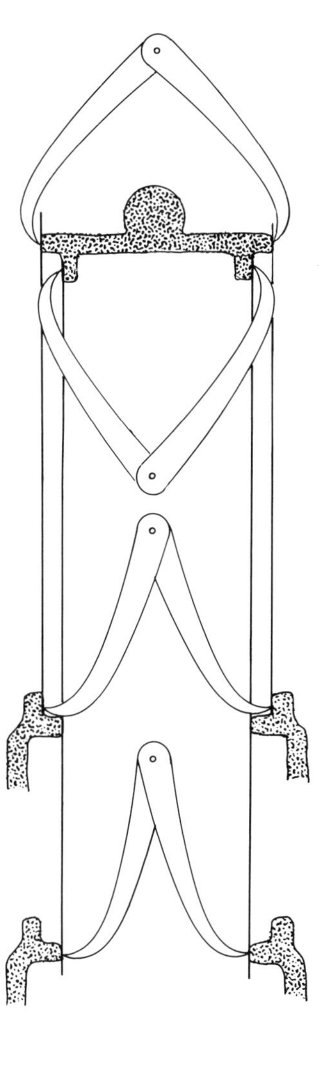

Dome lid

The diameter of the pot's *internal* rim must equal the diameter of the lid's *external* rim — remember, of course, that these measurements must be taken after all the trimming has taken place.

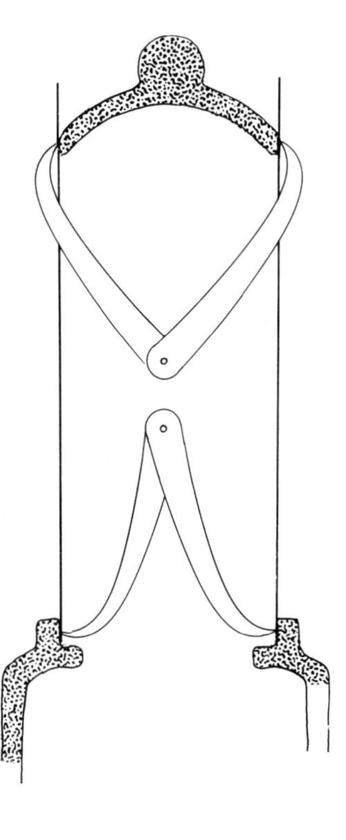

Flanged dome lid

The lid's *external* rim diameter must equal the diameter of the pot's *external* rim. The diameter of the lid's flange and the diameter of the pot's internal rim must be the same to ensure a neat fit.

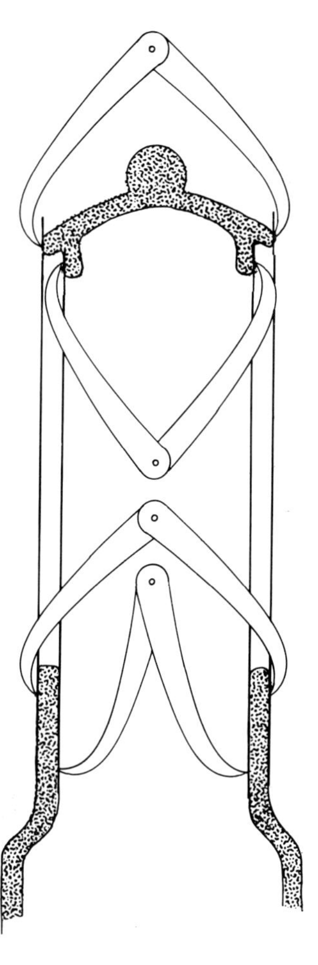

Joining thrown shapes

(see also pages 92-3)

Joining shapes will allow you to produce a greater variety of pot forms but do first make sure that the walls of each thrown section to be joined are of the same thickness. Although the internal differentiation may not be obvious when the pot is complete, an uneven join will be less secure and may well open up again or crack at the drying or the firing stage. So measure the diameters of both the external *and* internal rims and try to make the thicknesses of both sections match as closely as possible.

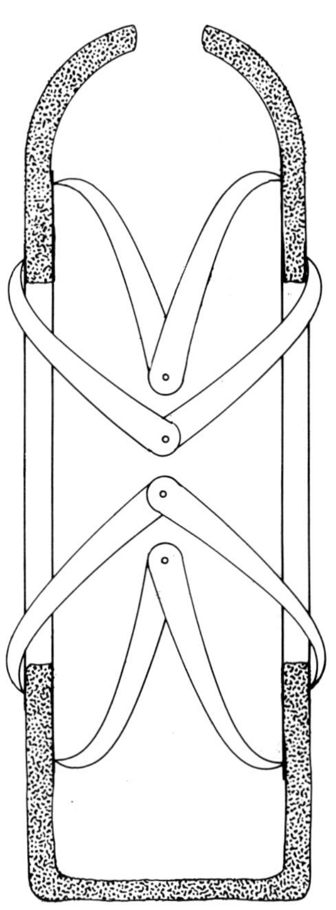

The chemistry of ceramics

Slip colorant	% of total	Result
chromium oxide	1-5%	green
cobalt carbonate	1-5%	light to dark blue
copper oxide	3-5%	green, charcoal
iron chromate	1-5%	light/medium gray
manganese	6-10%	brown/purple-brown
nickel oxide	2-5%	gray, tan, gray-green
red iron oxide	2-3%	tan
	4-10%	light/dark brown
	10-20%	dark brown/iron red
rutile	6-8%	tan, yellow
vanadium stain	8-10%	yellow
cobalt carbonate	1-2%	gray-blue
cobalt carbonate	1%	
red iron oxide	1-2%	turquoise
chromium oxide	½-1%	
cobalt carbonate	1-2%	
red iron oxide	3-5%	black
manganese dioxide	2-3%	
illmenite or granular manganese	2-3%	produces specks

Please note that kiln atmospheres will affect results considerably and that the colors in the table are likely to occur in oxidation rather than reduction firings.

Many of the chemical processes exploited by potters to achieve the effects that they require are actually brought about by fairly complicated chemical changes. It is not necessary to study these in depth — unless this is a particular area of interest for the potter concerned — but understanding a little of what goes on will help the average potter to anticipate and control results and to some extent reduce the trial-and-error aspect of ceramics.

All physical materials are combinations of different elements, some twenty of which are regularly used by potters while about the same number are employed occasionally. Often the elements are used as compounds when two or more elements have combined to form a useful substance such as sodium chloride (salt).

When an element combines with oxygen an oxide is created. This change takes place and can be altered again only at high temperatures, so in fact many of the effects which take place in the kiln are to do with such reactions between oxides. Firing can rearrange those compounds that exist already and can also create new ones — as indeed happens with a glaze .

Oxides can usefully lower or raise melting points or may help other elements to be better 'balanced' and so to unite more successfully. Atoms combine into compounds in the exact proportions which are indicated by the number which follows the symbol for that particular element. (Sometimes the letter R appears and this indicates an element that has combined with oxygen.)

The symbols stand for the molecular makeup of the material — which in the case of minerals can usually be only an approximation as minerals are seldom really pure. It is simply a convenient way to describe the chemistry of substances and the various changes that take place.

Slip colorants

The most straightforward way to make colored slips is to add coloring oxides or commercial stains to a basic white slip. Generally the percentages required are somewhat larger than for glazes: about 10 to 20 per cent of a commercial glaze stain is what would normally be required. The full color intensity of these slips is shown to its best advantage when covered by a simple clear glaze.

Some ideas for slip colorants are given in the table. Most of these colors can be adapted to cover a wide range of temperatures and will be effective whether fired in oxidation or reduction: they apply to any text discussion of slip mixing.

Ceramic raw materials

Material	Formula
antimony oxide	Sb_2O_3
barium carbonate	$BaCO_3$
bone ash (calcium phosphate)	$Ca_3(PO_4)_2$
boric acid	$B_2O_3 \cdot 3\ H_2O$
borax	$Na_2O \cdot 2\ B_2O_3 \cdot 10\ H_2O$
calcium borate (colemanite, gerstley borate)	$2\ CaO \cdot 3\ B_2O3 \cdot 5\ H_2O$
calcium carbonate (whiting)	$CaCO_3$
chromic oxide	Cr_2O_3
cobalt carbonate	$CoCO_3$
cobalt oxide, black	Co_3O_4
copper carbonate	$CuCO_3$
copper oxide, green (cupric)	CuO
copper oxide, red (cuprous)	Cu_2O
Cornwall stone	K_2O Na_2O CaO MgO CaF_2
cryolite	$Na_3 \cdot AlF_6$
dolomite	$CaCO_3 \cdot MgCO_3$
feldspar, potash	$K_2O \cdot Al_2O_3 \cdot 6\ SiO_2$
feldspar, soda	$Na_2O \cdot Al_2O_3 \cdot 6\ SiO_2$
flint (quartz, silica)	SiO_2
fluorspar (calcium fluoride)	CaF_2
iron chromate	$FeCrO_4$
iron oxide, red (ferric)	Fe_2O_3
iron oxide, black (ferrous)	FeO
kaolin (china clay)	$Al_2O_3 \cdot 2\ SiO_2 \cdot 2\ H_2O$
kaolin (calcined)	$Al_2O_3 \cdot 2\ SiO_2$
lead carbonate (white lead)	$2\ PbCO_3 \cdot Pb(OH)_2$
lead monosilicate	$3\ PbO \cdot 2\ SiO_2$
lead oxide (litharge)	PbO
lead oxide, red	Pb_3O_4
lepidolite	$LiF \cdot KF \cdot Al_2O_3 \cdot 3\ SiO_2$
lithium carbonate	Li_2CO_3
magnesium carbonate	$MgCO_3$
manganese carbonate	$MnCO_3$
manganese dioxide (black)	MnO_2
manganese oxide (greenish)	MnO
nepheline syenite	Na_2O K_2O CaO MgO
nickel oxide, green	NiO
nickel oxide, black	Ni_2O_3
petalite	$Li_2O \cdot Al_2O_3 \cdot 8\ SiO_2$
potassium carbonate (pearl ash)	K_2CO_3
pyrophyllite	$Al_2O_3 \cdot 4\ SiO_2 \cdot H_2O$
sodium bicarbonate	$NaHCO_3$
sodium carbonate (soda ash)	$Na2CO_3$
spodumene	$Li_2O \cdot Al_2O_3 \cdot 4\ SiO_2$
talc (steatite)	$3\ MgO \cdot 4\ SiO_2 \cdot H_2O$
tin oxide (stannic oxide)	SnO_2
titanium dioxide (rutile impure)	TiO_2
wollastonite	$Ca \cdot SiO_3$
zinc oxide	ZrO_2

Glazing recipes

The following recipes have been suggested by many of the potters who have contributed to this book. Some may suit your particular needs and should provide an interesting source of experimental ideas.

Where appropriate I have indicated known results. As different sources of materials vary, so will the results. It is all a matter of trial and error but the recipes should provide a basis for further exploration of glazing.

The basic recipe for a glaze is measured in parts: these may be grams, ounces, cupfuls or jugfuls! For simplicity, the total of the components for the basic glaze is usually considered to be 100 parts. Any additions to this basic glaze are calculated as a percentage of the base and are then added on to the total 100 parts.

Low-firing glazes 1560-2102 °F (850-1150 °C)

Raku

Basic glaze Cone 08 1560-1742 °F (850-950 °C)

Alkaline frit	58.3%
China clay	26.3%
Flint	15.4%
Green glaze	
to basic glaze add	2% copper oxide
Blue glaze	
to basic glaze add	0 · 5-1% cobalt oxide

Alkaline frit can be used over colored slips

Basic clear glaze

china clay plus oxides, used under and over glaze

Earthenware glazes

Clear glaze Cone 04 1900-1940 °F (1040-1060 °C)

Lead frit	71.5%
Ball clay	24.5%
Flint	4.0%
Honey glaze	substitute red clay for ball clay

Majolica - tin glaze

Basic clear glaze Cone 04 1940 °F (1060 °C)

Lead bisilicate	62.8%
Whiting	3.1%
China stone	5.6%
China clay	28.5%
White add	
Tin oxide	10%
Black add	
China clay	10%
Manganese dioxide	6%
Iron oxide	3%
Cobalt oxide	1%

Clear glaze	Cone 01	1868-2010 °F (1020-1100 °C)

Lead bisilicate	83%
China clay	17%
Honey glaze	
Lead bisilicate	83%
Red clay	17%

High-firing glazes 2100-2370 °F (1150-1300 °C)

Stoneware glazes

Celadon	Cone 9—10	2327-2372 °F (1280-1300 °C) reduced

Feldspar	21%
Whiting	22%
China clay	27%
Flint	20%
Iron oxide	2%

Cone 9 reduced

Cornish stone	60%
Dolomite	20%
China clay	20%
Blue/green add	
red iron oxide	1 • 5%

Oxidized	Cone 8	2268 °F (1260 °C)

Milky white	
China stone	50%
Whiting	25%
China clay	25%

Wood ash oxidized	**Cone 8**	**2268 °F (1260 °C)**

For a thin yellow/ocher coat

Mixed wood ash	40%
Feldspar	40%

For a thick coating of red/brown

China clay	20%
+ Iron oxide	10%

For a purple/brown semi-matte — looks very good over grogged clay

Wood ash	50%
Ball clay	40%
Crocus martis	10%

Wood-ash — dry

Oxididized or reduced

Cone 8 2268 °F (1260 °C)

Mixed wood ash	50%
China clay	50%
Add oxides for color	
Yellow ocher	5%
Blue	0.5% cobalt carbonate

Wood-ash glaze for raw glazing

Wood ash	50%
Plastic clay	50%

Basic clear Cone 8 2268 °F (1260 °C) oxidized

Wood ash	45%
Feldspar	33%
Dolomite	22%

Add percentages of color for tests

White opaque

Oxidized Cone 8 2268 °F (1260 °C)

Feldspar	54.6%
China clay	22.7%
Whiting	22.7%

Clear Cone 8 2268 °F (1260 °C) oxidized

Feldspar	40%
China clay	20%
Flint	20%
Whiting	20%

Porcelain glazes

Clear glaze Cone 10 2372 °F 1300 °C)

China stone	80%
Whiting	10%
China clay	10%

Smooth semi-matte Cone 9—10 2336 °F 1280 °C)

Feldspar	30%
Whiting	16%
China clay	24%
Ball clay	6%
Flint	12%
Talc	12%

Fault finding - firing problems

Problems	Cause	Result	Effect	What to do
Blow out or split out	Impurities such as sulphates/carbonates in the clay body	Contamination of the clay or the glaze	Craters form in the biscuit body	Use clays that are better refined
	Particles of plaster left on the surface of a mold		White powdery lump may be seen underneath	Check all plaster is properly removed from molds
Body explodes or fractures	Not dried correctly before firing	The body has air pockets trapped inside	Body explodes or fractures badly during firing	Dry body slowly before firing
	Fired too quickly	The clay has not dried thoroughly		Check thickness of work and create vents for air escape
Bubbles and bloating	Irregular or excessive firing	Trapped gases cause build-up of pressure so body expands	Bubbles form in the body when firing	Slower firing
	Carbon trapped in vitreous body			Add grog to make coarser open body Leave escape passages for air
Cracks	Overworking clay in manufacture	Surface stress	Cracks develop at biscuit stage	Handle clay less
	Using non-plastic clay			Use more plastic clay
	Blunt trimming tools			Use sharper tools
	Uneven drying			Dry more slowly
	Initial firing too fast			Slow down initial firing up to 212° F (100° C) per hour
Splitting or 'dunting'	Body has been overfired	Body is heated or cooled too quickly	Splitting due to silica inversion	Fire and cool the body gradually, especially at around 1065° F (575°C) and 435°F (225°C)

Fault finding - glazing problems

Problems	Cause	Result	Effect	What to do
Body will not absorb glaze	Biscuit firing too high	Glaze cannot penetrate body surface	Body rejects glaze	Fire to lower temperature at biscuit-firing stage Before glazing, warm the pot, and then fire to a higher temperature
Crawling	Too much handling of pots before glazing	Uneven distribution	Patches of surface left unglazed	Reduce handling of biscuitware
	Soluble salts in body		Glaze forms ruckled mounds	Add 1% barium carbonate to eliminate the soluble salts
	Glaze mix contains too much clay or is too thick (either in mix or application)			Reduce clay content of glaze and keep biscuit ware scrupulously clean Use less glaze
Crazing	Unsuitable mix of glaze and body types	Different levels of expansion when firing	Multiple fine cracks in the glaze surface	Increase the body expansion by firing to a higher temperature Reduce glaze expansion with a borax frit, silica or china clay
	Body too moist			Increase silica in body so it is less porous
	Glaze too thick			Use finer coat of glaze
Matte or milky finish	Glaze cools too slowly in early stages	Precipitation occurs during cooling stages	Crystals or a milky deposit spoil the final surface	Cool rapidly while glaze is still liquid Reduce lime content or add china clay to the glaze; use low-solubility glaze

Problems	Cause	Result	Effect	What to do
Peeling or flaking	The clay contracts more than glaze and so exerts pressure on the glaze	Poor adhesion of glaze	Glaze peels or flakes from surface, usually on edges of handles and rims	Fire at slightly lower temperature Reduce silica content of glaze Sponge rims and handles very carefully before biscuit firing
Pinholes in glaze surface	Underfired glaze or fired too fast or for too long	Air trapped in the clay	Glaze covered with pinholes after firing	Make sure you wedge clay more thoroughly Fire to recommended temperature Reduce silica content of glaze
Sulfuring	Sulfates in body	Sulfur gases or sulfates in the kiln react with the glaze	Surface of glaze is covered with a dull scum	Biscuit fire to higher temperature which releases any sulfur or carbon

Pyrometric cones

No	Large cones						Small cones		Seger cones	
	Approximate temperature rise per hour								Degrees	
No	60°C	140°F	100°C	212°F	150°C	302°F	300°C	572°F	C	F
022	576	1068			586	1086	630	1116		
021	602	1115			614	1137	643	1189		
020	625	1157			635	1175	666	1231	670	1238
019	668	1234			683	1261	723	1333	690	1274
018	696	1284			717	1323	752	1386	710	1310
017	727	1340			747	1377	784	1443	730	1346
016	764	1407			792	1458	825	1517	750	1382
015	790	1454			804	1479	843	1549	790	1454
014	834	1533			838	1540	870	1598	815	1499
013	869	1596			852	1566	880	1616	835	1535
012	866	1590			884	1623	900	1652	855	1571
011	886	1626			894	1641	915	1679	880	1616
010	887	1628			894	1641	919	1686	900	1652
09	915	1705			923	1693	955	1751	290	1714
08	945	1733	950	1742	955	1751	983	1801	940	1724
07	973	1783	978	1792	984	1803	1008	1846	960	1760
06	991	1815	995	1823	999	1830	1023	1873	980	1796
05	1031	1887	1036	1896	1046	1915	1062	1944	1000	1832
04	1050	1922	1055	1931	1060	1940	1098	2008	1020	1868
03	1086	1986	1092	1997	1101	2014	1131	2068	1040	1904
02	1101	2013	1110	2030	1120	2048	1148	2098	1060	1940
01	1117	2042	1127	2060	1137	2079	1178	2152	1080	1976
1	1136	2076	1145	2093	1154	2109	1179	2154	1100	2012
2	1142	2087	1150	2102	1162	2124	1179	2154	1120	2048
3	1152	2105	1160	2120	1168	2134	1196	2185	1140	2084
4	1168	2134	1176	2148	1186	2167	1209	2208	1160	2120
5	1177	2150	1186	2166	1196	2185	1221	2230	1180	2156
6	1201	2193	1210	2210	1222	2232	1255	2291	1200	2192
7	1215	2219	1227	2240	1240	2264	1264	2307	1230	2246
8	1236	2256	1248	2278	1263	2305	1300	2372	1250	2282
9	1260	2300	1270	2318	1280	2336	1317	2403	1280	2336

	Large cones						Small cones		Seger cones	
	Approximate temperature rise per hour								Degrees	
No	60°C	140°F	100°C	212°F	150°C	302°F	300°C	572°F	C	F
10	1285	2345	1294	2361	1305	2381	1330	2426	1300	2372
11	1294	2361	1306	2382	1315	2399	1336	2437	1320	2408
12	1306	2382			1326	2419	1335	2435	1350	2462
13	1321	2409			1346	2455			1380	2516
14	1388	2530			1366	2491			1410	2570
15	1424	2595			1431	2608			1430	2606
16	1455	2651			1473	2583				
17	1477	2691			1485	2705				
18	1500	2732			1506	2742				
19	1520	2768			1528	2782				
20	1542	2807			1549	2820				
23	1586	2886			1590	2894				
26	1589	2892			1605	2921				
27	1614	2937			1627	2960				
28	1614	2937			1633	2971				
29	1624	2955			1645	2993				
30	1636	2976			1654	3009				
31	1661	3021			1679	3054				
32	1706	3102			1717	3122				
33	1732	3149			1741	3166				
34	1757	3194			1759	3198				
35	1784	3243			1784	3243				
36	1798	3268			1796	3265				

In some cases these temperatures must be considered to be approximate only.

Melting points and color changes

Melting points of some of the minerals and compounds available to potters

Mineral or compound	°C	°F
Alumina	2050	3722
Barium carbonate	1360	2480
Bauxite	2035	3695
Borax	741	1365
Calcium oxide	2570	4658
Cobaltic oxide	905	1661
Copper oxide	1064	1947
Corundum	2035	3695
Cryolite	998	1830
Dolomite	2570-2800	4658-5072
Ferric oxide	1548	2818
Fireclay	1660-1720	3020-3128
Fluorspar	1300	2372
Kaolin	1740-1785	3164-3245
Lead oxide	880	1616
Magnesium carbonate	350	662
Magnesium oxide	2800	5072
Nepheline syenite	1223	2232
Nickel oxide	400	752
Potash	1220	2228
Potassium oxide	red hot	
Rutile	1900	3452
Silica	1715	3119
Silicon carbide	2220	3992
Sodium oxide	red hot	
Tin oxide	1130	2066
Titanium oxide	1900	3452
Whiting	825	1517
Zircon	2550	4622

Color and temperature guide

Color	°C	°F
Just turning red	475	885
Just turning red/dark red	475-650	885-1200
Dark red/cherry red	650-750	1200-1380
Cherry red/orange-cherry	750-815	1380-1500
Orange-cherry/orange	815-900	1500-1650
Orange/yellow	900-1090	1650-2000
Yellow/pale yellow	1090-1315	2000-2400
Pale yellow/white	1315-1540	2400-2804
White/brilliant white	1540 +	2804 +

Temperature conversion formula

Centigrade to Fahrenheit	×9 ÷5 +32
example: 100° C ×9/5 = 180	180 + 32 = 212° F
Fahrenheit to Centigrade	−32 ×5 ÷9
example: 212° F − 32 = 180	180 × 5/9 = 100° C

°C	°F	°C	°F	°C	°F	°C	°F
-273	-477	-18	0	0	32	10	50
50	122	100	212	120	248	200	392
225	437	250	482	300	572	350	662
400	752	450	842	500	932	550	1022
573	1063	600	1112	660	1220	700	1292
750	1382	800	1472	870	1598	900	1652
950	1742	1000	1832	1050	1922	1080	1976
1100	2012	1120	2048	1140	2084	1150	2102
1180	2156	1200	2192	1220	2238	1230	2246
1240	2264	1250	2282	1260	2300	1275	2318
1280	2336	1285	2345	1300	2373	1310	2390
1325	2417	1335	2435	1350	2462	1380	2516
1400	2552	1450	2642	1500	2732	1575	2867

The main tools used by any potter are of course his or her hands but many other manufactured tools can be made simply from wood, plastic and rubber or adapted from household articles. Manufacturers produce a wide range of useful tools and equipment and the potter can purchase everything from modeling and clay preparation tools to kilns.It all depends on your budget but an enormous outlay is not essential if you are prepared to improvise. The charts on the following six pages show a selection of tools used.

	Bamboo tools	Bow (harp)	Brushes	Calipers	Cardboard tube (former)	Cheese grater
Agate						
Burnishing						
Centering & pulling						
Clay preparation				77		
Clay testing						
Coiling			48			49
Combing & painting slip						
Extrusion						
Faceting		141				141
Firing						
Fluting	140					
Glazing			205, 212-14, 217			
Handles, lugs & knobs			102-4			
Incizing						
Inlay						
Impressed/applied & sprigs					142-45, 150	142
Joining shapes			92	91, 92-3		
Lamination			152			
Lids, lips & spouts				97-9		
Majolica			190			
Modeling						
Molds & casting			113		113	
Piercing						
Pinching			38			
Removing pots from wheel						
Repeat throwing/problems						
Resist, tiles & color			182-83, 186, 188			
Sgraffito			180			
Slab building		52	52, 54, 56	52	56	
Slab molding		63	63		63	
Slip trailing			165-69, 175			
Turning						
Wedging						

Combs	Curved blade hole cutter	Cut-out cards & flexible scrapers	Feather	Fork	Heat-proof gloves	Jugs	Kidney, rubber
						22	
		51		48-9, 51			
175		175		175			
					243		
	140						
						212, 214-16	
				138			
		142		142, 157			
							91-2
	101						
						112-13, 115	111
	134						
				37, 38-9, 41			37
						182	
				52			52
						63	63
169-70			168-69, 172			168-69, 171	165, 167, 174
		94					94-5
						24	

	Knives & blades	Lino & modeling tools	Long-handled tongs	Measure (ruler)	Metal scrapers	Metal tube
Agate						
Burnishing	132, 133					
Centering & pulling						
Clay preparation	77			77	77	
Clay testing	27			27		
Coiling	48-9, 51				49	49
Combing & painting slip						
Extrusion						
Faceting						
Firing			243			
Fluting				140		140
Glazing						
Handles, lugs & knobs						
Incizing	139					
Inlay	153				156	
Impressed/applied & sprigs	142, 145, 150	142, 145				
Joining shapes	91, 92-3	92			93	
Lamination						
Lids, lips & spouts	96, 98					
Majolica						
Modeling						
Molds & casting	113, 116	113			115	
Piercing	134					134
Pinching		37				
Removing pots from wheel					84	
Repeat throwing/problems						
Resist, tiles & color	186					
Sgraffito	180	180				
Slab building	52; 54-6			52, 57	52	
Slab molding						
Slip trailing	167	167				
Turning					94	
Wedging						

Needle in cork or hatpin	Paint scraper & square-ended tools	pebble,bone or stone	pencil or pen	pugmill	rolling pin	roulettes	saw blades
					154		
		132-3					
	77						
49, 51					48		48
				117			
		228					
				102			
	152				152, 56		
142			150	152	142-43	142, 150	142
91							
				152			
98-9, 101	97, 99						
			190				
					113		
37			37, 41				37
107							
183, 186							
52					52-3		
					63		
168, 172			165				
							94

	Scalpel blades	Scraper	Set-squares & T-square	Sieves	Sliptrailers	Sponges or sponge on stick
Agate						
Burnishing						
Centering & pulling						
Clay preparation		77		20, 23		76, 78
Clay testing						
Coiling	51	48				48-9
Combing & painting slip						
Extrusion						
Faceting						
Firing						
Fluting						
Glazing				200, 204, 212-13		205
Handles, lugs & knobs						105
Incizing	139					138
Inlay						
Impressed/applied & sprigs	150					
Joining shapes		93				91-2
Lamination						
Lids, lips & spouts						101
Majolica						190
Modeling						
Molds & casting	116	110-15				113, 115-16
Piercing						
Pinching						
Removing pots from wheel		84				84-5
Repeat throwing/problems						
Resist, tiles & color				149		182, 188
Sgraffito						
Slab building	56	52, 54, 56	52, 54			52, 54
Slab molding					63	
Slip trailing				174	165-67, 174	165, 167, 173
Turning	95					
Wedging						

	Spoon or scoop	Surform (clay plane)	Templates & stencils	Twist drills	Wash leather	Wire and wire with toggles	Wire loop tools	Wooden slat or paddle (batten/beater)
1								
2	132-33							
3					83-4, 86			
4	23	77			77	72, 76-7	77	76
5								
6	49			49		49	49	51
7								
8								
9		141				141		
10								
11							140	
12	212-23		205, 214					
13							103	
14						138		
15								
16	142						152	142-43, 157
17							92	
18								
19				101		101		
20			190					
21							64	
22		111-12						111
23				134				
24	37			37				40
25						84-5		
26								
27			182, 188					
28								
29		58	56, 58			52		53
30						63		
31						167		
32					94	94	94	
33						25		

Bibliography

Birks, T. *The Art of the Modern Potter* Van Nostrand Reinhold, New York 1977

Birks, T. *The Potter's Companion* Collins & Batsford, UK 1977

Cardew, M. *Pioneer Pottery* Longman, London 1969 and St Martins, New York 1976

Casson, M. *The Craft of the Potter* BBC Publications, UK

Casson, M. *Pottery in Britain Today* Transatlantic Arts, New York 1967

Ceramics Monthly Box 12448, Columbus, Ohio 43212, USA

Ceramic Review 17a Newburg Street, London W.1.

Charleston, R.J. *World Ceramics* Paul Hamlyn, London 1968

Clark, K. *The Potter's Manual* Macdonald, London 1983

Colbeck, J. *Pottery Techniques of Decoration* Batsford 1983

Cooper, E. *Handbook of Pottery* Longman, London 1970

Cooper, E. *History of World Pottery* Batsford, London

Cooper, E. *The Potter's Book of Glaze Recipes* Batsford, London 1980

Cooper, E. & Royle, D. *Glazes for the Potter* Scribner, New York 1978

Cosentino, P. *Creative Pottery* Ebury Press, London 1985

Craft Australia Craft Council of Australia, 100 George Street, Sydney, NSW

Craft International 24 Spring Street, New York, NY 10012 USA

Crafts Craft Council, 8 Waterloo Place, London SW17 4AT

Fraser, H. *Glazes for the Craft Potter* Pitman

Green, D. *Understanding Pottery Glazes* Faber & Faber

Hagger, R. *Pottery through the Ages* Methuen, London 1959

Hamer, F. *Dictionary of Material and Techniques*

Holden, A. *The Self-Reliant Potter* A & C Black

Honey, W.B. *Art of the Potter* Faber & Faber, London 1955

Honey, W.B. *Ceramic Art of China* Faber & Faber, London 1945

Lane, P. *Ceramic Form* Collins UK, 1988 & Rizzoli, USA

Lane, P. *Studio Ceramics* Collins UK, 1983 & Chilton, USA

Leach, B. *A Potter's Book* Faber & Faber, London 1945 & Transatlantic Arts, New York 1965

Leach, B. *Beyond East and West: Memoirs, Portraits and Essays* Faber & Faber

Lewenstein, E. and Cooper, E. *New Ceramics* Studio Vista, London 1974 & Van Nostrand Reinhold, New York 1974

Nelson, G. C. *Ceramics: A Potter's Handbook* Holt, Reinhart & Winston, New York 1978

Potter, The Craft Potters' Society of Ireland, Dublin

Radnor, P.A.(Ed) *Kilns, Design, Construction and Operation* Chilton, 1980

Radnor, P.A.(Ed) *Pottery Form* Chilton, 1977

Radnor, P.A.(Ed) *Stoneware and Porcelain: The Art of High-Fired Pottery* Chilton, 1959

Reigger, H. *Raku: Art and Technique* Van Nostrand Reinhold, New York 1970

Rhodes, D. *Clay and Glazes for the Potter* (Pitman 1959)

Rhodes, D. *Stoneware and Porcelain* Chilton Company

Rogers, M. *Pottery and Porcelain* Alphabooks/Watson-Guptill, New York

Shoki, H. *A Potter's Way and Work*

Caiger-Smith, A. *Tin Glazed Pottery* Faber and Faber

Wynne, R. *Wynne Pottery* Springfield, Sandyford Road, Co. Dublin, Ireland

Health and safety ◇

Mixing slips and glazes

Always wear a mask when weighing out dry ingredients and cover with water as soon as possible. Wipe up spilt powders with a damp sponge. Do not brush away into the atmosphere of the studio.

Using cutting wires

Attach toggles (pieces of wood, stone and so on) to ends of wire to help you retain a good grip on the wire and to prevent the wire from cutting into fingers. Fix wire to toggles carefully, making sure there are no loose ends to inflict damage on hands.

Clay

When handling powdered dry clay, treat this with great care. Try not to rub down pots when the clay is dry. Scrape them at the leather-hard stage. If dry clay has to be rubbed, wear a good filter mask over mouth and nose to prevent inhalation of dust particles.

Lifting and moving equipment

Most potters pride themselves on being relatively fit for their particular needs. However, do not try to be too independent; use a trolley to move sacks of clay and materials. Lift heavy objects carefully, with plenty of support, and keep the back straight. Use special lifting gear if moving machinery or enlist the help of any friends or callers when moving a heavy piece of equipment.

Electricity

Water is used constantly in a studio where inevitably there will be electrical equipment, wall plugs, switches and lights — so be careful!

Do not touch switches with wet hands. Wherever possible have switches changed to a string-pull variety. Make sure all wiring is done by an experienced qualified electrician. Electric kilns are heavy users of power; fit a door safety-switch and an isolator switch to main power.

Firing

Kilns produce fumes when in use so try to place them in a separate room to the studio. Allow plenty of ventilation for fumes to escape.

Do not fire salt-glaze kilns in enclosed spaces; the fumes given off are toxic. Enamel, luster bisque, sawdust, gas and oil all need plenty of ventilation in the kiln room and extractor fans to remove fumes which do not go up the flue.

Do not look into a kiln through the spyhole with the eye too close; hot air, or even a flame, may escape and damage the eye. Look into a hot kiln through protective dark lenses to prevent damage to your eyes by the bright glow.

Keep surfaces and equipment as clean as possible. Wipe up any dust or spillage with a damp cloth or sponge. If possible wash floors with a wet and dry vacuum cleaner to remove dust and water. Do not sweep up dust; mop the floors with a damp sponge.

Spraying

Glaze, oxides or slip sometimes require the use of a spray.

Use a spray booth with extractor fan and wear a mask.

Food and drink

Food or drink should not be consumed in the pottery studio environment. Harmful particles may be taken in through the mouth. Smoking while potting can also involve the inhalation of dust particles from the atmosphere and so should not be allowed.

Acknowledgments

I should like to thank all the potters who have so generously contributed examples of their work with details of their working methods and glazes. I am especially grateful to Peter Lane for his invaluable help throughout the production of the book.

Thanks are also due to Ray Scott of FordhamThermal Systems, Newmarket, Roland Curtis of Littlehope Potteries, Ripon, the Country Centre, Cambridge College of Further Education, the Fulham Pottery, London, and English China Clay, St Ives.

Potter	Country	Page	Photographer
G.E. Arnison	UK	164,176	
Oldrich Asenbryl	UK	4, 5, 120	
Paul Astbury	UK	2, 3, 4, 59, 118-9, 121	David Cripps
Rudy Autio	USA	45	
Maggie Barnes	UK	130, 134-5, 137	
Jenny Beavan	UK	46, 99, 154, 163, 178	
Tony Bennett	UK	17, 110, 114, 120, 193	
Maggie Angus Berkowitz	UK	187	
Sandra Black	Australia	116, 121, 136	
Betty Blandino	UK	46	G.O. Jones
Yvonne Boutell	USA	45, 62, 160	
Hilary Brock	UK	38, 119, 162	
Sandy Brown	UK	29, 56, 58, 208	
Sally Bowen-Prange	USA	123, 147, 203, 214	
Joan Campbell	Australia	209	
Virginia Cartwright	USA	57, 159	
Sally Cocksedge	UK	44, 65	David Playne
Marianne Cole	Australia	191	
Gordon Cooke	UK	191	
Delan Cookson	UK	59, 90, 99	
John S. Cummings	USA	203	
Greg Daly	Australia	74, 149, 161, 193, 211, 239	
Derek Davis	UK	209	
Sue Davis	UK	200	
Walter Dexter	USA	192	
Dorothy Feibleman	UK	152	
Ray Finch	UK	88	
Graham Flight	UK	29, 40-1, 58-9, 81, 128, 176-9	David Playne
Tessa Fuchs	UK	29, 161, 193, 207, 238	Ronald E. Brown
William Hall	UK	121, 147	
Frank Hamer	UK	100, 185, 223	
Jane Hamlyn	UK	74, 150, 196, 210, 240	
Robin Hopper	Canada	154-5	
William Hunt	USA	131, 203	
Neil Ions	UK	47, 120, 123	
Roberta Kaserman	USA	121	
Lisa Katzenstein	UK	104, 114, 116, 206, 223	
Ann Kenny	UK	210	
Peter Lane	UK	6, 29, 75, 86, 137, 139, 140	
Jennifer Lee	UK	29, 43	Haynes
Pauline Lurie	USA	223	
Jim Malone	UK	141, 213	
John Maltby	UK	148, 181	

Potter	Country	Page	Photographer
Andrew McGarva	UK	163, 195	John Coles
Peter Meanley	Ireland	124, 131, 160	
Eric James Mellon	UK	199	David Turner
Chris Myers	Australia	163	
Susan Nemeth	UK	158, 239	Kenneth Grundy
Sidig El Nigoumi	UK	132, 181	Andrew Dowsett
Alan Peascod	Australia	75	
David Pendell	USA	164	
Henry Pim	UK	59, 60, 122	
Ursula Morley Price	UK	32, 34, 44	Ces Thomas
David Roberts	UK	47, 238	
James Robison	UK	119, 146	
Ray Rogers	New Zealand	44, 192, 240	Ces Thomas & John Storey
Jerry Rothman	USA	76	
Jill Ruhlman	USA	44	
Gail Russell	USA	239	
Jan Schachter	USA	117, 147, 195, 205	
Karl Scheid	Germany	209	
Peter Simpson	UK	118	
Ian Sprague	Australia	89, 141, 147, 157	Mark Strizic
Petrus Spronk	Australia	79, 132	
Robyn Stewart	New Zealand	132	
Angus Suttie	UK	61	Tim Hill
Hiroe Swen	Sweden	159	
Geoffrey Swindell	UK	207, 223	
Bryan Trueman	Australia	73, 185, 191, 209	Hahn Tran
Sue Varley	UK	36, 43, 184	
Angela Verdon	UK	12, 136, 138, 160	John Coles
Monique Vezina	Canada	43, 131	
Sasha Wardell	UK	161, 149, 178	
Robert Washington	UK	60	
John Wheeldon	UK	208	
Mary White	Germany	42, 146, 202	
Lana Wilson	USA	10, 54, 184, 240	
Mollie Winterburn	UK	42, 50	
Andrew Wood	UK	122	

Photographs of historical ceramics have been reproduced by courtesy of:

C.C.E. of Antiques 292
American Museum in Britain 327
Arizona State Museum 301
Art & History Museum, Shanghai, China 264
Ashmolean Museum, Oxford 256
Bethnal Green Museum, London 308
Bridgeman Art Library 250, 295, 296, 306, 308
The Trustees of the British Museum 142, 169, 248-9, 252, 254-5, 256-7, 258-9, 261, 267, 275, 300, 305, 310-11
Cambridge Archaeology & Anthropology Museum 253
Christies Colour Library 265, 282-3, 285, 299
City Museum & Art Gallery, Plymouth 290
Dyson Perrins Museum 299
E.T. Archive 194, 299
Fitzwilliam Museum, Cambridge 263, 267, 270, 278-9, 287, 294, 297
Phillip Goldman Collection, London 303
Hanley Museum & Art Gallery, Staffordshire 296
Michael Holford 253, 291
The Maxwell Museum of Anthropology 303
Moravian Museum, Brno 251
Museum of London 260, 262-3, 293
National Museum of Anthropology, Mexico 302
National Museum of Archaeology, Madrid 276
National Palace Museum Taiwan 265
Ono Collection, Osaka 273
Peabody Museum, Mass. USA 301
Scarborough Museum 291
Cyril Staal, Cotehele House, Saltash 295
Stavenhagen Collection, Mexico 302
State Museum, East Berlin 250
Sudan Museum 133
The Victoria and Albert Museum 189, 194, 206, 253, 255, 266, 269, 270, 272, 276-7, 278, 280-1, 283, 284, 286-7, 289, 292, 295, 296
Villa Guilia Museum, Rome 261
Josiah Wedgwood & Sons Ltd. Stoke-on-Trent 153, 298
Wellington Museum, Apsley House 288
Werner Forman Archive 273, 274, 304, 311
David Whitelaw, St Davids 162, 192

Glossary of terms

Absorbency The term which describes the rate at which a body of material (clay unfired and fired, plaster, wood, stone etc.) will soak up moisture.

Ash Residue of burnt wood or vegetation, used in the preparation of high-fired glazes. It contains silica, alumina and flux, and forms a glaze combined with the body of the pot. More usually mixed with feldspar, clay, whiting, talc or oxides in varying proportions to give a wide variety of glaze effects and surfaces, depending on the type of ash used and its source.

Alkali The fluxes used in certain types of glazes. Soda and borax magnesia form the main part of low-firing glazes. They have the effect of heightening the color reaction to oxides, for example in Egyptian paste and Persian pottery.

Alumina One of the three main parts of a glaze together with silica and flux. It is the stabilizing neutral part of the glaze. It balances the effect of the silica and the flux and is usually represented in a glaze recipe as a clay.

Ball clay A fine, plastic secondary clay, containing organic material. It fires white or off white and can be used in glaze preparation, as an ingredient in porcelain and in colored slips.

Ball mill A revolving drum, mechanically driven, containing hard balls of flint or fired porcelain which grind powders finely in a mixture with water.

Basalt ware A high-fired unglazed body, colored with oxides and left unglazed. Used in the 18th century most famously by Josiah Wedgwood.

Bat A tile made from highly refractory material used for kiln shelves. A plaster disc used for throwing on the wheel. A piece of wood used as a throwing platform or to support pots when removed from the wheel.

Bentonite A clay-like material, used to add plasticity to clay, and added in small percentage to a glaze mixture to assist suspension of the particles.

Biscuit or bisque Clay which has been fired to a sufficient hardness, changing it from a dried clay to a hard porous state prior to glazing.

Body The term used for the mixture of clays which make up the basic material of the pot together with any added sand or grog.

Bone china A type of porcelain china used to imitate the oriental wares being imported into England during the 18th century. It contains calcined bones which impart hardness, whiteness and translucency to the body.

Burnishing Smoothing and polishing a clay or slipped surface with a smooth hard object such as a stone or the back of a spoon, to compact the clay so it forms a hard shell. Originally employed to make a pot more watertight. Generally used as a decorative process now.

Calcine Ceramic materials may be heated to drive off carbon-dioxide gases. The residue usually needs grinding to a fine powder before use.

Calcium carbonate Chalk in the form of whiting, limestone. Used as a flux in middle-to-high temperature glazes.

Casting (slip or plaster casting) *Plaster casting* means covering an object or model with a mixture of plaster of paris and water to form an exact copy of that object. The plaster hardens chemically forming a mold which can then be used to reproduce a form many times. *Slip casting* means filling a plaster mold with liquid clay (slip) to allow a build-up of clay as the water content is absorbed. The excess is poured out, leaving a coating or shell of clay in the form of the mold.

Celadon Stoneware and porcelain glazes containing small amounts of iron, which when fired in a reducing atmosphere produce soft gray/green glazes.

Chattering A pattern of uneven bumps on the surface of a turned pot, caused by irregular objects in the clay, blunt turning tools or not holding the turning tool firmly enough.

China clay A pure white primary non-plastic clay, usually used in combination with other clays to form a body and in glaze recipes.

Chuck A shaped object for holding leather-hard pots while they are being turned. They can be made in clay, wood, metal or plaster.

Clay body A general term to indicate a mixture of clays and minerals used for building pots.

Coiling A means of building by hand; a pot using ropes of clay, laid on top of each other and smoothed together.

Combing Scraping into a damp surface, either directly on to clay or through a coating of slip with an even-toothed tool to make a pattern of lines.

Cone A small conical object made from glaze materials which has a known melting point used in a kiln in front of the spyhole to indicate a given temperature.

Crank A heavily grogged and open clay mixture; its coarse-grained texture withstands shrinkage and thermal shock particularly well.

Crazing A fine line pattern appearing on the surface of a glaze caused by tension between glaze and body (when these have an uneven contraction after firing). Sometimes used as a decorative process when the pattern of lines is darkened by oxides or smoke.

Crawling Retraction of the glaze during firing to expose the clay body, caused by reaction of two or more glazes fired over each other or glazing a pot which is dusty or greasy.

Damp box or cupboard A metal or plastic-lined box or cupboard which will retain moisture, allowing the storage of unfinished work.

Deflocculent Alkaline mineral, usually sodium silicate, which is added to a clay mixture to separate the particles, so making the mixture more liquid without adding water.

Delft Tin-glazed earthenware pottery with onglaze painted decoration. Derives its name from *Delft* in Holland, but originates from Middle-Eastern majolica.

Dog-head wedging A way of preparing clay, wedging it into spirals that resemble a pointed dog's head.

Down-draft kiln A kiln in which the heat is forced upwards from the fire box and then down through the interior of the kiln through flues at the base of the kiln.

Dunting Cracking or breaking of pots caused by cold air being let into a hot kiln too quickly.

Earthenware Lower-fired pottery in which the body remains porous under the glaze.

Egyptian paste A body formed of clay and glaze materials and coloring oxides in which a glaze forms on the surface after the soda content migrates there. A non-plastic mixture, it was first used by the Egyptians to form small models and beads in molds.

Enamels Low-firing colors incorporating glaze which are painted on to an already glazed surface and refired to seal them into the glaze.

Engobe A slip mixture with parts of glaze materials incorporated to use on damp pots or on once-fired biscuit.

Faience The French name for tin-glazed earthenware. Also known as majolica or *Delft*.

Feathering A decorative process made by drawing the tip of a feather through various colored slips.

Fettle Smoothing and finishing damp pots to remove marks of tools or molding lines.

Fireclay A high-firing refractory clay used to make bricks for furnaces, kiln shelves and props — often used in clay bodies.

Flint silica Made by grinding pure flintstone to a fine powder.

Flux Glaze materials which reduce the melting point of a glaze or glass. Usually the most influential of the glaze materials, it dictates the gloss and color of a glaze.

Foot ring A circle of clay left while turning or trimming the base of a pot and which allows pot to stand evenly.

Frit A process of heating, rapid cooling and grinding materials which would normally be insoluble or toxic, to make them safe to use in a glaze as a flux.

Galena A lead ore, powdered and scattered on a damp pot by medieval potters to form a shiny glaze.

Glaze A mixture of silica, alumina and flux applied by brushing, pouring, dipping and spraying on to the surface of a pot to make it impervious to water and create at the same time a decorative pleasing quality when the glaze is fired.

Glost firing The firing of the kiln to sufficiently high temperatures to melt the glaze on a pot.

Green ware Pots which are hard but not completely dried or fired.

Grog Ground fired clay or sand which is calibrated in particle size from flour to coarse grit; it is added to clay to provide openness to the texture and to reduce shrinkage. It gives strength to smooth clays, and 'tooth' (slight resistance) to an otherwise creamy clay.

Gum arabic A natural gum mixed with glaze or oxide to prevent smudging, to bind the color to the body before firing.

Handbuilding Forming pots without using a wheel, by coiling, pinching and slabbing.

Hard paste True porcelain which is fired to temperatures in excess of 2370°F (1300°C).

Inglaze Oxides and colorants that are painted on to and then fired into a glaze to combine with it.

Iron oxide One of the main oxides used for coloring and decoration. Present in most dug clays except the pure kaolins.

Jigger and jolley A mechanical method of producing uniform shapes repeatedly. Clay can be pressed into a form to produce a shape on one side and a tin plate is pressed over the clay to shape the other side.

Kaolin Pure white clay, known as *china clay.*

Lead Red lead, white lead or galena used as a low-temperature flux in glazes. Highly toxic in its raw state, it is used as a lead frit in glaze composition.

Leather hard Stiff damp clay that can be turned, trimmed, cut and joined while it is still moist yet hard enough to handle without being damaged.

Levigation Repeated mixing and settling of coarser particles of clay so that the finer particles can be syphoned off to form a very fine slip.

Lug Handle or other applied shape which can be purely decorative or essential as a means of lifting the pot.

Luster Salts of metals applied in a thin film over glaze-fired pots and refired at a low temperature to give a metallic sheen to a surface.

Lute Joining leather-hard pieces of pottery together with slip or slurry.

Majolica Soft earthenware covered in a tin glaze with painted decoration. Derives from the island of Majorca from whence it was exported to other parts of Europe.

Mocha Decoration produced by brushing a solution of oxide and an alkaline liquid on to wet slip where it spreads quickly to form tree-like patterns.

Mold A form made from plaster of paris in which clay can be shaped. It absorbs moisture until the dried hardened piece can be removed.

Muffle A refractory container holding pots inside a kiln and preventing direct contact with the furnace.

Opacifier Chemical added to glaze which does not completely melt in a glaze and produces a whiteness. Tin oxide and zinc oxide are most commonly used and can form up to 10% of a clear glaze.

Oxidation Firing a kiln with an oxidizing atmosphere; a good supply of air produces a clear atmosphere in the kiln.

Peeling The separation of a slip or glaze from a body. (For example, slip may peel off after firing, taking the glaze with it.)

Plaster of paris Hydrate of calcium sulfate which hardens chemically, often mixed with water. Used widely in ceramic making for its absorbent qualities — for molds, bats, casting, jigger and jolley.

Porcelain Very fine high-fired mixture of clays and glaze materials, which is hard, white and translucent when thin.

Press molding Pressing sheets of rolled clay or clay pieces into or over a former to allow the clay to dry in that shape.

Primary clay Clays which have been formed on the site of the parent rock and have not been washed further away. They are not colored by impurities and are usually non-plastic. China clay or kaolin is the main form used in ceramics.

Pugmill A mixing machine for clay, which is then extruded either as a large round coil or in a shape formed by a preformed plate fixed over the mouth of a pug mill. Some pug mills will also extract the air bubbles from the clay in the process.

Pyrometer A means of measuring temperatures in a kiln.

Raku Low-fired earthenware, made of grogged open clay, covered with low-melting glaze and oxides. The pot is taken from the kiln when the glaze melts (using long-handled tongs) and reduced or treated with various substances to change the colors of the glaze and oxides.

Raw glazing Glazing a pot which is unfired and firing slowly through to the melting point of the glaze, thus eliminating the biscuit firing.

Reduction firing To fire a kiln with incomplete combustion so that the lack of oxygen creates a smoky atmosphere, thus changing the color of clay, glazes and oxides.

Refractory Resistance to heat, making a substance capable of withstanding high temperatures.

Saggar A container with a lid, made of fire clay, for holding a pot or pots to prevent damage to the pot by naked flames in the kiln.

Salt glaze The forming of a glaze by adding salt to a high-firing kiln where it vaporizes and then creates a coating on pots.

Secondary clay Clay which has been moved from its source of origin by erosion and has collected other minerals on the way.

Sgraffito decoration Scratching through a layer of slip, glaze or oxide to the layer below.

Short Term used to describe a clay which is dry and lacks plasticity.

Shrinkage Contraction of the clay during drying and firing. This varies from clay to clay; fine particled clays such as porcelain shrink the most. Grog added to clay will reduce shrinkage.

Slabs Clay sheets rolled out and allowed to harden when they can be cut and handled, perhaps being joined together with slurry to form flat-sided constructions. Soft slabs can be draped over or inside formers and left to harden in that shape.

Slip Liquid clay mixture with water, sometimes called an engobe, originally used as a light color over a dark body. Mixed with colorants for pouring, brushing, dipping and trailing to create a great variety of decorative processes.

Soaking Maintaining a temperature in a kiln by the control of the power or fire to allow the glaze to mature and settle.

Sprigging Plastic clay forms (sometimes made in a small press-mold) applied to a leather-hard pot.

Stilt A small pointed stand used to support pots in a kiln to prevent the glazed base sticking to the shelf.

Stoneware A hard stone-like ceramic material formed when the clay is fired into a semi-vitrified state at 2190°F (1200°C) and over.

Terracotta A term for clay or an object made in a high iron-content clay which is smooth and fires a rich red brown.

Terrasigillata A very fine slip used for coating pots; it is often fine enough to shine without being burnished.

Tenmoku A Japanese name for a glaze which has a high iron content giving a rich red/black/brown color.

T-material An English term for a beautifully plastic grogged clay ideal for large pieces of work.

Trailing Decorating by squeezing slip from a rubber bulb through a thin nozzle.

Turning or trimming Finishing off, by removing spare clay with a sharp tool from the base of a pot while it revolves on a wheel or lathe.

Vitrify To heat to a hard glass.

Volatilize To change into a vapor when at a high temperature, eg salt glazing.

Wax resist A method of decoration by painting wax on to a surface which will resist a water-bound mixture placed over it.

Weathering Clay left out in the open to mature in rain. The sun and ice breaks down the particles making the clay more plastic.

Wedging Kneading and mixing clay to remove air bubbles before commencing throwing or handbuilding.

Index